LABYRINTH AND PIANO KEY WEIRS IV

Practitioners of hydraulic structures engineering find nonlinear weirs attractive because they can provide a hydro-economical solution for rehabilitation projects. Their complex flow patterns, wide range of geometries and applications have also contributed to their appeal in academia. In a world increasingly defined by ageing infrastructure, diverse project goals and hydrological extremes, non-linear weirs, in particular labyrinth and piano key weirs, have emerged as a desirable option for construction.

This proceedings compilation presents the latest research results and innovations in the field on labyrinth and piano key weirs. Each contribution has been reviewed and selected by the International Scientific Committee of the 4th International Workshop on Labyrinth and Piano Key Weirs (7–9 October 2025, Atlanta, Georgia, USA). References to projects at the design and under-construction stage have also been included.

This publication is a useful guide for students, researchers and professionals interested in dams engineering, civil engineering and hydraulic structures engineering.

PROCEEDINGS OF THE FOURTH INTERNATIONAL WORKSHOP ON LABYRINTH AND PIANO KEY WEIRS (PKW 2025), ATLANTA, USA, 7–9 OCTOBER 2025

Labyrinth and Piano Key Weirs IV

Edited by

Sébastien Erpicum

Liege University, Belgium

Brian Crookston

Utah State University, USA

Greg Paxson

Schnabel Engineering, USA

CRC Press
Taylor & Francis Group
Boca Raton London New York

CRC Press is an imprint of the
Taylor & Francis Group, an **informa** business

Designed cover image: Lake Williams Dam, Pennsylvania, USA. York Water Company. © GFT Infrastructure, Inc.

First published 2026
by CRC Press/Balkema
4 Park Square, Milton Park, Abingdon, Oxon, OX14 4RN

and by CRC Press/Balkema
2385 NW Executive Center Drive, Suite 320, Boca Raton FL 33431

CRC Press/Balkema is an imprint of the Taylor & Francis Group, an informa business

British Library Cataloguing-in-Publication Data
A catalogue record for this book is available from the British Library

ISBN: 978-1-041-10527-5 (hbk)
ISBN: 978-1-041-10532-9 (pbk)
ISBN: 978-1-003-65554-1 (ebk)

DOI: 10.1201/9781003655541

Typeset in Times New Roman
by Integra Software Services Pvt. Ltd.

Labyrinth and Piano Key Weirs IV – Erpicum, Crookston & Paxson (Eds)
© 2026 The Editor(s), ISBN 978-1-041-10527-5

Table of contents

II.3 Construction

Labyrinth and Piano Key Weirs IV – Erpicum, Crookston & Paxson (Eds)
© 2026 The Editor(s), ISBN 978-1-041-10527-5

Preface

In the general field of hydraulic structures engineering, nonlinear weirs are attractive to practitioners as they can provide a hydro-economical solution for new and rehabilitation projects. To academia, they remain a topic of interest due to a combination of the complex flow patterns they generate and the wide range of relevant geometries and application scenarios in rivers, dams, and other water resources applications with research topics including discharge efficiency, aeration, energy dissipation, local scour, driftwood and ice, and flow instabilities. The usage of computational fluid dynamics continues to increase both for research and design while physical model studies remain the standard and are also commonly used, particularly for large dam projects.

In the general context of ageing infrastructure, diverse project goals, and increasing uncertainty on future hydrological extremes, nonlinear weirs, in particular labyrinth and piano key weirs, offer a solution often selected for construction. Bringing together the engineers and researchers interested in these non-conventional hydraulic structures is of utmost importance to discuss and share the latest knowledge and developments. Therefore, this workshop series was developed.

Following three successful editions held in Europe and Asia (Liège, Belgium in 2011; Paris, France in 2013 and Qui Nhon, Vietnam in 2017), the 4th International Workshop on Labyrinth and Piano Key Weirs moved to the United States where numerous Labyrinth and several Piano Key weir prototypes coexist. The fourth edition of the workshop has been co-organized by the United States Society on Dams, University of Liege, Electricité de France and Utah State University.

It has been the occasion to put special emphasis on planned and existing projects worldwide and also share the latest in research developments. The conference took place on 7–9 October 2025 in Atlanta, Georgia, USA, close to several labyrinth and piano key weirs in operation. The two-day workshop was followed by a technical tour visiting both types of structure in Georgia and Alabama.

This book is a compilation of high-level papers selected by the International Scientific Committee that were presented during the workshop. The contents herein present the state of the art of research, development, and lessons learned from the field on labyrinth and piano key weirs. Most of the main projects in design or under construction at the time of publication are also mentioned. It can thus be considered as an up-to-date reference document for students, practitioners and researchers.

Sébastien Erpicum, HECE – ULiege
Brian Crookston, UWRL – USU
Greg Paxson, Schnabel Engineering

Acknowledgements

The organization of this event is only possible with motivated contributors. We are therefore grateful to the practitioners and researchers who took up the challenge of writing papers for each of the four workshops in this series on the very specialized domain of Labyrinth and Piano Key Weirs. In order to achieve the high technical and scientific level of manuscript contributions, a rigorous single-blind review process by a minimum of two experts was carried out by the International Scientific Committee, in collaboration with the United States Society on Dams (USSD) and the Committee on Hydraulic Structures of the International Association for Hydro-Environment Engineering and Research (IAHR). All Reviewers merit our gratitude for the quality of their work and the respect of the event planning schedule.

The success of a workshop is also dependent of the support of professional associations that helped communicate and advertise this event for participants and attendees. Our thanks to the Comité Français des Barrages et Réservoirs (CFBR), the International Commission on Large Dams (ICOLD), USSD, and IAHR.

The edition of these proceedings has been made possible thanks to the contribution of all workshop sponsors. We give specific recognition to Schnabel Engineering (USA) and ISL Ingénierie (France) for their support.

The technical tour in Georgia and Alabama, USA allowed visiting the recently commissioned Peachtree Piano Key Weir and the XX labyrinth weir, designed by Schnabel Engineering. The tour was guided by competent persons in charge of management and safety of this hydraulic scheme. All our thanks go to each owner for making their facilities available during this event.

The involvement of five co-organizing institutions, USSD, University of Liège, Electricité de France, Schnabel Engineering, and Utah State University was fundamental for the preparation of the workshop, the field tour, and for the edition of the proceedings. Concerning these activities, the organizers want to express their recognition to the book publisher for his guidance and special thanks to Catrice Jones (Executive Director), Laura Wright (Director of Professional Development and Event Strategy) and the Board of Directors of USSD for backing this event as a USSD Fall Workshop and for their many contributions.

The Scientific Secretariat of PKW 2025

Labyrinth and Piano Key Weirs IV – Erpicum, Crookston & Paxson (Eds)
© 2026 The Editor(s), ISBN 978-1-041-10527-5

Organization

Members of the International Scientific Committee

Members of the Organizing Committee

Catrice JONES, *USSD, USA*
Laura WRIGHT, *USSD, USA*
Greg PAXSON, *Schnabel Engineering LLC, USA*
Sébastien ERPICUM, *University of Liège, Belgium*
Brian CROOKSTON, *Utah State University, USA*
Frédéric LAUGIER, *Electricité de France, France*
Julien VERMEULEN, *Electricité de France, France*

Labyrinth and Piano Key Weirs IV – Erpicum, Crookston & Paxson (Eds)
© 2026 The Editor(s), ISBN 978-1-041-10527-5

Editor Biographies

Sébastien Erpicum is a civil engineer and holds a PhD in Applied Sciences from Liege University, Belgium, where he is currently Associate Professor. He is in charge of the Hydraulic Engineering Laboratory – HECE, where he teaches and develops for more than 25 years activities related to hydraulics and hydraulic structures engineering. His expertise covers experimental and numerical modeling with application to more than 50 dams, hydropower and flood risk related projects worldwide. He is the Belgium representative to ICOLD Committee on Hydraulics for dam and has served as Chair of IAHR Hydraulic Structures Technical Committee. In recognition of his work, he was awarded the ASCE Hydraulic Structures Medal in 2020 and the 22nd IAHR Arthur Ippen Award in 2021.

Brian Crookston is a civil engineer and holds a PhD in Water Resources from Utah State University, USA where he is currently an Associate Professor. His research and consulting activities are focused on water sustainability and resiliency in the areas of hydraulic structures, fluvial hydraulics, and modeling and technology. Brian has particular interest in spillways, energy dissipators, river and canal systems, scour and erosion, public safety at hydraulic structures, physical modeling, computational fluid dynamics, and flooding with experience in over 300 dam and flood projects including the design of multiple labyrinth and piano key weirs in the USA. He is the USSD Spillways subcommittee chair, a member of National Dam Safety Review Board Research Workgroup,Vice Chair of the Joint Task Committee National Inventory of Low-head Dams, and has served as Chair of the IAHR Hydraulic Structures Technical Committee.

Greg Paxson is a civil engineer specializing in assessment and design of dams and levees. He currently serves as National Practice Leader for Dams at Schnabel Engineering, where he is responsible for the technical direction of Schnabel's dam engineering practice. He has expertise in the evaluation of and modifications to existing dams and in the design of new dams. Through his career, he has been involved in all aspects of dam engineering, with specific expertise in hydrologic and hydraulic analyses, hydraulic structures, dam safety, and risk analysis, with a focus on dam rehabilitation design. He has performed and led assessments, evaluations, analyses, and design for more than 100 dams and has been involved in the design of over 20 labyrinth spillways throughout the USA.

Labyrinth and Piano Key Weirs IV – Erpicum, Crookston & Paxson (Eds)
© 2026 The Editor(s), ISBN 978-1-041-10527-5

Organizing Institutions

Labyrinth and Piano Key Weirs 2025 was organized by the following institutions:

USSD
U.S. National Committee of ICOLD

US Society for Dams

eDF

Electricité de France
Hydro Engineering Center

LIÈGE université

Liège University
Hydraulics in Environmental and Civil Engineering

STATE U

Utah State University
Utah Water Research Laboratory

Schnabel ENGINEERING

Schnabel Engineering, LLC

Labyrinth and Piano Key Weirs IV – Erpicum, Crookston & Paxson (Eds)
© 2026 The Editor(s), ISBN 978-1-041-10527-5

Supporting Institutions

Labyrinth and Piano Key Weirs 2025 was supported by the following institutions:

Comité français des barrages et réservoirs

Commission Internationale des Grands Barrages
International Commission on Large Dams

Hydraulic Structures Technical Committee
International Association for Hydro-environment Engineering and Research

Sponsoring Companies

Labyrinth and Piano Key Weirs 2025 was sponsored by the following companies:

Schnabel ENGINEERING

Schnabel Engineering, USA

ISL Ingénierie

ISL Ingénierie, France

1 Research and development

I.1 Hydraulics

Labyrinth and Piano Key Weirs IV – Erpicum, Crookston & Paxson (Eds)

Remote sensing of the 3D water surface downstream of labyrinth weirs

P. Langohr and D.B. Bung
FH Aachen University of Applied Sciences, Aachen/Germany

B.M. Crookston
Utah State University, Logan/USA

ABSTRACT: Labyrinth weirs are routinely constructed at dams. However, flows exiting this weir type are aerated, three-dimensional and have a complex water surface. To investigate this water surface, an optical approach with high temporal and spatial resolution was used; the Intel® RealSense™ Depth Camera D455. Camera data was locally validated by using ultrasonic displacement sensors. Three different trapezoidal labyrinth weir geometries were tested for three downstream conditions. Results show that the selected camera was capable to produce consistent free surface profiles as well as time dependent fluctuations. An inverse relationship with increasing tailwater and decreasing water surface fluctuations were observed. It was also found that fluctuations reduced with decreasing labyrinth weir sidewall angle. It is anticipated that these results may be of interest to researchers, designers and others concerned with spillways and dam safety.

1 INTODUCTION

Labyrinth and piano key weirs are a modern type of hydraulic structure developed at the beginning of the twentieth century (Hager et al. 2015; Crookston et al. 2019). The geometries of these weirs increase weir crest along a given width and correspondingly the hydraulic efficiency increases significantly. Due to the wide range of applications, such structures are not only suitable for the rehabilitation of spillways but also for in-river applications. Despite numerous head-discharge studies (Hay and Taylor 1970; Crookston and Tullis 2012; Anderson and Tullis 2013; Tullis et al. 2020; Said and Ouamane 2021), the downstream region is gaining importance especially regarding submergence (Tullis et al. 2007; Belzner et al. 2017; Dabling and Tullis 2017; Carrillo et al. 2019), sedimentation (Gebhardt et al. 2019; Herbst et al. 2018) and energy dissipation (Lopes et al. 2006; Lopes et al. 2008; Erpicum et al. 2011; Merkel et al. 2018; Eslinger and Crookston 2020). These topics may be of great importance to ensure structural safety for spillway and in-river applications. These weirs provide additional discharge capacity considering limited space and revised design flow estimates or catchment hydrology. On the other hand, such weirs cause a three-dimensional flow field immediately downstream, yielding to complex aerated flow properties and turbulence, with significant fluctuations of the free surface (Shen 2023; Valero et al. 2024; Belay and Oertel 2025; KC et al. 2025). Consequently, the determination of relevant flow parameters via in-situ measurements can be challenging.

Independent of the developments in hydraulic engineering, the use of remote sensing techniques for 3D data acquisition is gaining significance in many fields. For instance, Giancola et al. (2018) highlights disciplines like medicine, geoscience and robotics. The distinction between reflective and transmissive approaches divides them into two general groups. A further separation inside the reflective group is non-optical (i.e. RADAR, SONAR) and optical methods. In contrast to passive optical methods, active methods use an artificial light source to create structured patterns on a surface that serve as additional reference (Giancola et al. 2018; Lachat et al. 2015).

DOI: 10.1201/9781003655541-1

The accuracy of various remote sensing technologies, often with compact dimensions and consumer friendly control features, makes such devices interesting in terms of hydraulic research. Sensors with high temporal and spatial resolution may provide a major advantage when comparing to classic point data acquisition methods such as point gauges, conductivity probes, or ultrasonic sensors. However, some optical sensors are not able to detect a water surface or struggle under clear water conditions, making reliable data collection difficult. Nonetheless, several authors such as Montano et al. (2018), Kramer et al. (2020), Bung et al. (2021) and (Li et al. 2021a; 2021b) conducted studies focused on detecting water surfaces of highly-turbulent and aerated flows such as hydraulic jumps and stilling basins.

Therefore, to consider the complex downstream water surfaces generated by labyrinth weirs, this study used the Intel® RealSense™ Depth Camera D455 for three laboratory scale trapezoidal labyrinth weirs of varying tailwater conditions. This study focused on detailed information about the free surface properties, including water surface profiles and fluctuations.

2 EXPERIMENTAL SETUP

For the laboratory work, the facilities of the Utah Water Research Laboratory (UWRL) of Utah State University (USU) were used. A 1.22 m wide (w), 0.9 m deep and 14.6 m long flume with smooth boundaries was selected for testing. Inside the flume, three trapezoidal quarter-round labyrinth weirs with two cycles (N) with a constant height of P = 30.5 cm and different crest length (L_c) to width ratios (L_c/w) were investigated (see Table 1). Each weir geometry was tested for three downstream conditions of stoplog height (δ = 0.00 m; 0.05 m; 0.10 m) and eight discharges Q (0.1 m³/s ≤ Q ≤ 0.45 m³/s). Note that the length of a labyrinth weir in flow direction (B), outside apex width (D) and sidewall angle (α) are also included in Table 1 for reference.

Table 1. Investigated weir parameters.

Test No. [-]	L_c [m]	L_c/w [-]	B [m]	D [m]	α [°]
G1	1.969	1.609	0.393	0.076	35
G2	3.095	2.529	0.705	0.088	20
G3	3.983	3.254	0.936	0.093	15

The supply pipe of the flume was equipped with a calibrated orifice flow meter and a pressure transducer for discharge measurements. Flows entering the flume were quiescent and uniform. Flow depths were measured 5P upstream of the labyrinth weir with a point gauge in a stilling well (Crookston 2010). To scan downstream flow conditions, the Intel® RealSense™ Depth Camera D455 (RGB-D) was installed along the flume center line with an installation height of 1.1 m. Further details of the camera specification can be found in KC et al. (2025).

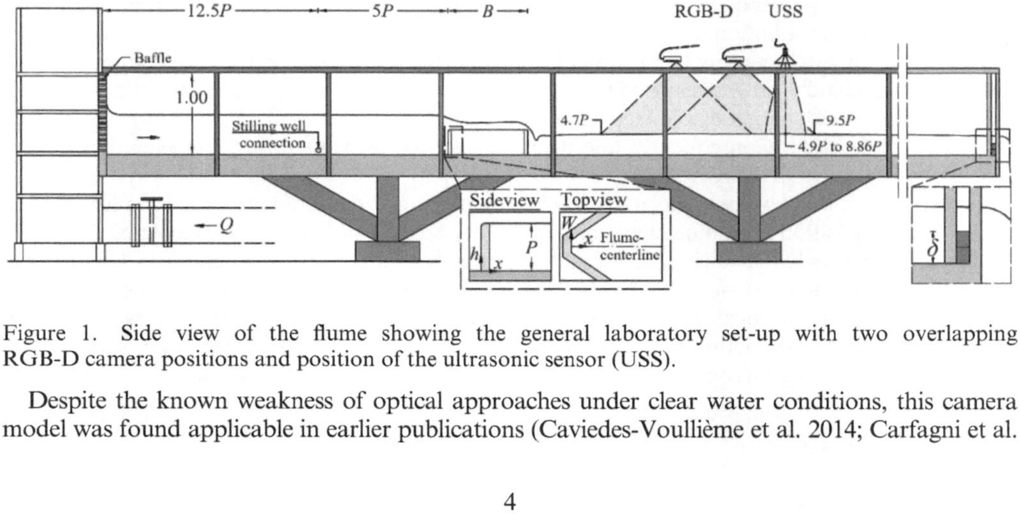

Figure 1. Side view of the flume showing the general laboratory set-up with two overlapping RGB-D camera positions and position of the ultrasonic sensor (USS).

Despite the known weakness of optical approaches under clear water conditions, this camera model was found applicable in earlier publications (Caviedes-Voullième et al. 2014; Carfagni et al.

2019; Bung et al. 2021; Crookston and Bung 2022; Panthi and Crookston 2024). The investigation area was between 4.7*P* and ~9.5*P* downstream of each weir. Besides the selected optical approach, ultrasonic displacement sensors (USSM, Microsonic™ mic+130/IU/TC; Zhang et al. (2018)) were installed for additional pointwise comparison at ±0.21 m from the centerline. The footprint size was estimated with ±7.5 cm around each specific position. The experimental set-up including the device positioning in longitudinal direction is illustrated in Figure 1.

All measurements were postprocessed according to the filtering approach proposed by Valero et al. (2020) for free surface turbulence characterization. Thus, filtering technique was applied pixelwise to each of the 480×848 pixels, considering the 90 Hz recording frequence over a sampling time of 60 seconds. Subsequently, each recorded depth map was averaged and converted to a metric point cloud by using the intrinsic camera parameters of the focal length and the principal offset point. Finally, all results were reworked into a uniform coordinate system by taking the relative device position into account.

3 RESULTS

3.1 *Free surface profiles*

Flow conditions downstream of labyrinth weirs are known as complex and three-dimensional. However, the recorded spatial data from the RGB-D camera were reduced to one longitudinal section. For comparison to the USS data, the longitudinal axis at $W = 0.21$ m was selected. Consequently, in Figure 2 only data from this section is shown (-0.21 m were found to be similar with neglectable differences).

Maximum detected water surface values from the camera data inside the estimated footprint were selected for comparison to USS data. Therefore, an estimated size of ±7.5 cm

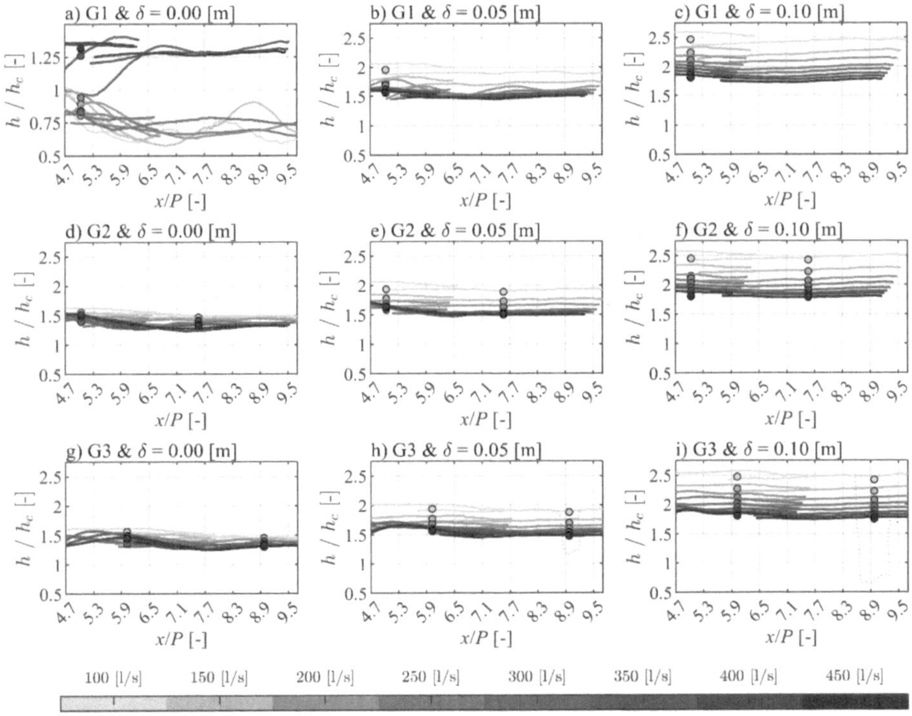

Figure 2. Comparison of the relative free surface as a ratio of the free surface (h) and the critical flow depth (h_c) along the relative downstream position (x/P). USS data are included as points. Local deviations due to false detection around 8,9P are highlighted with a dashed line.

perpendicular to the specific position was considered as footprint. All measurements are illustrated in Figure 2 as a ratio of the free surface (h) and the critical flow depth (h_c) alongside the relative downstream position (x/P).

When comparing the USS and camera data, generally consistent results across all test scenarios were found. Except for the first testing configuration (G1; δ = 0; Figure 2a) all h/h_c ratios were greater or around 1.25. However, the geometry G1 combined with a free outlet condition (δ = 0) was the only scenario with clear shockwaves, resulting in highly uneven free surface profiles.

Independent of the longitudinal section, the camera positioning (in flow direction) becomes more significant when investigating such highly uneven free surface profiles. Furthermore, a weaker quality is expected when the region of interest is located at the edges of the field of view (FoV). Consequently, all edges should be considered carefully, which generally applies to optical approaches. Throughout the measurements this leads to minor deviations when overlapping the results of two camera positions (i.e. Figure 2c, 2e, 2f). However, the deviations are found to decrease with increasing Q and δ.

When calculating the averaged instead of the maximum water surface profile along the selected width band of ± 7.5 cm, the results show negligible differences with one exception. The cases with Q = 100 l/s in Figure 2h and 2i show abrupt local changes at x/P = 8.86 (dashed line). This is caused by a combination of two factors. First, less aeration was present, which leads to a weaker contrast and detection by the RGB-D camera and second, through that weak turbidity the installed high contrast positioning markers (bright orange dots) stand out from the homogeneous relatively dark coloring of the flume (blue) resulting in a locally limited false depth detection. As a result, a homogeneous coloring of the background is recommended.

It is noteworthy that both geometries G2 and G3 in Figures 2d and 2g barely differ in their relative free surface amplitudes when comparing them under free outlet conditions. Furthermore, the increasing downstream boundary condition (δ) leads to smoother relative free surface profiles across all discharges and geometries.

Figure 3 exemplifies the observed shockwaves in a top (3a) and a side view (3b) for one selected testing configuration (G1; δ = 0; Q = 100 l/s). The gradients of the free surface at the selected positions along the upstream and downstream apex as well as the longitudinal sections along the USS positions are illustrated. Despite these longitudinal sections, Figure 3a highlights the equivalent depths measured by the USS within the camera field of view.

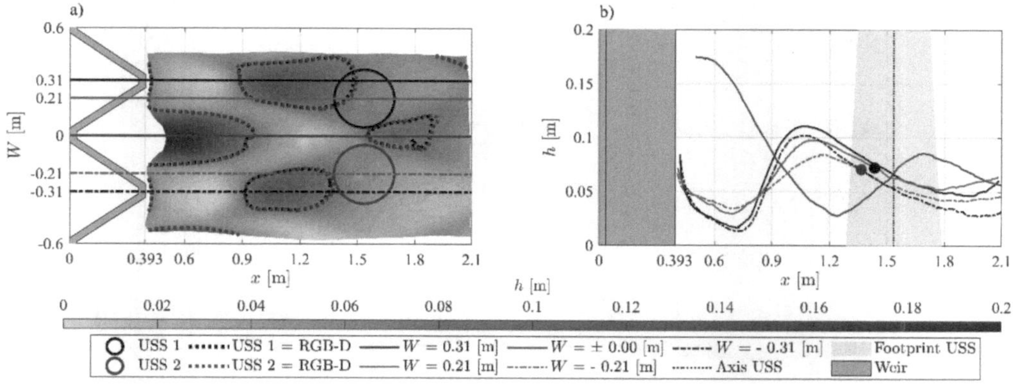

Figure 3. Influence of the longitudinal position under shockwave conditions (G1; δ = 0; Q = 100 l/s).

Regardless of the previously described observations and influences, generally consistent results were found. To conclude, the selected Intel® RealSense™ Depth Camera D455 seems applicable for measuring time averaged three-dimensional free surface profiles downstream of labyrinth weirs. Yet, data should be carefully validated with established measuring methods. This is generally recommended when a strong contrast in the background and less turbidity is observed.

3.2 *Free surface fluctuations*

The free surface fluctuations were used as an indicator of the turbulence downstream of labyrinth weirs. Therefore, time dependent fluctuations were decomposed into a time-averaged free surface (h) and the standard deviation of the timeseries for each pixel (h').

Figure 4 summarizes the recorded camera data of the relative free surface fluctuations ($h'/\bar{h}$) along the relative downstream position (x/P). To reduce averaging effects, the selected longitudinal width band was reduced from ±7.5 cm (Figure 2) to ± 5 mm. The longitudinal section at 23.5 cm from centerline was selected, which is located inside the USS footprint but neglects the influence of the false depth detection caused by the positioning markers. Furthermore, the observed shock waves as well as the vertical amplitudes of the smallest discharges are less relevant. Consequently, these lowest discharges were excluded in each testing configuration.

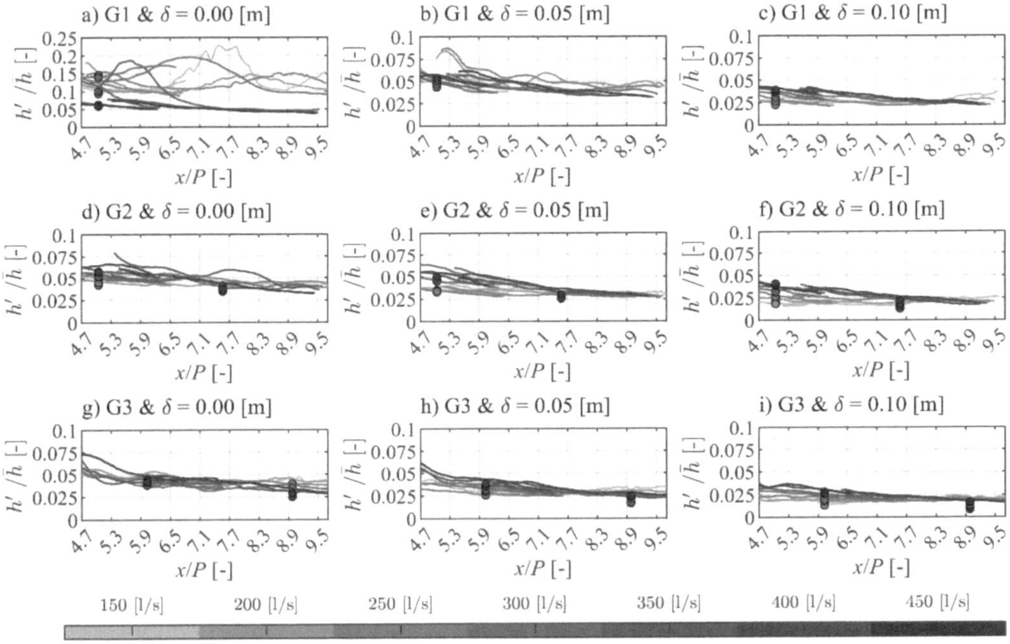

Figure 4. Comparison of the relative free surface fluctuations expressed as a ratio of $h'/\bar{h}$ along the relative downstream position (x/P). Position of the longitudinal section $W = 23.5$ cm.

Except for the first testing configuration shown in Figure 4a, generally decreasing relative free surface fluctuations were observed downstream of the investigated weirs. However, this testing configuration shows the highest changes of $h'/\bar{h}$, especially at lower discharges. These differences were caused by the high changes in the free surface rather than the relatively small amplitudes of the fluctuating component at the crest of the shock waves. Consequently, the analysis of the free surface fluctuations under shock wave conditions seems not appropriate.

When comparing the results for each tested labyrinth weir geometry (each column) and different downstream conditions (each row) both influence the relative free surface fluctuations. The comparison of Geometry G1 in Figure 4a, 4b and 4c shows significant changes in the downstream flow pattern. For instance, the free outlet condition (4a) can be summarized by differently strong pronounced shockwaves with $h'/\bar{h}$ ratios up to ~ 0.225. On the other hand, the downstream condition with δ = 0.1 m shows much smaller $h'/\bar{h}$ ratios. Such a decrease is visible across all geometries when increasing the downstream conditions. However, when comparing different geometries under free flow conditions, major differences are visible as well. For example, the largest fluctuations at the geometries G2 (4d) and G3 (4g) were generally lower than 0.075 and regardless of the exact value, this is approximately three times lower than the $h'/\bar{h}$ ratios from geometry G1 (4a). On the other hand, the results of higher downstream conditions show a convergence along x/P with values mostly lower than 5% of the free surface. To conclude, decreasing sidewall angles (α) as well as increasing downstream boundary conditions decrease the relative free surface fluctuations. In this context, the influence of α seems smaller than the influence of δ which is primarily caused by the increase of $\bar{h}$.

To build on these results, a deeper analysis for two instantaneous measurements was conducted. For each investigated configuration, one pixel at the intersecting centerline of the USS was extracted. Figure 5 compares both the instantaneous absolute free surfaces (h_{inst}) as well as the fluctuations around the median value. A comparison of the USS and the camera data in Figure 5a shows an offset of around 11.8 mm when comparing the averaged results. Considering the flow conditions, an overestimation by the USS is not surprising. Due to the nature of reflective approaches as well as the USS footprint size thus results are biased towards the foreground. This averaging effect shifts the results towards the crest of the shock wave. For completeness, the estimated footprint is illustrated in Figure 3b.

In contrast, the free surface measured by the USS shows a 5.8 mm lower free surface than the camera (Figure 5b). While a turbulent surface is crucial for good results of the optical approach, the USS is known to penetrate aerated flows and provide results at a level of 60 to 85% void fraction. This may be an explanation for thus difference (Bung et al. 2021; Bung 2013; Kucukali and Chanson 2008; Zhang et al. 2023).

Both devices show higher standard deviations under shock wave conditions. Regardless of the investigated configuration, the flow calms down sufficiently downstream of the weir. This is notable in Figure 5d with a harmonization of the histograms and decreasing standard deviations. Apart from the relatively good agreement between the fluctuations in Figure 5c and 5d, the limited response time and the smoothening effect caused by the footprint should be considered carefully. This smoothing effect is visible when comparing other results in Figure 4 and it is accompanied by the tendency to slightly underestimate the free surface fluctuations in such three-dimensional conditions. However, the results of the USS may serve as a validation for the free surface fluctuations. A further comparison with phase detection probes may be helpful to reference the camera data to a specific void fraction. However, such validation is beyond this manuscript.

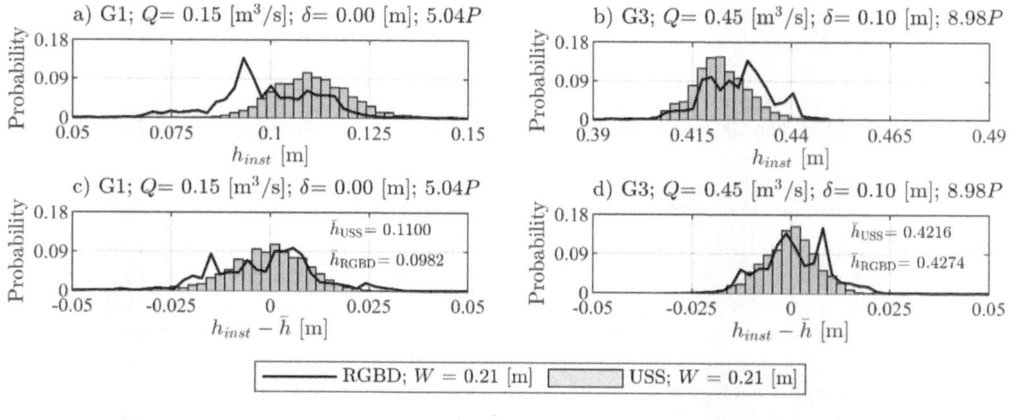

Figure 5. Normalized histograms (probability distributions) of recorded data from two different configurations. With a) and b) as absolute flow depth and c) and d) as fluctuations around the median value.

4 CONCLUSIONS

A laboratory-scale study with three trapezoidal labyrinth weirs was conducted. Each geometry was investigated considering three downstream conditions ($0 < \delta < 0.10$ m) and eight discharges ($0.1\ m^3/s \leq Q \leq 0.45\ m^3/s$). The Intel® RealSense™ Depth Camera D455 as well as the mic+130/IU/TC ultrasonic sensors were used for data acquisition up to a position of 8.9P downstream. The recorded spatial data from the RGB-D camera were reduced to longitudinal sections inside the footprint of the USS. Averaged free surfaces as well as instantaneous fluctuations were quantified. A comparison of the devices showed that the selected optical approach is capable to produce similar results. However, special care must be taken throughout the data acquisition process. It was found that a sufficient contrast provided by the free surface as well as a homogeneous background color is crucial for consistent results. When observing weak turbidity or high contrasts in the background, a synchronized validation with other measuring methods is recommended. Furthermore, the results of optical approaches are highly sensitive to misalignments. Such deviations increase towards the edges of the field of view and must be added to the general weakness of optical approaches at the edges. In this context a careful positioning of the camera is essential. One solution to increase the confidence of results, especially at the edges, may be the use of multiple synchronized cameras with a large overlapping field of view. However, an attempt was made to describe the flow features downstream of labyrinth weirs by analyzing the free surface profiles and their instantaneous fluctuations. For both, a general harmonization along the downstream position was observed. Thus, harmonization was mainly influenced by the weir geometry and the downstream boundary condition. Despite some minor deviations, generally consistent results were observed throughout this study.

REFERENCES

Anderson, R. & Tullis, B. 2013. Piano Key Weir Hydraulics and Labyrinth Weir Comparison. *J. Irrig. Drain Eng.* 139 (3): 246–253.

Belay, B. & Oertel, M. 2025. Void Fraction Distribution Downstream of Nonlinear Weirs: An Experimental Investigation Using Trapezoidal A-Type Piano Key Weirs, a Labyrinth Weir, and a Linear Weir. *J. Hydraul. Eng.* 151 (3): 04025006.

Belzner, F.; Merkel, J.; Gebhardt, M. & Thorenz, C. 2017. Piano Key and Labyrinth Weirs at German waterways: Recent and future research of the BAW. Labyrinth and Piano Key Weirs III – PKW 2017: 167–174. CRC Press: London.

Bung, D. 2013. Non-intrusive detection of air–water surface roughness in self-aerated chute flows. *J. Hydraul. Res.* 51 (3): 322–329.

Bung, D.; Crookston, B. & Valero, D. 2021. Turbulent free-surface monitoring with an RGB-D sensor: the hydraulic jump case. *J. Hydraul. Res.* 59 (5): 779–790.

Carfagni, M.; Furferi, R.; Governi, L.; Santarelli, C.; Servi, M.; Uccheddu, F. & Volpe, Y. 2019. Metrological and Critical Characterization of the Intel D415 Stereo Depth Camera. Sens. 19 (3).

Carrillo, J.; Matos, J. & Lopes, R. 2019. Numerical modeling of free and submerged labyrinth weir flow for a large sidewall angle. *Environ. Fluid Mech.* 20: 357–374.

Caviedes-Voullième, D.; Juez, C.; Murillo, J. & García-Navarro, P. 2014. 2D dry granular free-surface flow over complex topography with obstacles. Part I: experimental study using a consumer-grade RGB-D sensor. *Comput. & Geosci.* 73 (0098-3004): 177–197.

Crookston, B. 2010. Labyrinth Weirs. All Graduate Theses and Dissertations. Paper 802. Logan: USU Library.

Crookston, B. & Bung, D. 2022. Application of RGB-D cameras in hydraulic laboratory studies. 39th IAHR World Congress: 5127–5133. IAHR: Spain.

Crookston, B.; Erpicum, S.; Tullis, B. & Laugier, F. 2019. Hydraulics of Labyrinth and Piano Key Weirs: 100 Years of Prototype Structures, Advancements, and Future Research Needs. J. Hydraul. Eng. 145 (12).

Crookston, B. & Tullis, B. 2012. Labyrinth Weirs: Nappe Interference and Local Submergence. *J. Irrig. Drain Eng.* 138 (8): 757–765.

Dabling, M. & Tullis, B. 2017. Modifying the downstream hydrograph with staged labyrinth weirs. J. Appl. Water Eng. Res. 6 (3): 183–190.

Erpicum, S.; Machiels, O.; Archambeau, P.; Dewals, B. & Pirotton, M. 2011. Energy dissipation on a stepped spillway downstream. Labyrinth and Piano Key Weirs I – PKW 2011: 105–111. CRC Press: London.

Eslinger, K. & Crookston, B. 2020. Energy Dissipation of Type a Piano Key Weirs. Water 12 (5): 1253.

Gebhardt, M.; Herbst, J.; Merkel, J. & Belzner, F. 2019. Sedimentation at labyrinth weirs – an experimental study of the self-cleaning process. *J. Hydraul. Res.* 57 (4): 579–590.

Giancola, S.; Sala, R.; Valenti, M. 2018. *A Survey on 3D Cameras: Metrological Comparison of Time-of-Flight, Structured-Light and Active Stereoscopy Technologies*. Switzerland: Springer.

Hager, W.; Pfister, M. & Tullis, B. 2015. Labyrinth weirs: Developments until 1985. 36th IAHR World Congress IHE Delft: Delft.

Hay, N. & Taylor, G. 1970. Performance and Design of Labyrinth Weirs. J. Hydraul. Div. 96 (11): 2337–2357.

Herbst, J.; Gebhardt, M.; Merkel, J.; Belzner, F. & Thorenz, C. 2018. Sediment Transport over Labyrinth Weirs. 7th IAHR International Symposium on Hydraulic Structures: 489–497. USU Library: Logan.

KC, M.; Crookston, B.; Flake, K. & Felder, S. 2025. Enhancing flow aeration on an embankment sloped stepped spillway using a labyrinth weir. J. Hydraul. Res. 63 (1): 32–47.

KC, M.; Crookston, B. & Bung, D. 2025. Capturing Free Surface Dynamics of Flows over a Stepped Spillway Using a Depth Camera. Sens. 25 (8).

Kramer, M.; Chanson, H. & Felder, S. 2020. Can we improve the non-intrusive characterization of high-velocity air–water flows? Application of LIDAR technology to stepped spillways. J. Hydraul. Res. 58 (2): 350–362.

Kucukali, Serhat & Chanson, Hubert 2008. Turbulence measurements in the bubbly flow region of hydraulic jumps. Exp. Therm. Fluid Sci. 33 (1): 41–53.

Lachat, E; Macher, H.; Landes, T. & Grussenmeyer, P. 2015. Assessment and Calibration of a RGB-D Camera (Kinect v2 Sensor) Towards a Potential Use for Close-Range 3D Modeling. Remote Sens. 7 (10): 13070–13097.

Li, R.; Splinter, K. & Felder, S. 2021a. Aligning free surface properties in time-varying hydraulic jumps. Exp. Therm. Fluid Sci. 126: 110392.

Li, Rui; Splinter, Kristen D. & Felder, Stefan 2021b. LIDAR Scanning as an Advanced Technology in Physical Hydraulic Modelling: The Stilling Basin Example. Remote Sens. 13 (18): 3599.

Lopes, R.; Matos, J. & Falcao de Melo, J. 2006. Discharge capacity and residual energy of Labyrinth weirs. 1st International Junior Researcher and Engineer Workshop on Hydraulic Structures: 47–55. The University of Queensland: Brisbane.

Lopes, R.; Matos, J. & Melo, J. 2008. Characteristic depths and energy dissipation downstream of a labyrinth weir. 2nd International Junior Researcher and Engineer Workshop on Hydraulic Structures Pisa University Press: Italy.

Merkel, J.; Belzner, F.; Gebhardt, M. & Thorenz, C. 2018. Energy Dissipation Downstream of Labyrinth Weirs. 7th IAHR International Symposium on Hydraulic Structures: 508–517 USU Library: Logan.

Montano, L.; Li, R. & Felder, S. 2018. Continuous measurements of time-varying free-surface profiles in aerated hydraulic jumps with a LIDAR. EFT science 93: 379–397.

Panthi, Manisha & Crookston, Brian M. 2024. Self-Cleaning Ability of the Piano Key Weir with Driftwood Blockages. J. Hydraul. Eng. 150 (4): 04024022.

Said, M. & Ouamane, A. 2021. Performance of rectangular labyrinth weir – an experimental and numerical study. Water Supply: ws2022005.

Shen, Xiaoyang 2023. Piano-Key-Weirs as in-channel applications: Flow characteristics and geometry optimizations. Hamburg: Helmut-Schmidt-University.

Tullis, B.; Young, J. & Chandler, M. 2007. Head-Discharge Relationships for Submerged Labyrinth Weirs. J. Hydraul. Eng. 133 (3): 248–254.

Tullis, B. P.; Crookston, B. M. & Young, N. 2020. Scale Effects in Free-Flow Nonlinear Weir Head-Discharge Relationships. J. Hydraul. Eng. 146 (2): 4019056.

Valero, D.; Felder, S.; Kramer, M.; Wang, H.; Carrillo, J.; Pfister, M. & Bung, D. 2024. Air–water flows. J. Hydraul. Res. 62 (4): 319–339.

Valero, Daniel; Chanson, Hubert & Bung, Daniel B. 2020. Robust estimators for free surface turbulence characterization: a stepped spillway application. Flow Meas. Instrum. 76: 101809.

Zhang, G.; Valero, D.; Bung, D. B. & Chanson, H. 2018. On the estimation of free-surface turbulence using ultrasonic sensors. Flow Meas. Instrum. 60: 171–184.

Zhang, L.; Zhang, F.; Xu, W.; Bo, H. & Zhang, X. 2023. An innovative method for measuring the three-dimensional water surface morphology of unsteady flow using light detection and ranging technology. Ocean Eng. 276:114079.

Labyrinth and Piano Key Weirs IV – Erpicum, Crookston & Paxson (Eds)
© 2026 The Author(s), ISBN 978-1-041-10527-5

Interfacial velocity characteristics downstream of a piano key weir versus a linear weir in a horizontal channel

B.S. Belay and M. Oertel
Helmut-Schmidt-University, Hydraulic Engineering Section, Hamburg, Germany

ABSTRACT: The downstream flow characteristics of piano key weirs (PKWs) are highly complex, manifested by significant localized self-aeration due to nape collisions and plunging jets impacting the downstream free-surface. For PKWs installed in horizontal channels, the interfacial velocity characteristics have received limited attention in previous research. In this experimental study, air-water flow measurements were conducted using a double-tip conductivity probe downstream of a PKW and a reference ogee-crested linear weir to evaluate the velocity trends under comparable hydraulic boundary conditions. The findings reveal the existence of two distinct velocity profiles downstream of piano key weirs: quasi-wall-jet and uniform profiles. The occurrence of these profiles depends on both the longitudinal position relative to the weir crest and the inflow magnitude. At higher discharges, wall-jet profiles dominate near the jet impact zone, transitioning into uniform flow profiles further downstream. Conversely, at lower discharges, uniform velocity distributions persist throughout the aerated region. Notably, these quasi-wall-jet profiles exhibit moderate similarities with hydraulic jump characteristics observed downstream of the linear weir, particularly in the vertical positioning of the maximum velocity, which resides close to the channel bed. Furthermore, evaluation of the longitudinal decay of maximum time-averaged velocities indicates a comparable exponential trend for both weir models. However, the PKW exhibits lower velocity magnitudes due to enhanced energy dissipation mechanism by the plunging jets and nappe interactions, distinguishing them from the hydraulic jump behavior of the linear weir. Overall, these insights contribute to the understanding of PKW hydraulics, particularly on the downstream interfacial velocity properties in horizontal channel applications.

1 INTRODUCTION

1.1 *Self-aeration mechanisms in piano key weirs*

Piano key weirs feature a highly complex flow field in the downstream region due to their nonlinear geometric attributes. A subcritical approaching flow distributes over their nonlinear crest, which transitions into a supercritical flow regime entering the tailrace channel. Such flow characteristics make piano key weirs effective in energy dissipation, particularly at low heads (Eslinger and Crookston 2020). The overall flow behavior regarding discharge capacity and energy dissipation is also reported to be influenced by the shape of the piano key weir crest, which should be considered (Shen and Oertel 2023). The interaction between side nappes within the outlet keys of piano key weirs and the overfalling plunging jets from their inlet keys results in the localized self-aeration in the downstream channel. A recent experimental study by Belay and Oertel (2025) depicted that the combined effect of weir geometry and the hydraulic boundary conditions greatly dictate the mechanism of air entrainment downstream of nonlinear weirs. The authors also comparatively analyzed the scale of air entrainment for piano key weirs against a reference ogee-crested linear weir. Their results outlined the prevalence of air entrainment by the hydraulic jump downstream of the linear weir over the plunging jets of the piano key weir.

DOI: 10.1201/9781003655541-2

As part of the nonlinear weir family, piano key weirs are being implemented as contemporary hydraulic structures instead of linear weirs due to their advantage in space constrained environments. Thus, understanding the resulting velocity characteristics for in-channel installations is crucial to supplement current design practices. Comparative evaluation of their velocity trends relative to a linear weir can be a vital input to enhance the current understanding and fill the missing gaps, particularly in the aerated region. Recent developments analyzed the overall aeration performance and velocity characteristics of a labyrinth weir placed as a control structure upstream of a stepped chute (Crookston et al. 2024, Megh et al. 2025). Previous experimental studies have also illustrated the velocity characteristics downstream of piano key weirs in a horizontal channel beyond the aerated region (Besser 2024). However, given the inherent nature of the downstream macro-scale turbulence induced by the plunging jets and the resulting shockwaves, characterization of the interfacial velocity distributions in the case of in-channel applications was not sufficiently researched.

1.2 *Interfacial velocity characteristics in hydraulic jumps*

Contrary to piano key weirs, or nonlinear weirs in general, the air entrainment mechanism downstream of linear weirs is governed by the hydraulic jump phenomenon. Such a process features an abrupt transition of a rapidly varied flow from supercritical to subcritical regime (Hager 1992). The formation of flow recirculation (rollers) due to the interaction of the impinging jet and the downstream tailwater derives a strong shear region that facilitates localized air entrainment (Chanson and Brattberg 2000). This event plays a crucial role in the process of energy dissipation downstream of sluice gates and linear weirs. The corresponding velocity characteristics in the jump toe vicinity are properly represented by a wall jet profile where maximum longitudinal velocities reside close to the channel bottom (Rajaratnam 1965). In horizontal channels, such wall jet profiles transition into an open channel flow type and to uniform flow as the distance further increases (Wu and Rajaratnam 1965, Liu et al. 2004).

A significant number of research has already been carried out that addressed a wide range of flow characteristics ranging from the basic to the advanced air-water flow characteristics. Particularly, several studies showcased the self-similar nature of the downstream interfacial velocity properties. A wall-jet analogy was derived by Rajaratnam (1965) that reproduced interfacial velocity profiles below the recirculating regions at jet impingement zones of classic hydraulic jumps.

$$\frac{u}{U_{\max}} = \left(\frac{z}{Z_{u,\max}}\right)^{\frac{1}{N}}, \text{for } zZ_{u,\max}^{-1} < 1 \tag{1}$$

where u denotes the time-averaged velocity at a vertical position z, $U_{\max}$ = the maximum time-averaged velocity at a vertical position of $Z_{u,\max}$, and based on previous studies N takes a value between 6 and 10 in open channels (Chanson and Brattberg 2000, Wang 2014). In addition, other modified wall-jet solutions of dimensionless velocity profiles were also suggested by previous research in which the velocity in the recirculation zone was considered (Ohtsu et al. 1990, Chanson and Brattberg 2000).

$$\frac{u}{U_{\max}} = \exp\left(-\frac{1}{2}\left[1.765\left(\frac{z - Z_{u,\max}}{Z_{0.5}}\right)\right]^2\right), \text{for} 1 < zZ_{u,\max}^{-1} < 4 \tag{2}$$

where $Z_{0.5}$= the vertical position where $u = 0.5 \times U_{\max}$.

For in-channel applications, piano key weirs (including other types of nonlinear weirs) are recently being implemented in several rehabilitation projects, replacing aged linear weirs. In such cases, understanding the downstream complex velocity characteristics plays a vital role in enhancing design practices, particularly related to energy dissipation and downstream protection works. Thus, this experimental investigation evaluates the resulting velocity distribution characteristics downstream of a piano key weir and compares them with a linear weir (an equivalent dimension) case under similar hydraulic conditions.

2 EXPERIMENTAL METHODS

2.1 *Test layout and model preparations*

This experimental study was conducted in a horizontal flume with dimensions of 20 m in length, 1 m in depth, and 1 m in height. The inflow discharges were monitored by a magnetic inductive flow meter (MID: Krohne™ Optiflux) with an accuracy of ±0.3 %. The downstream tailwater was controlled using a tailgate with adjustable flaps. Two ultrasonic sensors (microsonic®, mic+130/IU//tc, measurement range between 200 mm - 1300 mm) were utilized to measure the flow depths at positions in the upstream ($5\times P$, where P is the weir height) and the downstream at $15\times P$. Due to the inherent differences in their crest geometry, which derives the resulting hydraulic performance in discharge capacity, the flow depth upstream of a piano key weir is always smaller than the linear weir for the same inflow discharge or Reynolds number. As a result, it was fundamentally necessary to maintain similar hydraulic conditions by maintaining comparable relative flow depths between the upstream and downstream boundaries; thereby, the desired comparison of interfacial velocities between these two entirely different hydraulic structures is reasonable. Both weir models were prepared through 3D-printing techniques (BigRep One 3D-printer) using polylactide-based PLX filaments extruded layer by layer (layer height = 0.6 mm). These weir models were an A-type trapezoidal piano key weir with symmetric key width ratios (parapet walls with a half-round crest shape) and an ogee-crested standard linear weir of equivalent dimension.

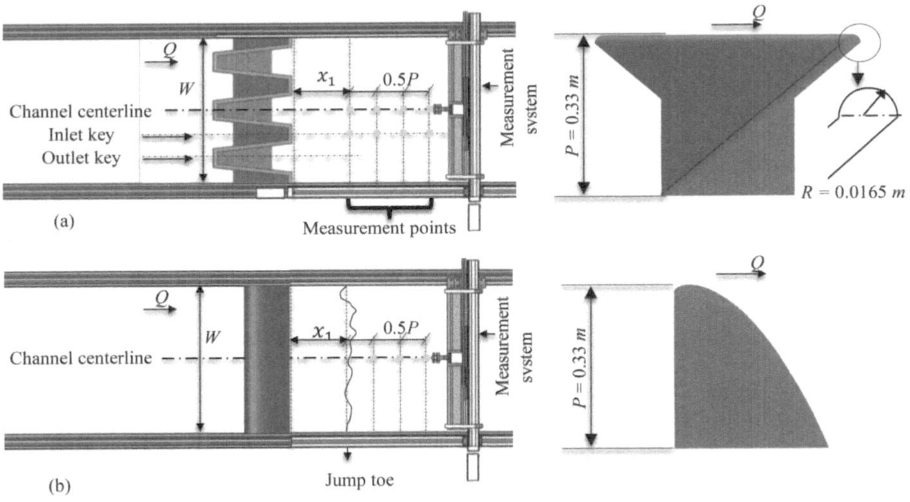

Figure 1. Experimental setup and weir models implemented.

For the piano key weir model, the location of x_1 was set at $1\times P$ (where P is the weir height). This location is at an optimum position from the impact of the plunging jets of the inlet keys. Based on a preliminary investigation, it was noted that the plunging jet impact position for each discharge could also be used as the reference location (x_1). However, the variation in the jet impact position for the inflows investigated in this experiment was negligible. Thus, a constant starting position of $x_1 = 1\times P$ was selected with the downstream weir crest as a reference for simplicity. For the case of the linear weir, after a comparable hydraulic condition is created, the approximate position of the jump toe is used as the relative reference location.

2.2 *Instrumentation and sampling characteristics*

The downstream two-phase flow characteristics were measured using a double-tip conductivity probe (Manufacturers of the probe and the measurement system: VAW, ETH-Zurich based on the design developed at WRL, UNSW (Felder et al. 2019)). The probe tip dimensions consist of an outer electrode diameter of 0.5 mm and a longitudinal spacing of 3.45 mm between the tips. A longitudinal array of measurements was conducted along the channel centerline for the linear weir case; for the piano key weir, two centerlines along the inlet and outlet keys were considered due to inherent cross-sectional differences caused by the complex flow (Figure 1). Before the main measurement campaigns, the optimum sampling parameters were selected to enable the acquisition of a statistically representative data set in the shortest time possible. This was achieved by comparing the results of four different sampling durations (45 s, 60 s, 90 s, and 120 s) at a sampling rate of 50 kHz. Based on this evaluation, the characteristics of the vertical velocity profiles were determined to be adequately represented with a sampling duration of 60 s. Despite the extreme turbulence near the channel bottom close to the plunging jet impact region, the scatter of the time-averaged vertical profiles near the channel bed did not improve, even with a sampling duration of 120 s. The pseudo-instantaneous (time-series) velocities were obtained from the raw conductivity probe data using the Adaptive Window Cross Correlation (AWCC) technique (Kramer et al. 2019). The freely available MATLAB code (https://github.com/MatthiasKramer/Phase-detection-signal-processing-toolbox) was utilized, which included recent developments for velocity bias correction against the intrusive nature of the conductivity probe (Hohermuth et al. 2021).

3 RESULTS AND DISCUSSION

3.1 *Time-averaged velocity profiles*

The macro-scale nature of turbulence downstream of the piano key weir, particularly near the jet impact zone, considerably induced a highly scattered velocity distribution across the flow depth (Figure 2a). This influence was reflected in the data processing stage because the accuracy of velocity estimations is highly dependent on the strength of signal cross-correlation between the leading and trailing tips of the conductivity probe. The intensity of transversal flow disturbance due to the shockwaves induced by the plunging jets influenced the interaction between air bubbles and the horizontally aligned conductivity probe. This effect was pronounced along the outlet key centerline ($yW^{-1} = 0.5$), where the formation of standing waves from sideways spreading shockwaves of adjacent plunging jets is dominant. Such lateral shockwave propagation forced the probe tips to interact with air bubbles obliquely, affecting signal correlation. The results demonstrated that the interfacial velocity downstream of the piano key weir has two modes of vertical profiles along the flow. A quasi-wall-jet profile (Eq. 3) was observed for a longitudinal position ranging between the jet impact zone and $xP^{-1} = 2.2$ (Figure 2a). As the intensity of three-dimensional turbulence and the shockwave influence reduced further downstream, the vertical distributions moderately resembled a uniform velocity profile (Figure 2c). An attempt was also made to compare these far field velocity profiles ($xP^{-1} > 2.2$) based on the classic power-law profile (Eq. 4) with $N = 15$, however, the results (Figures 2c, d) illustrated that a vertical scaling factor ($\alpha \neq 1$) is required to describe the profiles in such relationship, depicting the profiles are not adequately following a true power-law relationship. Thus, uniform velocity trend is considered as a better descriptor in such cases.

$$\frac{u}{u_c} = 1.25\left(\frac{z}{h_c}\right)^{\frac{1}{12}} - 0.4\left(\frac{z}{h_c}\right)^{1.2}, \text{for the piano key weir}: \ xP - 1 \leq 2.2 \qquad (3)$$

$$\frac{u}{u_c} = \alpha\left(\frac{z}{h_c}\right)^{\frac{1}{N}}, \text{for the piano key weir}: \ xP - 1 > 2.2 \text{and} xP - 11.7 \text{ (at low discharges)} \qquad (4)$$

$$\frac{u}{u_c} = 2\left(\frac{z}{h_c}\right)^{\frac{1}{12}} - 1.4\left(\frac{z}{h_c}\right)^{1.2}, \text{for the linear weir}: \ 1 \leq xP - 1 \leq 2.7 \qquad (5)$$

where u = time-averaged interfacial velocity at a vertical position z, u_c = critical velocity, and h_c = critical flow depth. α is a vertical scaling factor, where a value of 1 represents a true power-law relationship. Equations 3 and 4 represent empirical approximations of the two modes of vertical velocity profiles observed downstream of a piano key weir. Scaling based on critical flow depth and critical velocity allowed for a reasonable comparison between the linear weir and the piano key weir. It is important to note that the scaling based on other flow parameters, such as, the maximum velocity ($U_{\max}$ at $Z_{\max}$) and the characteristic flow depth (U_{90} at Z_{90}, where the void fraction C is 0.9) did not consistently allow comparing the dimensionless velocity profiles of both the linear weir and the piano key weir.

The comparable hydraulic conditions were maintained for both weir models based on the dimensionless relative flow depth $\Delta h_{\text{rel}} = 1\text{–}h_2/h_1$, where h_1 and h_2 denote the flow depth upstream ($x = 5\times P$) and downstream ($x = 15\times P$) of the weir, respectively. For the piano key weir considered in this study, the distinction between the near-field quasi-wall-jet velocity profiles and the far-field uniform profiles is dependent on the longitudinal position for larger inflows. However, at low discharges, the impact of plunging jets on the formation of such wall-jet-like profiles is reduced, and the velocity distributions take a moderately uniform trend throughout the downstream channel. Although their three-dimensional flow characteristics between the linear weir and piano key weir are inherently different, the wall-jet trend of the time-averaged vertical velocity profiles downstream of the piano key weir showed some degree of similarity with the linear weir for $1 \le xP^{-1} \le 2.7$ (Figures 2a, b). This quasi-wall-jet profile downstream of the linear weir, scaled with the critical velocity and critical flow depth, showed a consistent representation of the dimensionless velocity profile for $zh_c^{-1} \le 1$. The results illustrated that such a trend conforms to the region below the flow recirculation or shear flow region ($zh_c^{-1} \le 1$ where $uu_c^{-1} \ge 0$).

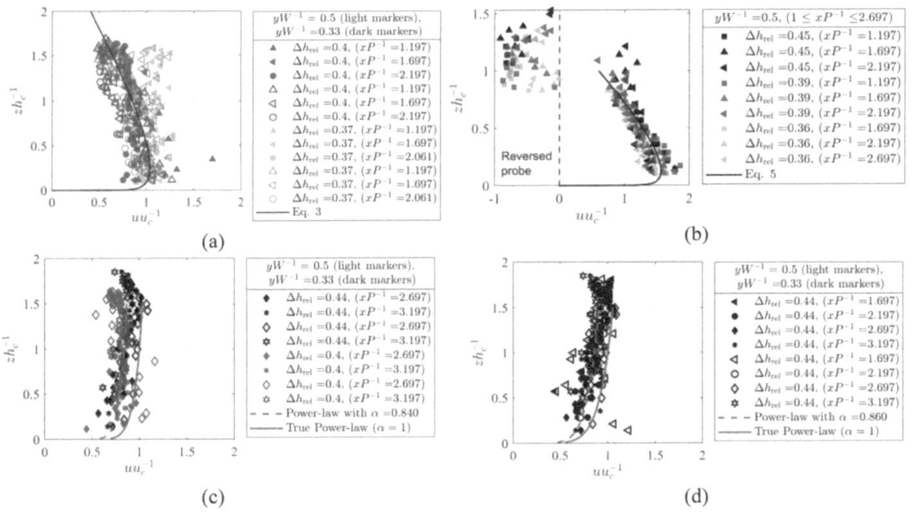

Figure 2. Vertical velocity profiles: (a) near-field quasi-wall-jet trend for the piano key ($xP^{-1} \le 2.2$), (b) quasi-wall-jet profile for the linear weir along the channel centerline ($1 \le xP^{-1} \le 2.7$), (c) far-field uniform flow trend for the piano key weir ($xP^{-1} > 2.2$), and (d) uniform flow trend along the flow for the piano key weir at low discharge.

Additionally, below the recirculation zone, the velocity profiles downstream of the linear weir were reproduced with a previously developed relationship (Figure 3) in which good fits were obtained with the wall jet solution proposed in literature (Eq. 2). This outcome is also relatively consistent with previous remarks on classic hydraulic jumps (Wang 2014) where good representation of velocity profiles with a wall jet solution is reported. Furthermore, the average dimensionless recirculation velocities ($U_{\text{rec}}\ U^{-1}_{\max}$) in the shear flow regions downstream of the hydraulic jump were within 0.26 to 0.48. Although the current measurements feature inflow Froude numbers ranging from 2.8 to 3.2, the results also depicted the presence of self-similarity with previously observed hydraulic jump flows (Chanson 2010), where an average value of 0.4 was reported for a hydraulic jump case with inflow Froude numbers of 9.2 and 10. On the other hand, the velocity profile characteristics of the piano key weir could not be well described with previously established relationships, mainly the wall jet solution presented in Equation 2. This dissimilarity is mainly attributed to the presence of highly three-dimensional flow due to the complex jet interactions caused by the inlet and outlet keys, as a result, the cross-sectional turbulence properties remain significant, affecting the nature of vertical velocity profiles and the corresponding longitudinal trends.

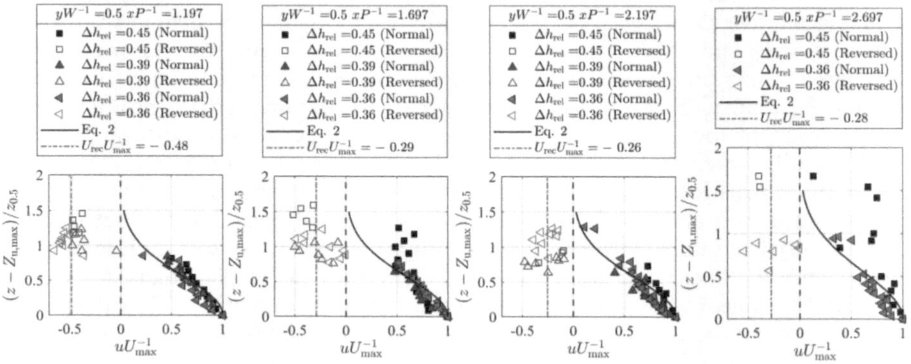

Figure 3. Comparison of vertical velocity profiles downstream of the linear weir with previous approximations in literature. Data includes measurements with normal probe alignment (facing upstream) and reversed alignment (facing downstream).

3.2 *Decay of maximum velocity*

For the linear weir, the flow parameters before (but close to) the jump toe (d_1: measured depth and u_1: calculated inflow velocity) were used to normalize the maximum depth-averaged interfacial velocity (U_{max}) and the corresponding longitudinal position $x - x_1$, where x_1 is the approximate position of the jump toe from the weir toe. However, for the piano key weir, because of the absence of clear hydraulic jump, the downstream measured flow depth (h_2) at $15 \times P$ and the upstream inflow discharge were used to calculate the equivalent inflow velocity (u_1) and depth (d_1) based on the momentum balance (Belanger equation) assuming negligible friction. It is imperative to note that these pre-jump properties (u_1 and d_1) do not physically exist in the same way as in the case of the linear weir due to the inherent differences in flow characteristics; however, these values are idealized as equivalent parameters to enhance a direct comparison of the longitudinal interfacial velocity trends between these two different weirs. Downstream of both weir models, the longitudinal decay of the maximum time-averaged velocity in a horizontal channel depicted an exponential trend along the flow (Figures 4a, b). Small differences (yet closely similar trends) were observed regarding the reduction rate between these two cases considered in the current investigation. This is illustrated in the coefficients of the empirically derived relationships based on the experimental data (Eqs. 6, 7). Because of the enhanced energy dissipation by the plunging jets and nape interactions, the maximum interfacial velocity near the jet impact zone of the piano key weir ($x = x_1 \sim 1 \times P$) was about 74% of the calculated equivalent inflow velocity (u_1). This corresponds to smaller magnitudes of velocity throughout the downstream channel compared to the classic hydraulic jump past the linear weir. Comparatively, the maximum interfacial velocity at the jump toe of the linear weir ($x = x_1$ ~jump toe location) was close to 90% of the nonaerated (monophase) depth-averaged inflow velocity upstream of the jump (u_1).

$$\frac{U_{max}}{u1} = 0.74 \exp\left[-0.051 \frac{(x - x_1)}{d_1}\right], \text{for the piano key weir} \quad (6)$$

$$\frac{U_{max}}{u_1} = 0.9 \exp\left[-0.043 \frac{(x - x_1)}{d_1}\right], \text{for the linear weir} \quad (7)$$

where x = absolute location along the flow from the downstream weir crest for the piano key weir and from the weir toe for the linear weir, x_1 = first measurement point from the weir crest for the piano key weir, and from the jump toe for the linear weir. The results in Figure 4 also depict that the longitudinal decay of the maximum velocity downstream of the piano key weir is significantly scattered (as shown by the small correlation with the empirical relationship). This outcome is related to the presence of strong three-dimensional flow compared to the linear weir, where the cross-sectional nonuniformities relative to the channel centerline are minimal.

Furthermore, a qualitative comparison of the downstream flow characteristics is shown in Figure 5 using side photos taken at the same hydraulic conditions for both weir models. The pictures depict the key differences in the overall longitudinal profile of the aerated region. It must be noted that, to impose comparable hydraulic conditions at the same inflow discharge (or Reynolds number), the tailwater downstream of the linear weir needs to be raised such that the dimensionless relative flow depths Δh_{rel} remain the same in both models. However, suppose the comparison is directly based on similar inflow Reynolds numbers and similar downstream flow depths. In that case, the hydraulic jump position significantly moves further downstream of the linear weir toe, leading to an extended section of the supercritical pre-jump region. This characteristic further informs the prevalence of energy dissipation downstream of the piano key weir compared to the linear weir.

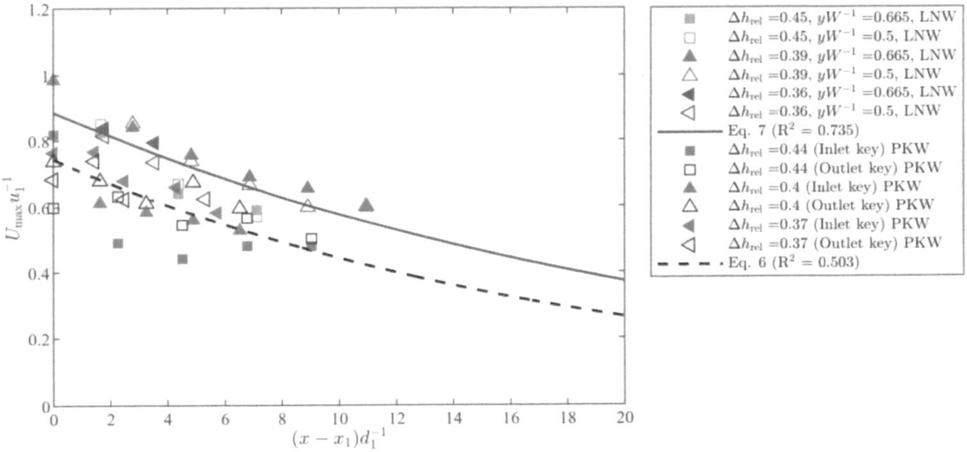

Figure 4. Longitudinal trend of the maximum velocity along the flow for the linear weir (LNW) and the piano key weir (PKW).

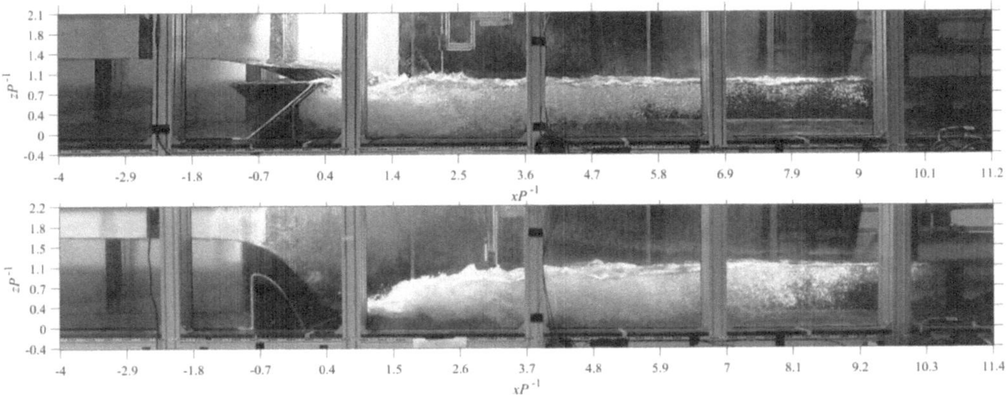

Figure 5. Picture presentation of comparable hydraulic conditions at q = 0.265 m^2/s, Δh_{rel} = 0.37: piano key weir (top) and linear weir (bottom).

4 CONCLUSIONS

The two-phase flow characteristics downstream of nonlinear weirs, particularly piano key weirs installed in horizontal channels, have not been extensively studied in previous research. Understanding this complex aerated flow field is crucial for improving future design practices in in-channel applications. This experimental study comparatively examines the downstream interfacial velocity characteristics of piano key weirs relative to an ogee-crested linear weir with equivalent dimensions. The results indicate that two distinct modes of velocity profiles persist downstream of piano key weirs: quasi-wall-jet profiles and uniform profiles. The type of trend depends on the location relative to the weir crest and the inflow discharge magnitude. At higher discharges, wall-jet profiles emerge near the weir and continue over a limited distance before transitioning to uniform profiles. In contrast, at lower discharges, uniform profiles remain dominant throughout the downstream aerated region. Additionally, scaling these velocity profiles based on the critical flow depth and critical velocity seems to depict better normalization across all flow conditions when comparing the linear weir and piano key weir. The observed quasi-wall-jet profiles resemble those found in hydraulic jumps downstream of linear weirs, particularly in the vertical positioning of maximum velocity, which occurs near the channel bed. The velocity profiles downstream of the linear weir also depicted self-similarity with previous hydraulic jump studies. The regions below the recirculation zone were sufficiently represented by the modified wall-jet solution. Additionally, the longitudinal decay of maximum time-averaged velocity follows a similar exponential trend for both weir models. However, velocity magnitudes differ, with piano key weirs exhibiting lower values due to enhanced energy dissipation from plunging jets and nape interactions compared to the hydraulic jump case of the linear weir.

REFERENCES

Belay, B. S. & Oertel, M., 2025. Void Fraction Distribution Downstream of Nonlinear Weirs: An Experimental Investigation Using Trapezoidal A-Type Piano Key Weirs, a Labyrinth Weir, and a Linear Weir. *Journal of Hydraulic Engineering*, *151*(3), 04025006.

Besser, L. (2024). Tailwater influence on downstream flow conditions of piano key weirs. In *Proceedings of the 10th International Symposium on Hydraulic Structures (ISHS) 2024*. ETH Zurich.

Chanson, H. & Brattberg, T. 2000. Experimental study of the air–water shear flow in a hydraulic jump. *International Journal of Multiphase Flow*, *26*(4), 583–607.

Chanson, H. (2010). Convective transport of air bubbles in strong hydraulic jumps. *International Journal of Multiphase Flow*, *36*(10), 798–814.

Crookston, B. M., Flake, L. K., & Felder, S., (2024). Flow nonuniformity and energy dissipation in moderate-sloped stepped chutes with a labyrinth crest. *Journal of Hydraulic Engineering*, *150*(5), 04024024.

Eslinger K.R. & Crookston, B. M,. 2020. Energy dissipation of type a piano key weirs. *Water*, *12*(5), 1253.

Felder, S., Hohermuth, B. and Boes, R.M., 2019. High-velocity air-water flows downstream of sluice gates including selection of optimum phase-detection probe. *Int. Journal of Multiphase Flow*, 116: 203–220.

Hager, W. H. (1992). *Energy dissipators and hydraulic jump* (Vol. 8). Springer Science & Business Media.

Hohermuth, B., Kramer, M., Felder, S. and Valero, D., 2021. Velocity bias in intrusive gas-liquid flow measurements. *Nature communications*, *12*(1), p.4123.

Kramer, M., Valero, D., Chanson, H. and Bung, D.B., 2019. Towards reliable turbulence estimations with phase-detection probes: An adaptive window cross-correlation technique. *Experiments in Fluids*, *60*(1).

Liu, M., Rajaratnam, N., & Zhu, D. Z. (2004). Turbulence structure of hydraulic jumps of low Froude numbers. *Journal of Hydraulic Engineering*, *130*(6), 511–520.

Megh, K.C., R., Crookston, B., Flake, K., & Felder, S., 2025. Enhancing flow aeration on an embankment sloped stepped spillway using a labyrinth weir. *Journal of Hydraulic Research*, *63*(1), 32–47.

Ohtsu, I., Koike, M., Yasuda, Y., Awazu, S., & Yamanaka, T., 1990. *Free and submerged hydraulic jumps in rectangular channels*.

Rajaratnam, N., 1965. The hydraulic jump as a well jet. *Journal of the Hydraulics Division*, *91*(5), 107–132.

Shen, X. & Oertel, M., 2023. Influence of piano key weir crest shapes on flow characteristics, scale effects, and energy dissipation for in-channel application. *Journal of Hydraulic Engineering*, *149*(6), 04023010.

Wang, H., 2014. Turbulence and air entrainment in hydraulic jumps.

Wu, S., & Rajaratnam, N. (1996). Transition from hydraulic jump to open channel flow. *Journal of Hydraulic Engineering*, *122*(9), 526–528.

Labyrinth and Piano Key Weirs IV – Erpicum, Crookston & Paxson (Eds)
© 2026 The Author(s), ISBN 978-1-041-10527-5

Model-prototype upscaling of discharge capacity of piano key weir

J. Yang
R&D Hydraulic Laboratory, Vattenfall AB, Älvkarleby, Sweden
Civil & Architectural Engineering, KTH Royal Institute of Technology, Stockholm, Sweden

P. Nilsen and M. Billstein
R&D Hydraulic Laboratory, Vattenfall AB, Älvkarleby, Sweden

L. Johnsson
Sydkraft Hydropower AB, Sundsvall, Sweden

S.C. Li
Civil & Architectural Engineering, KTH Royal Institute of Technology, Stockholm, Sweden

ABSTRACT: In Sweden, many dams are undergoing upgrades to comply with the revised dam safety guidelines, which mandate higher spillway discharge capacities. As part of an ongoing project, a piano key weir (PKW) is proposed to replace an embankment dam and simultaneously increase the facility's discharge capacity. The PKW configuration was evaluated in a 2 m wide flume at two geometric scales (1:35 and 1:12) to assess potential scale effects. The former comprises eight units, while the latter includes three, with all components produced using 3D printing. Results indicated that the 1:35 model exhibits a slightly lower discharge capacity, likely due to scale effects. To further investigate, a 1:35 scale floor model was constructed using CNC milling to replicate the river bathymetry, the existing spillway with three tainter gates, and the proposed PKW consisting of 24.5 units. When scaled to prototype discharge levels, this model demonstrated a somewhat lower capacity than both flume models due to real approach conditions, though the differences between them were minor. In addition, the CNC milling technique yielded greater manufacturing accuracy compared to 3D printing. The findings offer insights for future PKW designs and scale model testing in similar applications.

1 BACKGROUND

Based on the criteria for design-flood determination, many existing spillways in Sweden are found to have lower discharge capacity than the updated design floods. The main reason is that they were built many decades ago with inadequate hydrological modeling procedures. A partial survey shows that, without considering effects of the climate changes, the current design floods become typically 20–50% higher (Yang *et al.* 2002, Yang *et al.* 2019). This implies that most of the existing spillways are under-dimensioned, only with a limited number of exceptions. Discharge safety is often one reason for dam upgrades. Dams are also rehabilitated due to aging, function deterioration, foundation settlement, etc.

The dam-safety guidelines stipulate that a dam should discharge and withstand the revised design flood without jeopardizing its structural integrity. The hydraulic measures include modification of the existing spillways, construction of new spillways, etc. The former refers to opening width enlargement, lowering of spillway crests, modification of bottom outlet to surface gated opening, etc. It is common that several rebuilding measures are combined to achieve a technically feasible and cost-effective solution. Owing to their high reliability, low

DOI: 10.1201/9781003655541-3

engineering costs etc., the labyrinth and piano key weirs (PKW) have become the major alternative in a few dam upgrade projects in Sweden (Yang *et al.* 2024, 2025).

The dam facility, for which the current study was performed, consists of a gated spillway, a 30 m long gravity dam to its left, and an embankment dam to its right. The max. structural height is ~10 m. The dam-safety evaluations show that the embankment dam is, due to aging in poor conditions and substantial rebuilding measures are needed to meet the structural safety standard. In addition, the existing spillway is insufficient to accommodate the design flood as prescribed by the dam-safety guidelines, leading to the need for additional capacity. The three existing gates discharge ~560 m^3/s at the full reservoir water level (FRWL), which is compared with the design flood ~1000 m^3/s. The upgrade option finally chosen is a PKW construction ~55 m downstream of the embankment dam that is partially removed upon its completion.

As shown in Figure 1, the PKW is designed with a L-shaped planform, connecting to the rightmost pier of the existing spillway to its left and extending to the right riverbank with a short buttress dam. Its total length amounts to ~100 m. The PKW is placed on a low concrete dam, and the crest elevation is set the same as the FRWL. It is composed of 23 regular units, a half unit at each end and a hybrid unit in the middle. Upon completion, the embankment dam is excavated to an elevation that does not unfavorably affect the approach flow. The excavated height is finalized through hydraulic model tests.

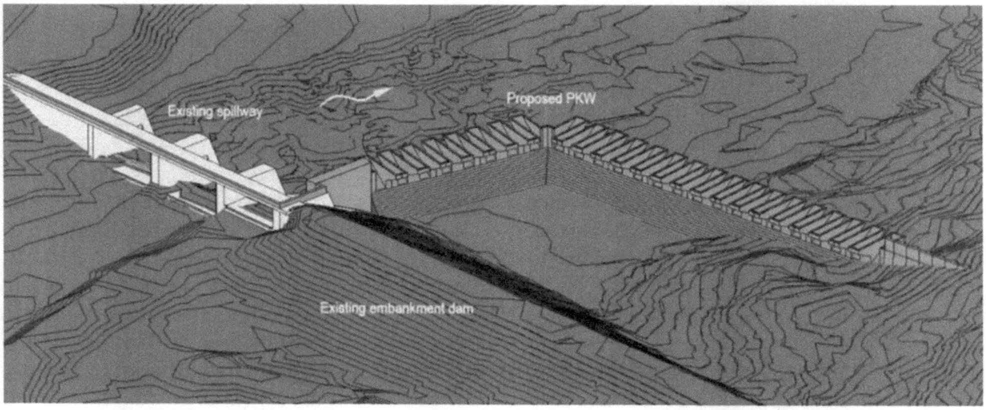

Figure 1. Adopted PKW layout to replace the existing embankment dam.

The major PKW parameters is defined in Figure 2, in which T_s = wall thickness at crest, W_i, W_o = inlet and outlet key width, P_i, P_o = inlet and outlet key height, P = weir height, B_i, B_o = crest length of inlet and outlet key overhang (excluding T_s), B_b = weir base length, S_i, S_o = parapet wall height of inlet and outlet key, α_i, α_o = angle of inlet and outlet key floor with the horizontal plane, and subscripts i and o denote the inlet and the outlet key, respectively.

The width of a single weir unit along the dam axis, and its crest length (inclusive of the crest walls) in the flow direction are

$$W = W_i + W_o + 2T_s \tag{1}$$

$$B = B_i + B_o + B_b + 2T_s \tag{2}$$

The angles of the key floors with the horizontal plane are

$$\tan(\alpha_i) = (P_i - S_i)/(B_b + B_i) \tag{3}$$

$$\tan(\alpha_o) = (P_o - S_o)/(B_b + B_o) \tag{4}$$

Measured along the crest centreline, the developed crest length of one weir cycle is

$$L = 2B + W_i + W_o \tag{5}$$

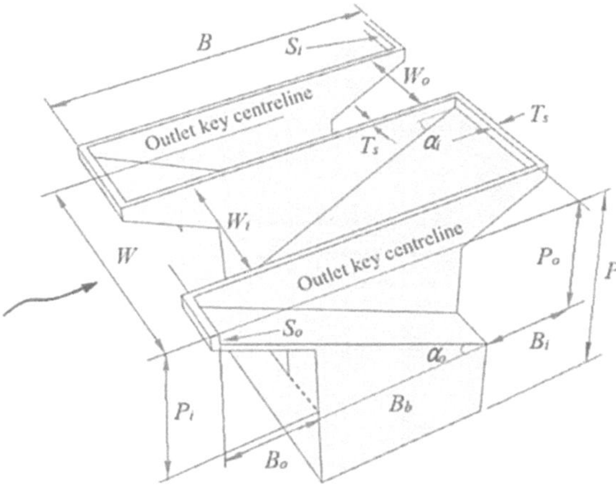

Figure 2. Parameter definition of PKW with both up- and downstream overhangs.

For the upgrade project, the finalised prototype PKW dimensions are summarised in Table 1. This gives $W = 3.90$ m and $L = 19.90$ m. A parapet wall is assigned to the outlet key, but not to the inlet key, a decision made by the dam owner. Partially due to this, the two floors have somewhat different angles. As well understood, a thinner wall leads to a higher discharge at a given reservoir water level. In this design, the wall thickness is $T_s = 35$ cm because of the reinforcement bars. The wall crest is chamfer-edged on both sides, each with a 5 cm chamfer.

Table 1. Prototype dimensions of the PKW.

W_i (m)	W_o (m)	T_s (m)	B_i (m)	B_b (m)	B_o (m)	B (m)
1.80	1.40	0.35	1.45	3.75	2.45	8.35
S_i (m)	S_o (m)	α_i (°)	α_o (°)	P_i (m)	P_o (m)	P (m)
0	0.50	33.9	25.8	3.50	3.50	3.75

Model tests of discharge capacity are performed both in a flume with straight approach flow conditions and in a conventional floor model with natural reservoir bathymetry. In the former, two different scale models (denoted as 1:12 and 1:35 flume models) are compared. In the latter, the scale is 1:35 and the model (denoted as 1:35 floor model) is used even for evaluation of other aspects related to the dam upgrade. The primary objective is to determine whether significant scale effects exist that may influence the PKW design.

2 FLUME EXPERIMENTS

The tests are performed in a flume. Its length, width and height are 11.00, 2.00 and 2.00 m. In the middle, a 4.50 m straight channel is set up, resulting in satisfactory approach flow conditions. In the prototype, the PKW is placed on a concrete dam of varying heights. For the tests, a 2.40 m base height is chosen. To provide reliable data to the design, two scales are evaluated, i.e., 1:12 and 1:35. The embankment dam is not included in the flume. The models are placed flush with the flume's downstream end, so that the overflow is aerated and does not suffer from any negative pressure behind the jets (Figure 3). More description of the rig is found in Yang *et al.* (2023) and Yang and Li (2024).

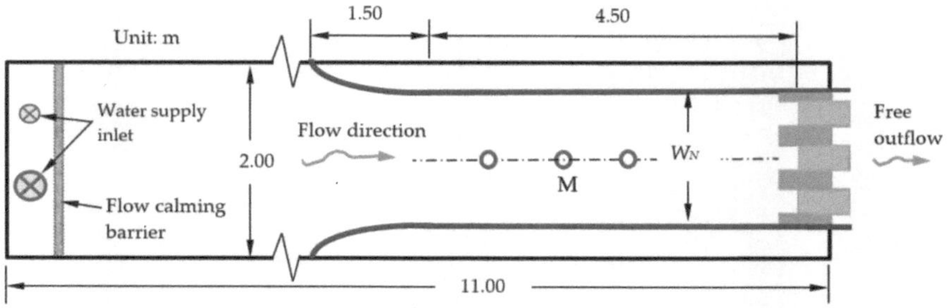

Figure 3. Experimental layout in a 2-m wide flume, with a central channel for PKW tests.

Both PKW models are scaled according to the dimensions in Table 1 and manufactured using 3D printing. The manufacturing process leaves a layered structure with a stair-stepping effect on surfaces that depend on its orientation. The surface finish is governed by the printing nozzle diameter and layer height, which affects to great extent the manufacture costs. The printing polymer is a recycled ABS plastic. To attain relatively smooth finishing and also to be affordable, the nozzle diameter and the layer height used are 6.0 and 1.1 mm for the 1:12 model, and 0.4 and 0.2 mm for the 1:35 model. Figure 4 shows the surface finishing of both flume models.

Figure 4. Surface finish of the 3D-printed flume models: (a) 1:12 flume model; (b) 1:35 flume model.

The 1:12 flume model comprises three identical PKW units, with a width of W_N = 97.50 cm ($3W$ = 11.70 m in the prototype) and a height of 51.25 cm (corresponding to 6.15 m in the prototype, i.e. P = 3.75 m + base height 2.40 m). The 1:35 flume model is composed of eight identical units, with a width of W_N = ~89.14 cm ($8W$ = 31.20 m in the prototype) and a height of ~17.57 cm (the same prototype height 6.15 m). The channel width is individually adjusted to accommodate the models.

Water stages in the channel are monitored at a few points (M) along its axis, 3.00–4.00 m upstream of the PKW for the 1:12 model and 1.00–2.00 m for the 1:35 model (Figure 3). For the capacity determination, the water head (H), i.e., the difference between the undisturbed upstream water stage and the weir crest elevation, is based on the reading at the position 35 m (prototype size) upstream of the vertical weir front for both weirs.

A Ø200 mm inlet pipe with a magnetic flow meter records flow rates in the flume, which is sufficient for the 1:12 model, requiring up to ~175 liter/s. For the 1:35 model, its maximum flow rate does not exceed 35 liter/s. At such low flow rates, the large flow meter would lead to significant measurement errors. To compensate this, a smaller inlet pipe (Ø100 mm), also equipped with a magnetic flow meter, is installed. Both the flow meters are calibrated in advance (Figure 5).

For each weir, tests are made at a few upstream water levels. Overflow patterns and flow rates are monitored. Figures 6 and 7 show the models and their overflows at selected flow rates. At given heads below H = 1.0 m (prototype size), no obvious difference in overflow between them is observed. Up to H = 1.0–1.2 m (prototype size), the outflow in the outlet key is still free, without any submergence effects. Beyond this limit, there is no interest anymore for the project, and no further tests are made.

The PKW is a hydraulic structure with free-surface flow. Defined by L, the flow rate over a single PKW unit (Q) is expressed as

$$Q = \frac{2}{3} CL\sqrt{2gH^3} \tag{6}$$

in which C = discharge coefficient and g = gravitational acceleration. Figure 8 compares the change of Q as a function of H. Obviously, the smaller model exhibits, at the same reservoir water stage, somewhat lower discharge capacity. The relative errors are below 3–5%, which suggests that some scale effects do exists. However, the two lines collapse onto each other if $H > \sim 1.00$ m. This limit corresponds to about 3 cm in the 1:35 model, which is consistent with the findings by Erpicum *et al.* (2016). After comparing the prototype data with the capacity results from three scale models (1:7, 1:15 and 1:25), they stated that a minimum head of 3 cm was required in the model tests to avoid the scale effects due to the surface tension and viscous forces. Pfister *et al.* (2013) came to similar conclusions in their studies.

Figure 5. Two separate inlet pipes (Ø200 mm and Ø100 mm) with magnetic flow meters and control valves for monitoring the flow in the 1:12 and 1:35 models, respectively.

Figure 6. The 1:12 flume model with three PKW units, at $Q = \sim 25$ m^3/s.

Figure 7. The 1:35 flume model with eight PKW units, at Q = ~30 m^3/s.

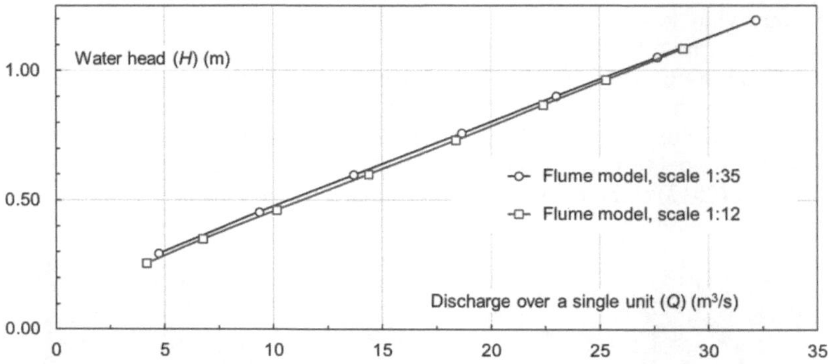

Figure 8. Relationship between H and Q for the 1:12 and 1:35 flume models. Discharge Q refers to the flow capacity of a single weir unit (width W = 3.90 m).

3 FLOOR MODEL (SCALE 1:35)

A floor model over the dam facility is constructed, the chosen model scale is 1:35. With the whole river width, a 150 m reservoir length and a 100 m river reach downstream are included. It measures ~10 m long, ~6 m wide and 0.6 m high. To achieve high modeling accuracy, the echo-sounded reservoir bathymetry, the laser-scanned river terrain downstream, the PKW and the existing spillway are all manufactured with the machine milling technique. The surface finishing is smooth and uniform, which is essential taking into consideration the relatively small model dimensions. For the natural topography and the concrete structure, Styrofoam (density 300 kg/m^3) and Ureol material (density 700 kg/m^3) are used. To compare with the flume test results, the embankment dam is not included in the floor model. Looking from the left bank downstream, Figure 9 shows the view over the model. A calibrated magnetic flow meter (Ø200 mm) measures flow rates into the model. The water head used for the discharge relationship is at a location 1.00 m (35.0 m in the prototype) upstream of the weir, which is consistent with the flume tests.

Figures 10 and 11 plot the experimental results of H-Q and C-H/W. Apparently, at the same water head, the floor model provides somehow lower flood discharge capacity than both the flume models. The deviations among them also get larger as H increases, which is probably due to the growing flow velocity of the approach flow.

There are also differences in discharge coefficient C. For either model, there is always a peak – C goes first up and then down with an increase in H. This is the inherent property of the weir crest, which is seemingly related to the crest width and shape in cross-section. The peak value for the floor model occurs at a larger H than in the flume models. The trend is consistent with the findings from the studies by Anderson and Tullis (2012) and Yang *et al.* (2023).

Figure 9. The 1:35 floor model incorporating the PKW and the existing spillway.

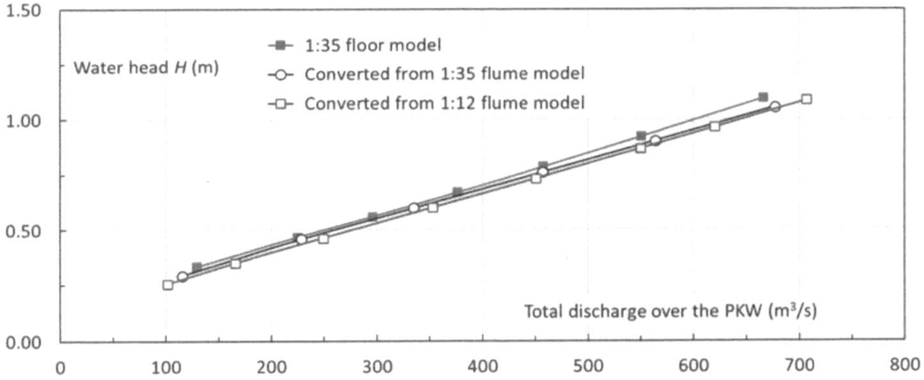

Figure 10. Comparisons of discharge over the PKW with 24.5 units and its discharge coefficient C.

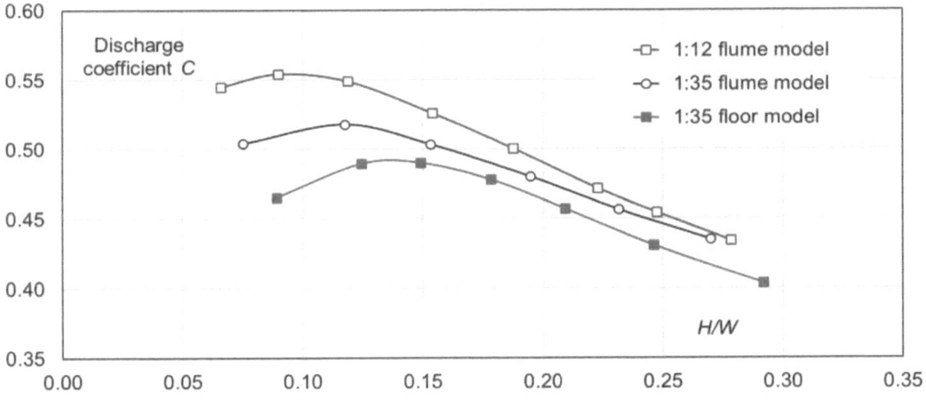

Figure 11. Comparisons of discharge over the PKW with 24.5 units and its discharge coefficient C.

4 RESULTS AND DISCUSSIONS

In the flume tests, the two PKW models are given the same prototype base height (2.4 m). The measurements of water stages (heads) are read at the same prototype location (35.0 m) upstream of the weirs. Two magnetic meters of different measurement ranges provide calibrated accuracy for flow rate determination. Except for small manufacture errors, the difference in discharge capacity between them is mainly due to scale effects, as the operating water heads are below 3.0 cm.

In the 1:35 floor model, the factors that affect the discharge capacity include the following aspects.

(1) The PKW model is aligned at a right angle with the flow in the flume. With its L-shaped planform, the weir axis of the left PKW section (~33 m long) is approximately at a 20° angle with the approach flow direction in the reservoir.
(2) Beneath the PKW in the flume, the base is 2.4 m in height, which can be compared with the varying height 0.5–7.0 m in the prototype.
(3) On the right riverbank, the reservoir is shallow, and its bed roughness is not small in relation to the flow depth, which results in head losses, especially at high flow velocity.
(4) The maximum H value is below the 3 cm limit, indicative of certain scale effects in the results.

As the PKW is complex in geometry, certain errors exist in the manufacture. For a gated overflow opening, deviations in the crest elevation and the opening width are the main sources of errors, which can be compensated when the experimental results are analyzed. For the PKW, this is not straightforward. Therefore, it is essential that the PKW be manufactured with 3D printing or machine milling.

Quality control show that the 3D-printed 1:12 model suffers from certain crest unevenness and side-wall deformations also appear, which are manually amended. Of course, a smaller printing nozzle diameter and layer height can be used to mitigate this. However, this would lead to a significant increase in costs. For the machined-milled PKW in the floor model, Figure 12 compares the dimensions of the nominal and the constructed weir. Obviously, for both the sidewalls and the keys, the deviations are below 0.3 mm, which is satisfactory. It also results in smooth surface and uniform model quality, which is more satisfactory than the 3D printed surfaces (Figure 13). Our experiences have demonstrated that the machine milling technique provides better geometrical accuracy.

5 CONCLUSIONS

As part of dam upgrade projects in Sweden, the PKW is being adopted in some cases as the preferred reconstruction alternative. In this project, it involves replacing the existing embankment dam in poor condition, while simultaneously increasing the facility's discharge capacity. The flume experiments assess two physical models at scales of 1:12 and 1:35. To reduce measurement inaccuracy, two magnetic flow meters of different pipe diameters (measuring ranges) are used. The tests indicate that the latter tends to underestimate the discharge capacity, likely due to scale effects. The floor model at scale 1:35 exhibits somewhat lower capacity than both flume models. The reason for the discrepancy involves a few factors such as scale effects, weir planform, shallow water and bed roughness, etc.

The models in this study are produced by either 3D-printing or machine milling. Due to the small PKW sizes (e.g. at the 1:35 models), manufacture deviations should be kept as low as possible. It is advisable to use the machine milling technique as it leads to higher accuracy in geometry and satisfactory surface finish.

Further model tests indicate that excavating the embankment dam to an elevation 3.0 m below the full reservoir water level (FRWL) is sufficient to ensure acceptable approach flow conditions without significantly compromising the PKW's discharge capacity. The project aims to safely convey the design flood—approximately 1000 m^3/s—while keeping the reservoir level within

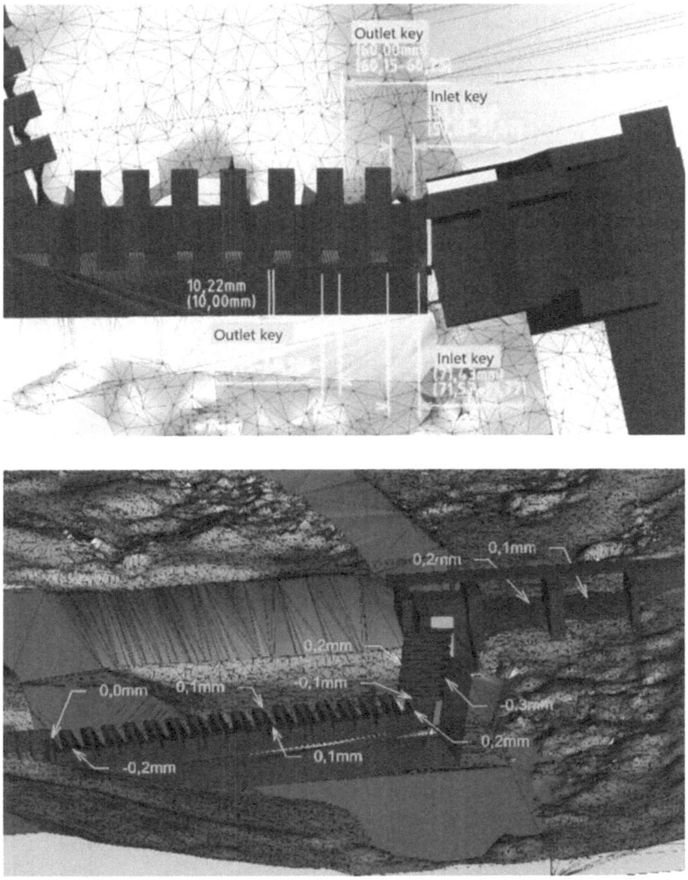

Figure 12. Machine milling of the PKW in the 1:35 floor mode – manufacture deviations.

Figure 13. Satisfactory smooth surface finish from machine milling technique.

0.80 m above the FRWL. Although the exact flood magnitude has not been finalized, at this target water level, the PKW and the existing spillway are expected to discharge approximately 500 m^3/s and 600 m^3/s, respectively.

From both hydraulic and structural perspectives, the implementation of the PKW represents a cost-effective solution for this project. Since the PKW is located downstream of the embankment dam, its construction does not impact the facility's operation or necessitate the construction of a cofferdam. Given its location in a cold climate, post-construction field monitoring and testing are expected to yield valuable insights for similar future projects. Further research should incorporate numerical modeling to complement physical testing. Additionally, investigations into sediment transport and ice-related effects in cold environments are crucial to understanding and optimizing PKW performance in such contexts.

ACKNOWLEDGEMENTS

The study was carried out for Uniper Sydkraft Hydropower AB; the tests were made at Vattenfall's Hydraulic Laboratory, Älvkarleby. Erik Skepparkrans, Anna Helgesson, Anders Sjödin, Magnus Lagerkvist, Jennifer Persson and Patrik Lidberg, all at Vattenfall R&D, Älvkarleby, were acknowledged for CAD, model manufacture, assembling, model testing and instrumentation. James Yang prepared and revised the manuscript, and all the other authors consented to the submitted version.

REFERENCES

Anderson, R.M. & Tullis, B.P. 2012. Piano key weir: reservoir versus channel application. *Journal of Irrigation and Drainage Engineering* 138: 773–776.

Erpicum, S., Tullis, B.P., Lodommez, M., Archambeau, P., Dewals, B.J. & Pirotton, M. 2016. Scale effects in physical piano key weirs models. *Journal of Hydraulic Research* 54(6): 692–698.

Pfister, M., Battisacco, E., De Cesare, G. & Schleiss, A.J. 2013. Scale effects related to the rating curve of cylindrically crested piano key weirs. *Proc. 2nd Int. Workshop on Labyrinth and Piano Key Weirs, Paris, France*, 20–22 November 2013.

Yang, J., Andreasson, P. & Cederström, M. 2002. Floods, new design-flood criteria and dam refurbishment in Sweden. *Proc. 2nd International Symposium on Flood Defense, Beijing, China*, September 10–13. 2002.

Yang, J., Andreasson, P., Teng, P. & Xie, Q. 2019. The past and present of discharge capacity modeling for spillways—a Swedish perspective. *Fluids* 4(1): 10.

Yang, J., Li, S.C., Helgesson, A., Skepparkrans, E. & Ansell, A. 2023. Geometric modification of piano key weirs to enhance hydraulic performance and discharge capacity. *Water* 15(23): 4148.

Yang, J. & Li, S.C. 2024. Piano key weir (PKW) – improvement in conventional geometry for augmented discharge capacity. *Water* 16(23): 3375.

Yang, J., Nordström, E., Li, S.C. & Kocahan, H. 2024. Labyrinth and piano key weirs in cold climate – experiences from prototype facilities and laboratories. *Proc. 92nd ICOLD Annual Meeting & International Symposium, New Delhi, India*, 29 September–3 October *2024*.

Yang, J., Frisk, A., Billstein, M., Mattiasson, P. & Johnsson, L. 2025. Dam upgrade with piano key weir for discharge of extreme floods. *Proc. 26th ICOLD Congress, Chengdu, China*, May 16–23, *2025*.

Labyrinth and Piano Key Weirs IV – Erpicum, Crookston & Paxson (Eds)
© 2026 The Author(s), ISBN 978-1-041-10527-5

Physical model investigations on nappe instabilities at labyrinth weirs

M. Gebhardt and F. Belzner
Federal Waterways Engineering and Research Institute, Karlsruhe, Germany

ABSTRACT: Free overflow hydraulic structures operating at low heads tend to develop nappe instabilities when there is no ventilation. A detailed investigation has been undertaken with a relatively large-scale labyrinth weir in the hydraulic laboratory of the BAW. A quarter-round crest was selected, because nappe instabilities mainly occur at sharp-crested weirs, which favor the development of an air pocket behind the nappe. The headwater ratios for which nappe instabilities could be observed were determined. Pressure fluctuations were measured and the frequencies determined. Adding splitters to the crest is one way to vent the air pocket and to avoid vibrations. Another possibility is an artificial ventilation to stabilize the pressure behind the nappe. This paper presents the results and observations and discusses them against the background of the existing literature.

1 INTRODUCTION

The Federal Waterways Engineering and Research Institute (BAW) has carried out studies to evaluate the suitability of fixed weirs for the waterway network (Gebhardt & Belzner, 2024). Head-discharge relationships for labyrinth weirs were evaluated experimentally for free and submerged flow conditions (Belzner *et al.*, 2017). Based on the experimental results, a feasibility study on the Ilmenau River was carried out in order to determine the most appropriate weir type (Gebhardt *et al.*, 2017).

Flow-induced vibrations, commonly classified as air-water-interface instabilities, can be an issue for light or elastic supported structures. This does not typically apply to solid and rigid labyrinth weirs. However, nappe vibrations can produce noise, which affects the quality of living environment in close proximity. Sound waves can also cause window glass to vibrate (Casperson, 1993). It is therefore not a question of structural safety for the labyrinth weir, but rather a question of suitability for use. The purpose of this study is to gather experiences and data on nappe instabilities and nappe vibrations at labyrinth weirs. In addition to visual observations, pressure measurements underneath the nappe are being carried out. The number and location of breakers or flow splitters were tested to determine their effectiveness in counteracting nappe instability.

2 EXCITATION DUE TO INTERFACE INSTABILITY

According to Naudascher and Rockwell (1994) nappe vibrations are classified as instability-induced excitation (IIE) and can be observed at flow systems involving one or more liquid-gas interfaces such as a water nappe with an air volume underneath. Like all IIEs, the excitation is associated with self-sustained flow oscillations induced by an instability and controlled by either fluid-dynamic, fluid-elastic, or fluid-resonant feedback mechanisms (Naudascher & Rockwell, 1994). One process involves the nappe beginning to oscillate in response to perturbations at the origin, the resulting alterations in the instantaneous profiles of the nappe induce volume changes in the air pocket beneath it. These volume fluctuations lead to pressure variations within the air pocket, which subsequently cause deflections of the entire nappe. This process establishes a feedback loop.

DOI: 10.1201/9781003655541-4

Two types of instabilities or oscillations may occur through inadequate aeration. The first is "flow surging" as reported by Falvey (2003) and was observed among others at the Boardman spillway in Oregon, United States. Crookston and Tullis (2013) described this phenomenon as "nappe instability". The process can be characterized as follows: As air becomes entrained, the volume of the air pocket decreases and the pressure in it falls below atmospheric pressure. The process continues until the nappe collapses when the pressure difference becomes sufficiently large, resulting in an abrupt pressure equalization. The nappe swings back to its initial state and the process starts again. The process appears to repeat regularly, but without a discernible pattern or fixed interval, suggesting an element of randomness.

In contrast to "flow surging" a second type of nappe instabilities without ventilation of the nappe could be detected in the physical model, which is not mentioned in literature according to the authors knowledge. The process is characterized by a clear oscillation of the nappe with a significant higher frequency than "flow surging", which can be measured in the pressure underneath the nappe. During the oscillation process the air pocket remains closed and is not ventilated even for a short period of time (Chapter 4.2).

Finally, a third type of nappe vibration can be observed under certain circumstances which was described by Casperson (1993) as "fluttering fountains". These vibrations have been observed for relatively small low-head discharges resulting in a very thin nappe. It is characterized by a horizontal banding on the nappe and an excessive noise which occurs even though the nappe is fully aerated. Examples of nappe vibrations with horizontal banding and undulations are described by Crookston *et al.* (2014). According to Lodomez *et al.* (2017a) a lack of consensus exists on the causes and source of the oscillation development. Partenscky and Khloeung (1967) provide an empirical correlation for the nappe oscillation frequency in terms of a Strouhal number, which is defined by the frequency, the fall height and the initial nappe velocity and nappe thickness. As these nappe instabilities could not be observed in the model, they are not discussed further below. Thus, for further details on this phenomenon it is referred, for example, to Lodomez *et al.* (2018b).

3 EXPERIMENTAL SETUP

3.1 *Testing apparatus*

Systematic experimental tests were performed by Linzen (2023) and Voges (2024) on a physical model in the laboratory of the BAW. The tests were performed in a 0.60 m wide flume with a length of 18.00 m and a height of 1.20 m. The upstream approach section was approx. 11 m (~12 *P*). Additional measures to help with approach flow uniformity included floating wave suppressors and a honeycomb flow conditioner near the flume inlet. Magnetic inductive flow meters of nominal diameters DN100 and DN300 were used to quantify discharge (precision ±0.2 %). Floating gauges in Plexiglas cylinders to measure the water depths were connected to the flume by using the principle of communicating vessels (±0.1 mm). The gauge for determining the water depth y_u was located approx. 0.70 m upstream of the weir.

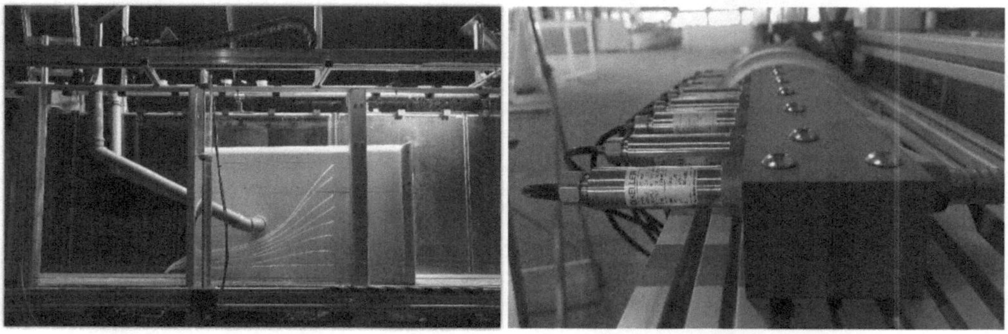

Figure 1. Labyrinth weir: pressure tubes and artificial aeration (left) and pressure transmitters (right).

Seven pressure sensors were employed to measure the air pressure (±0.05 %) beneath the nappe. The number was selected based on initial uncertainty concerning the location of nappe reattachment. These sensors were connected to ports in the wall through plastic tubing with a diameter of 1 mm (Figure 1). To ensure that there was only air in the pipes, lockable drainage pipes were provided. The presence of water would adversely affect the accuracy of the measurements.

3.2 *Geometry and crest shapes tested*

A trapezoidal labyrinth weir was tested consisting of one weir cycle with one full outlet key. A sketch of the labyrinth weirs is given in Figure 2. The geometric parameters are given in Table 1.

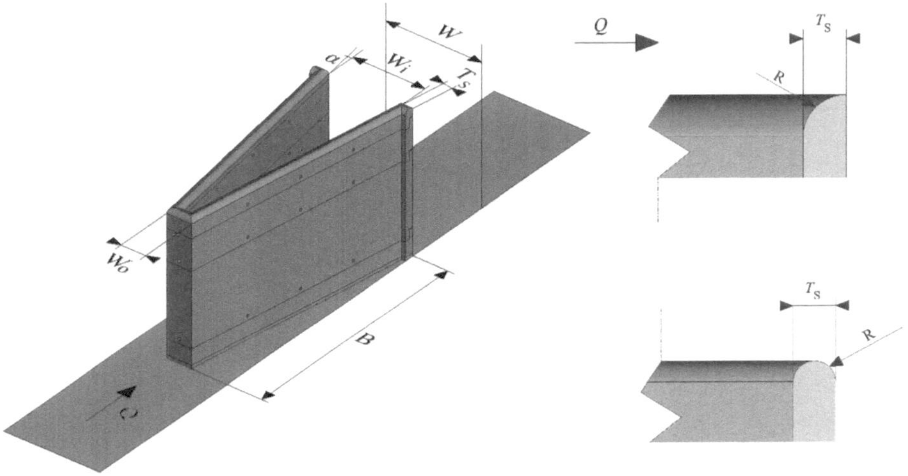

Figure 2. Isometric illustration of the labyrinth weir.

Table 1. Geometric parameters of the labyrinth weir.

Parameter			
Number of cycles	N	[m]	1
Weir height	P	[m]	0.925
Width of channel	W	[m]	0.600
Outlet key width	W_o	[m]	0.152
Inlet key width	W_i	[m]	0.448
Streamwise length	B	[m]	1.446
Sidewall thickness	T_S	[m]	0.060
Sidewall angle	α	[°]	8.21
Developed crest length	L	[m]	2.994
Radius of half-round crest	R_{HCR}	[m]	0.030 = 0.5 T_S
Radius of quarter-round crest	R_{QCR}	[m]	0.060 = T_S

3.3 *Experimental program*

Experimental data were collected under steady-state conditions. Therefore, the discharge Q was varied between 0.006 $m^3 \cdot s^{-1}$ and 0.300 $m^3 \cdot s^{-1}$ in steps between 0.001 $m^3 \cdot s^{-1}$ and 0.020 $m^3 \cdot s^{-1}$ depending on the discharge. Overall, 40 head-discharge data points were collected for each crest geometry. This included a number of checks, wherein more than 20 % of the tests were repeated to ensure accuracy and determine measurement repeatability. The measurements for discharge, water depth (sample rate 1 Hz) and pressure (sample rate 50 Hz) were recorded for about 300 s to ensure stationary conditions and to determine average values. States in which nappe instabilities were observed or expected were approached several times, from a higher to a lower discharge and vice versa.

4 EXPERIMENTAL RESULTS

4.1 *Discharge capacity*

Poleni´s Formula is used in this study to describe the head-discharge relationships of the labyrinth weir.

$$C_d = \frac{Q}{2/3 \cdot L \cdot \sqrt{2g} \cdot H_u^{3/2}} \tag{1}$$

where C_d = discharge coefficient, Q = discharge, L = developed crest length, g = gravity constant and $H_u = h_u + v_u^2/2g$ = upstream energy head, v_u = average cross-sectional velocity at the gauging location and h_u = piezometric head upstream of the weir relative to the crest (Figure 3).

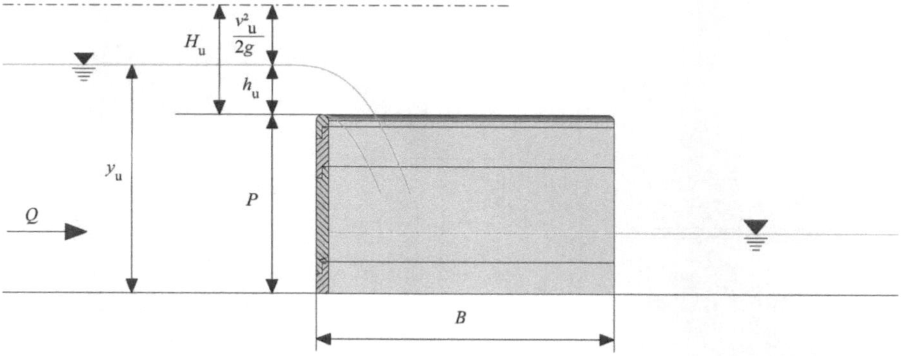

Figure 3. Side view of the weir with the indication of the main hydraulic and geometric parameters.

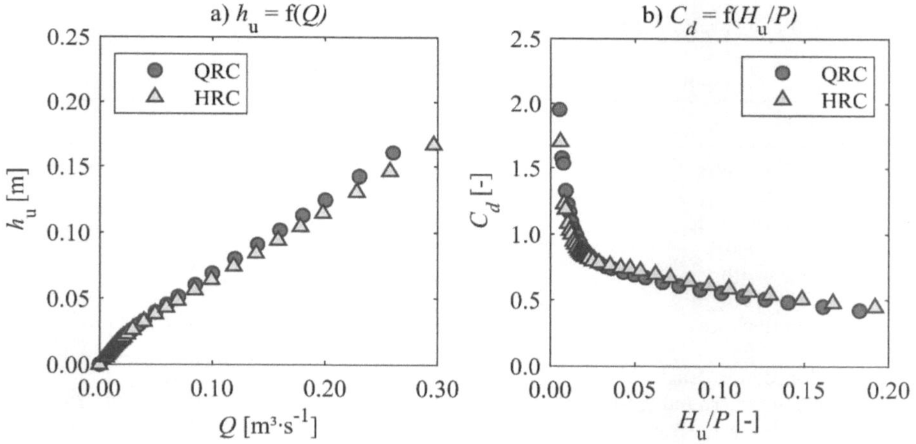

Figure 4. a) Discharge relationship and b) discharge coefficient C_d vs. H_u/P for the quarter-round (QRC) and the half-round crest shape (HRC).

Figure 4 a) shows the discharge rating curves for the quarter-round (QRC) and the half-round crest shape (HRC). C_d data are presented in terms of H_u/P in Figure 4 b) for the non-vented trapezoidal labyrinth weir. It is noticeable that the discharge coefficient C_d strongly increases for $H_u/P < 0.03$ and thus, raises the question of scale effects. This will be discussed in Chapter 5. In general, the data indicate that the half-round shape is slightly more efficient for $H_u/P > 0.03$, which can be explained by the fact, that the crest is rounded and the flow stays attached to the weir wall increasing flow efficiency. This trend fits in well with the observations of Crookston (2010) although the ranges for H_u/P are very different.

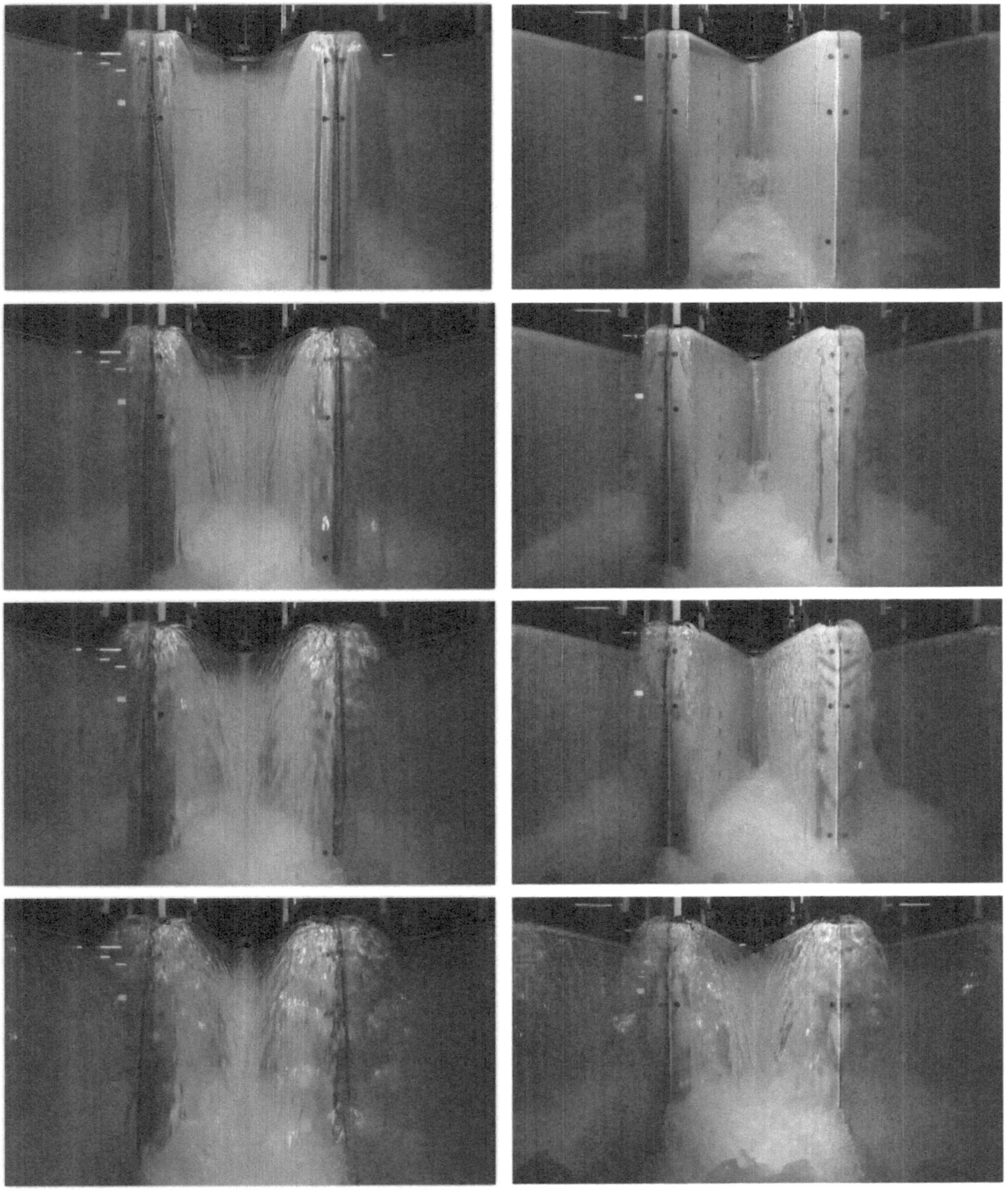

Figure 5. Photographic overview for Q = 0.02, 0.04, 0.06, 0.08 m³/s (from top to bottom); labyrinth weir with quarter-round crest (left) and half-round crest (right).

Figure 5 shows a comparison of the quarter-round and the half-round crest for discharges between 0.02 and 0.08 m³/s. This corresponds to a head of H_u/P = 0.02 and 0.06. The nappe is not vented. According to Crookston (2010) three nappe conditions can be expected: clinging, aerated and partially aerated. Drowned conditions were not the subject of this study. While the nappe detaches from the quarter-round crest even at low discharges, one can observe "clinging" for Q = 0.02 m³/s and "partially aerated" for Q = 0.06 m³/s, both conditions at the half-round crest. It should be noted that a certain randomness could be observed in whether the nappe broke up at the transition between the downstream crest and the edge of the flume, leading to ventilation. The nappe conditions at the quarter-round crest were characterized by the formation of an air cavity.

4.2 *Nappe vibrations and nappe instabilities*

Nappe vibrations could be observed in the physical model only for a very small range of discharges for $Q = 0.01$ m³/s (±10 %) which corresponds to a head of $H_u/P = 0.01$. However, the oscillations did not always occur, even though the same discharge was set. A dependency of gradually reduced or increased discharge was observed. When the discharge was reached, it took often a few minutes until the nappe vibrations started, but then with rapidly increasing amplitudes. The volume of the air cavity did not visibly seem to change but the trajectories and the vibrations were accompanied by a pulsating noise. Measured pressure fluctuations had an amplitude of about $\Delta p = 25$ Pa and a frequency of about $f = 2.0$ Hz (Figure 6).

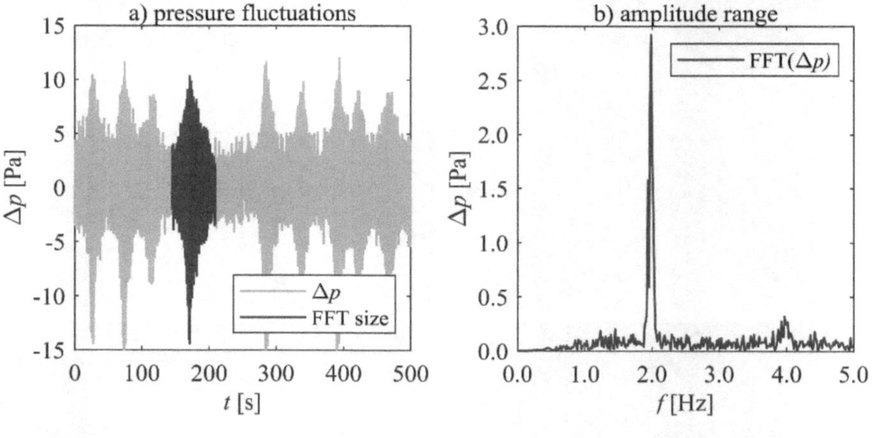

Figure 6. Nappe vibrations for $Q = 0.01$ m³/s ($H_u/P = 0.01$): a) pressure fluctuations underneath the nappe; b) amplitude range of pressure fluctuations obtained by fast Fourier transform (FFT).

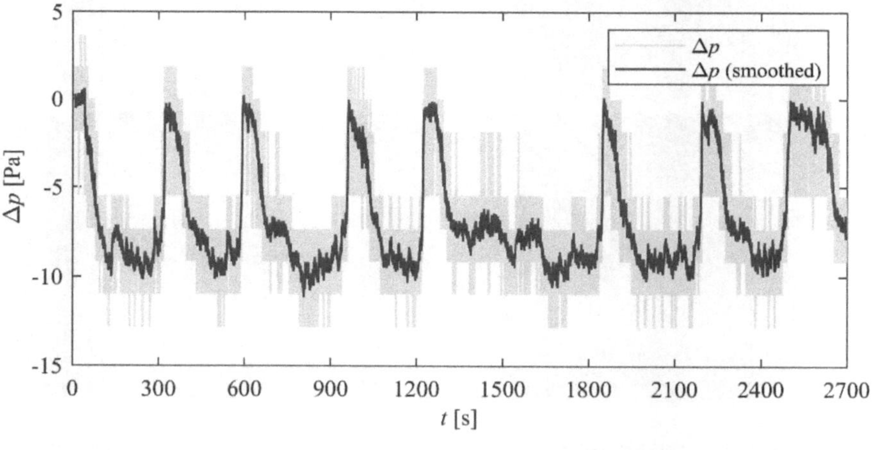

Figure 7. Nappe instabilities "flow surging" for $Q = 0.02$ m³/s ($H_u/P = 0.02$): pressure fluctuations underneath the nappe.

Nappe instabilities could be observed for slightly larger discharges of about $Q = 0.02$ m³/s (±10 %) which corresponds to a head of $H_u/P = 0.02$. The pressure fluctuations are shown in Figure 7. The pressure beneath the nappe reaches $\Delta p = 10$ Pa below atmospheric pressure, which is sufficient to induce deformation of the nappe and draw it towards the weir wall. After approximately $\Delta t = 150$-200 s, pressure equilibration occurs, with the pressure returning to atmospheric levels, causing the nappe to return to its initial position. This cycle typically repeats every 300 seconds, although it may occasionally take longer. No instabilities could be observed with larger discharges, which can be attributed to the increasing nappe thickness. The greater mass of the jet becomes too substantial to be deflected by the relatively small negative pressure.

4.3 *Artificial nappe aeration*

Artificial aeration can be accomplished with nappe breakers on top of the crest or with vents. Two circular conduits DN85 were tested in the physical model for the quarter-round crest (Figure 8), effectively maintaining atmospheric pressure beneath the nappe. With aeration, no nappe instabilities or vibrations were observed.

Various geometries of nappe breakers and positions on the crest were initially tested and evaluated with regard to split the water jet and ensure that the underside of the nappe is exposed to atmospheric pressure under low flow conditions (Linzen, 2023). The final geometry and positions, which were examined in the model, are shown in Figure 8. The model investigations demonstrate that the nappe breakers have no significant effect on the discharge capacity for $H_u/P > 0.04$ and a negligible impact for $H_u/P < 0.04$. This can be attributed to the fact that the configuration behind the crest does not substantially shorten the effective crest length, with the nappe being disrupted only after the jet deflection, rather than at the crest as observed with other solutions.

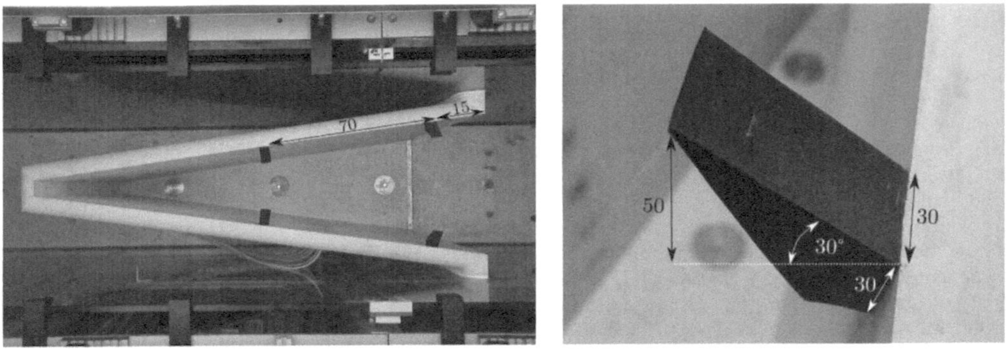

Figure 8. Nappe breakers: position on the quarter-round crest (left) and geometry (right), dimensions [mm].

5 DISCUSSION

It is known, that for absolute upstream head values lower than $h_u = 0.03$ m significant scale effects affect the head-discharge relation (Erpicum *et al.*, 2016) due to the fact that the relevance of surface tension and viscous forces increase as the flow depths decrease. When scale effects are present, the discharge coefficient is typically lower in the model than in the prototype. In this study it was observed that the discharge coefficient was increasing with decreasing upstream head, which seems to be contrary to the authors expections. This can be explained by the characteristic of the head-discharge relationship for $H_u/P < 0.5$ which is here better represented by a root function rather than a linear function (Figure 4 b)). For this case, the discharge coefficient must become infinitely large if H_u/P tends to zero. For further investigation it would be beneficial to compare results of different model scales. However, the wall thickness T_S must be considered as the characteristic length instead of the weir height P as Pfister *et al.* (2013) has shown.

For the quarter-round crest nappe instabilities, also known as "flow surging", could be observed for $H_u/P = 0.02$, as already described by Falvey (2003) or Crookston and Tullis (2013). The pressure under the nappe jet was only 10 Pa below atmospheric pressure, but large enough to deflect the nappe. In addition, nappe vibrations could be observed for $H_u/P = 0.01$ accompanied by a pulsating noise. Pressure fluctuations could be measured with an amplitude of about $\Delta p = 25$ Pa and a frequency of about $f = 2.0$ Hz (Figure 6). To the authors knowledge, this vibration type is new in the context of labyrinth weirs. In contrast to "flow surging" the pressure in the air cavity even exceeds atmospheric pressure, indicating a fluid-resonant feedback mechanism.

Nappe instabilities and nappe vibrations could no longer be observed, if ventilation of the nappe is ensured either by nappe breakers on the crest or with vents. Recommendations can be found in literature, for example in Falvey (2003). Crookston and Tullis (2013) recommend designing nappe breakers with a triangular floor plan and positioning them at the downstream apexes. This could not be confirmed in this study due to the experimental set-up. Here, two breakers per side wall proved to be very effective. At the same time the influence on the discharge capacity was found to be negligible relative to the non-vented nappe conditions.

Nappe vibrations, also known as "fluttering fountains", described by Falvey (2003) and intensively examined by Lodomez *et al.* (2018b), were expected but not observed. This could be due to the height of the labyrinth weir, but it is more likely that the interaction of the nappes prevents the occurrence. For the half-round crest no nappe instabilities and vibrations could be observed. Overall, this appears to be the more robust crest shape.

ACKNOWLEDGEMENTS

The authors thank udo pfrommer and timo krebs for their great support in designing and constructing the physical model, sara linzen and jost voges for the structured execution of the physical experiments and for producing high quality data.

REFERENCES

Belzner, F., Merkel, J., Gebhardt, M. & Thorenz, C. (2017) Piano Key and Labyrinth Weirs at German waterways: Recent and future research of the BAW. In: Erpicum, S., Laugier, F., HoTaKhanh, M. & Pfister, M. (eds.) *Labyrinth and Piano Key Weirs III*. Boca Raton, CRC Press. pp. 167–174.

Casperson, L.W. (1993) Fluttering fountains: Simplified models. *J. Appl. Phys.*, 74 (8), 4894–4898.

Crookston, B.M. (2010) *Labyrinth Weirs*. [Online]. Logan. Available from: https://digitalcommons.usu.edu/etd/802.

Crookston, B.M., Anderson, A.A., Shearin-Feimster, L. & Tullis, B.P. (2014) Mitigation investigation of flow-induced vibrations at a rehabilitated spillway. In: *Hydraulic structures and society - Engineering challenges and extremes:*, 5th IAHR International Symposium on Hydraulic Structures. The University of Queensland. pp. 1–8.

Crookston, B.M. & Tullis, B.P. (2013) Hydraulic Design and Analysis of Labyrinth Weirs. II: Nappe Aeration, Instability, and Vibration. *Journal of Irrigation and Drainage Engineering*. [Online] 139 (5), 371–377. Available from doi:10.1061/(ASCE)IR.1943-4774.0000553.

Erpicum, S., Tullis, B.P., Lodomez, M., Archambeau, P., Dewals, B.J. & Pirotton, M. (2016) Scale effects in physical piano key weirs models. *Journal of Hydraulic Research*, 1–7.

Falvey, H.T. (2003) *Hydraulic Design of Labyrinth Weirs*. Reston, Va, ASCE Press (American Society of Civil Engineers).

Gebhardt, M. & Belzner, F. (2024) Perspective for the Use of Labyrinth Weirs on Waterways. In: The World Association for Waterborne Transport Infrastructure (ed.) *Proceedings of the 35th PIANC World Congress 2024:*, 29.04–03.05.2024. Brussels. pp. 1601–1607.

Gebhardt, M., Merkel, J., Belzner, F. & Thorenz, C. (2017) A comparison of side weirs and labyrinth weirs at Ilmenau river. In: Erpicum, S., Laugier, F., Khanh, M.H.T. & Pfister, M. (eds.) *Labyrinth and Piano Key Weirs III: PKW 2017*. Boca Raton, CRC Press. pp. 221–228.

Linzen, S. (2023) *Hydraulische Untersuchungen zur Belüftung bei Labyrinth-Wehren*. Karlsruhe.

Lodomez, M., Pirotton, M., Dewals, B., Archambeau, P. & Erpicum, S. (2017a) Could Piano KeyWeirs be subject to nappe oscillations? In: Erpicum, S., Laugier, F., Khanh, M.H.T. & Pfister, M. (eds.) *Labyrinth and Piano Key Weirs III: PKW 2017*. Boca Raton, CRC Press. pp. 135–144.

Lodomez, M., Pirotton, M., Dewals, B., Archambeau, P. & Erpicum, S. (2018b) Nappe Oscillations on Free-Overfall Structures: Experimental Analysis. *Journal of Hydraulic Engineering*. [Online] 144 (3). Available from doi:10.1061/(ASCE)HY.1943-7900.0001420.

Naudascher, E. & Rockwell, D. (1994) *Flow-induced Vibrations: An Engineering Guide*. Rotterdam, A.A. Balkema.

Partenscky, H.-W. & Khloeung, I.S. (1967) Oscillations de Lames Déversantes non Aérées. *Proc. XII IAHR Congr.*, 6.1–7.1.

Pfister, M., Battisacco, E., De Cesare, G. & Schleiss, A. (2013) Scale effects related to the rating curve of cylindrically crested Piano Key weirs. In: Erpicum, S., Laugier, F., Pfister, M., Pirotton, M., Cicero, G.-M. & Schleiss, A.J. (eds.) *Labyrinth and Piano Key Weirs II*: CRC Press/Balkema. pp. 73–82.

Voges, J.C. (2024) Hydraulische Modelluntersuchungen an einem Labyrinth-Wehr – Einfluss des Unterwasserstandes auf die Belüftung und die Strömungsinstabilitäten des Überfallstrahls. Karlsruhe.

I.2 Design and application

Labyrinth and Piano Key Weirs IV – Erpicum, Crookston & Paxson (Eds)
© 2026 The Author(s), ISBN 978-1-041-10527-5

Theoretical and experimental studies of non-rectilinear weirs: From labyrinth to piano key weir

A. Ouamane
Laboratory of Hydraulic Planning and Environment, University of Biskra, Algeria

F. Lempérière
Hydrocoop, France

ABSTRACT: Free-flow rectilinear spillways, while simpler and safer than gated or shaft spillways, have a low specific discharge, requiring a significant water head and resulting in a storage loss of more than 20% for some small reservoirs. To overcome this drawback, labyrinth weirs provide an effective alternative, allowing high discharge under a reduced head while maintaining lower construction and maintenance costs. For several decades, global research has focused on optimizing their design. In this context, a research project was initiated in 1995 at the University of Biskra to enhance the understanding of the hydraulic behavior of these structures. This project was carried out in two phases: the first (1995-2000) was dedicated to theoretical and experimental studies on labyrinth weirs, serving as a key step in the development of the PK-Weir; the second (2001-2003) involved close collaboration with Hydrocoop-France, culminating in 2003 with the final design of the Piano Key Weir (PK-Weir). The present study focuses on the first phase, which led to this innovation.

1 INTRODUCTION

The spillway is essential for dam safety, preventing overtopping, which is a leading cause of failure. Given the risks associated with insufficient discharge capacity, ICOLD recommends their adaptation. The labyrinth weir, with its non-rectilinear crest (triangular, trapezoidal, or rectangular), offers an effective solution, ensuring improved flood discharge under a reduced head and optimized cost, making it particularly suitable for sites with limited spillway width. However, the complexity of the flow and the geometric variability remain challenges in its design. This article outlines the research conducted at the University of Biskra since 1995, combining theoretical studies and experimental investigations to analyse the flow behavior and optimize the hydraulic performance of non-rectilinear weirs. These efforts led, in 2003, to the development of the Piano Key Weir (PK-Weir), an innovative structure that enhances hydraulic efficiency, optimizes storage, and improves dam safety.

2 LABYRINTH WEIR

The labyrinth weir is characterized by a crest with a non-rectilinear axis, increasing the developed length for a given width to enhance discharge under a given hydraulic head. Designed to operate under low heads, these structures optimize hydraulic efficiency while limiting dam height. They adopt various plan shapes (trapezoidal, triangular, rectangular, curvilinear, or mixed), with the symmetric trapezoidal configuration being the most

DOI: 10.1201/9781003655541-5

common due to its simplicity and performance (Hay and al., 1970). Their geometry is defined by several parameters, including the length (L) and width (W) of a cycle, the upstream and downstream heights (P and D), the number of cycles (n), and the angle (α') formed by the sidewalls and the flow direction (Kathleen and al., 1982).

The L/W and W/P ratios are commonly used to characterize the geometry of labyrinth weirs, representing respectively the crest elongation and the vertical aspect, which expresses the horizontal width relative to the height of the upstream wall. Due to its non-rectilinear shape, the flow over a labyrinth weir is complex and three-dimensional. Studies, notably those by Indlekofer (1975) and Lux III et al. (1985), have shown that this flow can be divided into two zones: an area near the corners, where the flow is three-dimensional, and a downstream area, dominated by two-dimensional flow.

The flow over a labyrinth weir evolves through four phases depending on the water head: an aerated phase with a free nappe, a partially aerated phase with stable air pockets, a transitional unstable phase, and a depressed phase where the flow becomes non-aerated, reducing efficiency (Hinchliff and al., 1984; Lux III and al., 1985). Several studies confirm that performance decreases with increasing head, particularly for weirs with an extended crest length (Hinchliff and al., 1984; Afshar, 1985; Yildiz and al., 1996; Zerrouk and al., 1995).

The concept of labyrinth weir performance, introduced by Hay and Taylor (1970), is assessed using the discharge between a labyrinth weir (Q_L) and an equivalent rectilinear weir (Q_N), expressed as $q = Q_L/Q_N$. Darvas (1971), Magalhães (1985), and Houston and al. (1982) demonstrated that at low heads, Q_L/Q_N tends toward L/W, whereas higher heads reduce performance, confirming optimal efficiency under low-head conditions.

Performance depends on dimensionless parameters such as the crest length ratio L/W, the corner angle α, the number of cycles n, and the ratios H/P and W/P (Hay and al., 1970). These parameters define the geometric configuration of the weir and influence its efficiency.

The discharge coefficient, a key performance indicator, varies based on hydraulic head and geometry. Darvas (1971) developed experimental design curves, later revised downward by Magalhães (1985). Research by the USBR (Houston, 1982–1983; Hinchliff, 1984) highlighted variations depending on head definition (total vs. piezometric), with findings corroborated by Zerrouk and al. (1995). Hay and Taylor (1970) reported the highest values, while Magalhães obtained more conservative results.

Tullis (1995) proposed an adaptation of the standard rectilinear weir equation, incorporating the effective crest length and the total head. These studies indicate that the crest coefficient is influenced by geometry and nappe aeration and its values cannot exceed those of a rectilinear weir.

Finally, while discharge capacity is primarily governed by H/P, W/P, L/W, α, and n, secondary factors may also play a role. Their impact was studied at the University of Biskra between 1995 and 2001.

3 WORK ON THE LABYRINTH WEIR CARRIED OUT FROM 1995 TO 2001

The literature review on labyrinth weirs has highlighted the need for further research to deepen the understanding of their hydraulic behavior and improve their design. In this context, a research program dedicated to non-linear weirs was launched in 1995 at the University of Biskra (Algeria).

3.1 *Effect of the discontinuity of the labyrinth weir crest on the flow*

To deepen the understanding of flow dynamics over labyrinth weirs, experimental tests were conducted on six triangular weir models with crest corner angles ranging from 45° to 180° (rectilinear model). During the experiments, the formation of two opposing nappe flows was observed, converging as the hydraulic head increased. This convergence generates a central jet and localized disturbances at the weir corner, whose intensity increases with the head, highlighting the influence of weir geometry on the flow (Figure 1).

Figure 1a shows the convergence of the nappe flows upstream, capped by a central jet whose horizontal length (l_{jc}) increases in parallel with the extension of the disturbed area ($l_d/2$) along the walls. Figure 1b shows this dynamic schematically, with concentric circles illustrating the expansion of the disturbed area and ellipses symbolising the evolution of the central jet. This configuration was partially explored by Indlekofer and Rouvé (1975), who established a relationship between the discharge and the length of the disturbed area l_d.

In this study, the central jet is defined by its horizontal length l_{jc}, measured using a precision ruler, while the disturbance length l_d represents the sum of the two disturbance lengths ($l_d/2$) measured along the two side walls of the triangular labyrinth weir. The total length of the weir crest is l_c.

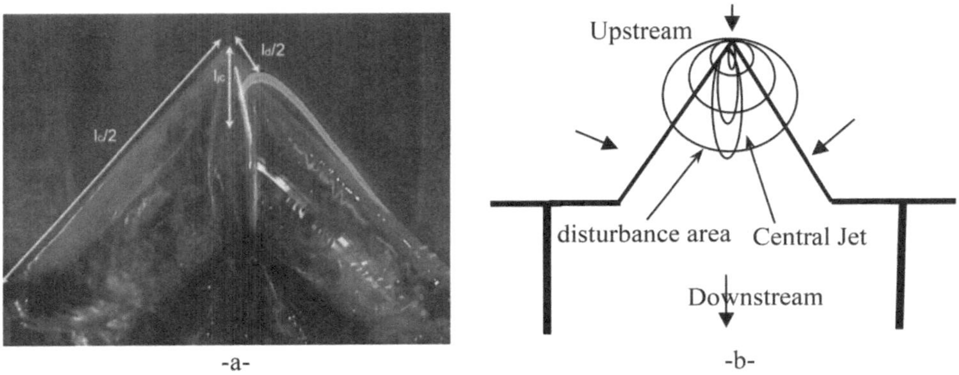

Figure 1. Flow over a labyrinth weir in the presence of disturbance.

In order to examine the relationship between the disturbance length, the opening corner and the hydraulic head, it was appropriate to plot the relative disturbance length l_d/l_c as a function of the relative head H^*/P and the opening corner.

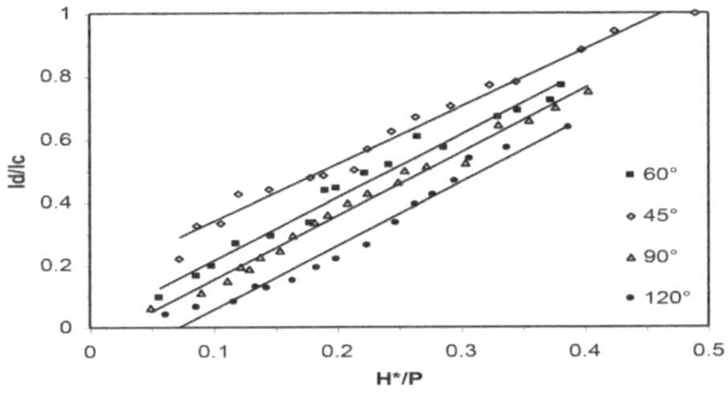

Figure 2. Relative disturbance length l_d/l_c as a function of relative head H^*/P.

From Figure (2), which shows the variation of l_d/l_c as a function of H^*/P and the corner, it can be seen that the relationship between l_d and H^*/P for each opening corner of the triangular weir follows a linear trend, of the form $y=ax+b$, where a and b are angle-dependent constants. Consequently, the relative disturbance length can be expressed by the following equation:

$$l_d/l_c = a \cdot H*/P + b \tag{1}$$

Table 1. Values of constants a and b in relation (1).

α	*a*	*b*
45°	1.824	0.1560
60°	2.012	0.0115
90°	2.033	-0.0514
120°	2.040	-0.1480

The straight lines representing the different corners are almost parallel, indicating similar slopes. The intersection of these lines with the y-axis shows that b can take on positive or negative values. The constants a and b were determined by analysing the pairs of values (l_d/l_c, H^*/P).

Table (1) shows that the slope (*a*) remains practically constant for different angles (α) (with a ≈ 2), except for α = 45°. In contrast, the constant (*b*) varies as a function of α can take on positive or negative values. This suggests that the constant (*a*) is independent of the corner, whereas the constant (*b*) is strongly influenced by α.

It is noteworthy that the discharge coefficient decreases with the increase in the relative perturbation length and the relative length of the central jet. This decrease is particularly pronounced for the opening angle of 45°, where the convergence of the spilling nappe occurs more rapidly due to the narrowing of the sidewalls, generating intense perturbation. This perturbation gradually extends over the entire crest, especially under high head conditions (Figure 3).

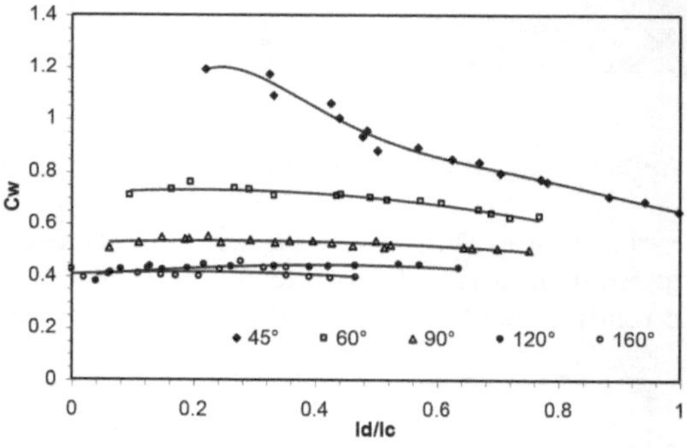

Figure 3. Discharge coefficient as a function of the angle and the relative perturbation length l_d/l_c.

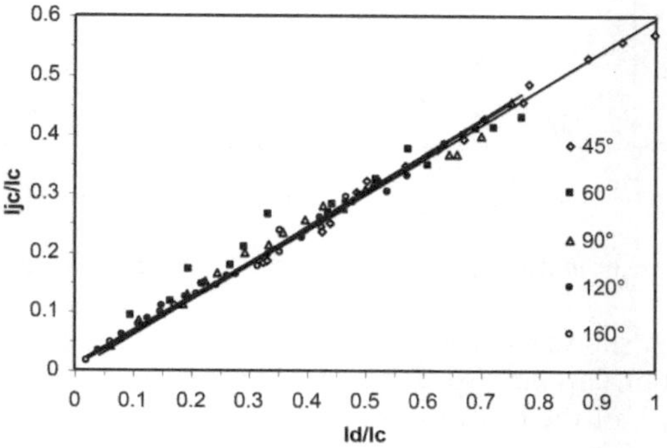

Figure 4. Relative central jet length as a function of relative perturbation length and angle.

Figure (4) illustrates the relationship between the relative central jet length l_{jc}/l_c and the relative perturbation length l_d/l_c, regardless of the opening corner. Although the curve appears as a single straight line, this does not imply that the central jet length l_{jc} is independent of the angle α. The relationship between l_d/l_c and l_{jc}/l_c is linear and unique, taking the form $y = a{\cdot}x$.

Using the least squares method to analyse the pairs of values (l_{jc}/l_c, l_d/l_c), the constant (a) was determined to be 0.6. Thus, the relative central jet length can be expressed by the linear relation:

$$l_{jc}/l_c = 0.6 \cdot l_d/l_c \tag{2}$$

This relationship demonstrates that the central jet length is closely linked to the evolution of the perturbation, which influences the flow over a labyrinth weir.

From a practical standpoint, the horizontal length of the central jet (l_{jc}) is easier to measure than the disturbance length, which tends to be unstable, particularly under high hydraulic heads. Therefore, the central jet length can serve as a reliable indicator of the degree of weir submergence.

3.2 *Impact of the relative length L/W and the relative width W/P*

The performance of labyrinth weirs primarily depends on the relative length L/W, while the impact of the vertical aspect ratio W/P remains minimal. Tests on ten trapezoidal models have shown that increasing L/W enhances performance due to the extended crest length. However, beyond a certain head ($H^*/P>0.5$), the convergence of the nappe flow reduces discharge capacity, thereby limiting this advantage.

Moreover, the effect of the vertical aspect ratio W/P on performance is significant up to a relative head H^*/P of approximately 1, after which it becomes negligible. Additionally, while increasing L/W enhances discharge capacity, its impact becomes marginal beyond a certain value for a given head. Thus, an excessive increase in L/W ($L/W>6$) may not be justified due to additional construction costs.

3.3 *Impact of upstream and downstream weir height*

Studies on labyrinth weirs have mainly been conducted with equal upstream and downstream heights ($D/P=1$). It is therefore relevant to examine the effects of height differences while keeping other geometric parameters constant. Three cases are considered: $P=D$, ($D/P=1$), $P<D$, ($D/P>1$), and $P>D$, ($D/P<1$).

The evaluation of the effect of downstream height on flow was conducted while keeping the upstream height of the weir constant. Tests with D/P ratios of 0.33, 0.5, and 0.67 showed that a decrease in downstream height can reduce the weir's performance.

Comparing these results with the reference configuration ($D/P=1$), it was observed that for $H^*/P < 0.3$, the discharge coefficients remain similar. However, beyond this value, a divergence appears, attributed to the rise in the downstream water level, which may reach or exceed the weir crest level, thereby modifying the upstream flow.

Conversely, tests conducted with $D/P>1$(1.5, 2.0, and 3.0) indicate an increase in discharge coefficients compared to the reference, suggesting that a higher downstream height enhances the weir's discharge capacity.

3.4 *Influence of the planform shape of the labyrinth weir*

Labyrinth weirs exhibit various planform shapes, characterized by a broken line following a specific geometry. To analyse their impact on flow behavior, four configurations were studied: triangular, trapezoidal, rectangular, and trapezoidal with an upstream rounded crest. The trapezoidal shape, commonly used in weir design, was chosen as the reference model for comparing the performance of the other configurations.

Trapezoidal shape   Triangular shape   Trapezoidal shape(n=2)   Rounded trapezoidal shape

Figure 5. Planform labyrinth weir of different shapes.

Figure 6a shows that triangular and trapezoidal labyrinth weirs exhibit similar flow behavior, with comparable variations in the discharge coefficient. The triangular model demonstrates a slight advantage at low heads, which diminishes as the head increases. However, its adoption is not justified, as it requires a larger footprint and leads to higher construction costs.

Graph 6b highlights the relationship between the discharge coefficient and the trapezoidal and rectangular plan shapes. For relative heads (H^*/P) below 0.5, the discharge coefficient is higher for the rectangular shape. However, beyond this value, the coefficients tend to equalize or even become slightly lower for the rectangular model. This trend is due to the early convergence of lateral nappe flows in the trapezoidal configuration, which limits its hydraulic efficiency.

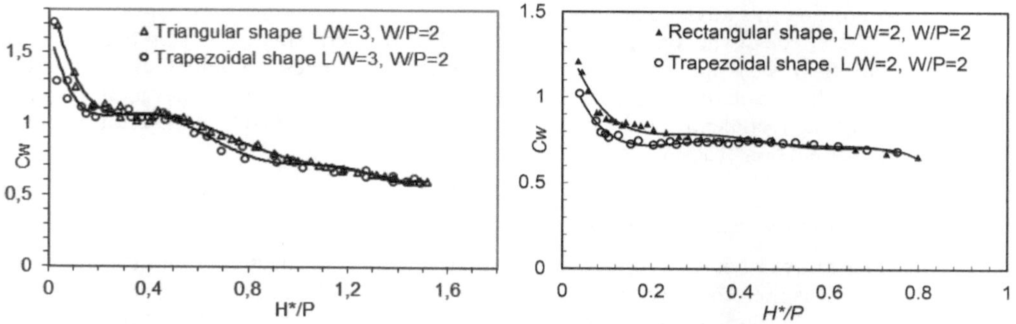

a   b

Figure 6. Discharge coefficient as a function of the plan shape of the labyrinth weir.

The upstream-rounded trapezoidal labyrinth weir is an economical solution for specific discharges below 15 m³/s.m. It is suitable for nappe depths ranging from 1 to 2.5 m and wall heights between 2 and 4 m, with an average wall thickness of approximately 10% of the wall height. This design allows for a 50 to 70% reduction in weir width compared to a rectilinear weir while maintaining the same nappe depth and doubling the discharge.

A hydraulic comparison between labyrinth weirs with upstream-rounded trapezoidal and planar trapezoidal shapes was conducted using tests on two models with identical dimensions. The results are presented in Figure 7.

Graph 7 shows that the labyrinth weir with an upstream-rounded trapezoidal shape has a higher discharge coefficient than the conventional trapezoidal model. This improvement is attributed to its rounded profile, which stabilizes the flow and optimizes discharge capacity. For relative heads below 0.6, its efficiency is significantly higher, while beyond this value, the performance of both configurations becomes similar.

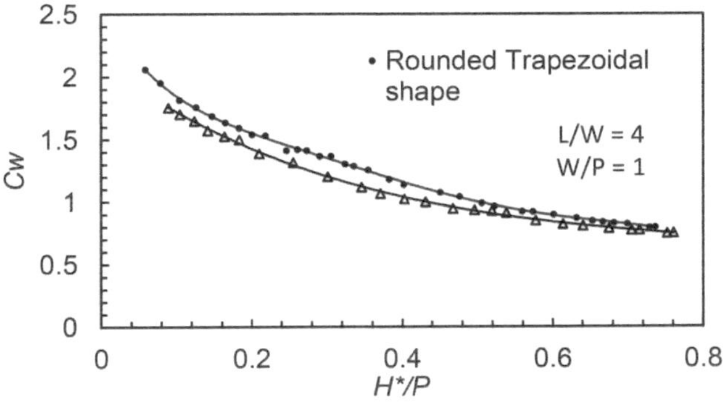

Figure 7. Discharge coefficient as a function of the labyrinth weir plan shape (Trapezoidal shape and upstream-rounded trapezoidal shape).

3.5 *Approach conditions*

The performance of a labyrinth weir is highly dependent on approach flow conditions, particularly flow direction. Two configurations were studied: without lateral contraction (constant-section channel) and with lateral contraction (reservoir outlet).

Tests conducted on a triangular-shaped model show that, without lateral contraction, the flow remains parallel and perpendicular to the weir width, whereas with lateral contraction, the water is directed perpendicularly to the sidewalls. Moreover, the absence of lateral contraction promotes a higher flow velocity, thereby reducing the head for a given discharge. This configuration allows for higher discharge rates under lower heads, providing improved hydraulic efficiency.

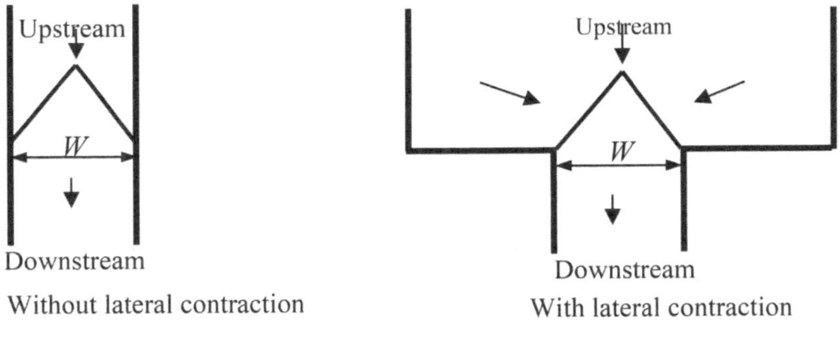

Figure 8. Diagrams of the approach flow conditions tested for the labyrinth weir.

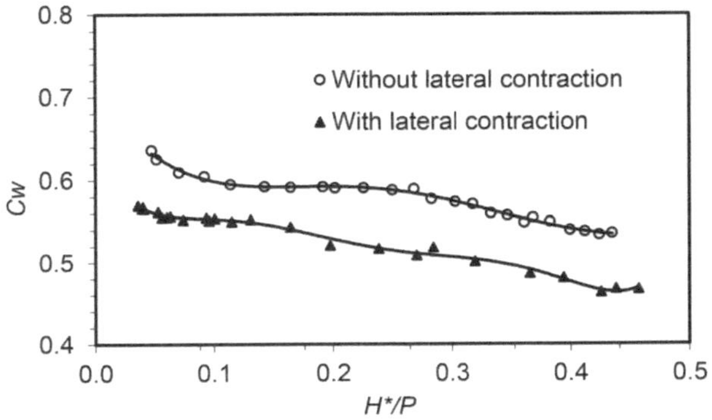

Figure 9. Discharge coefficient as a function of relative head and contraction.

The experimental tests showed that lateral contraction significantly affects the discharge coefficient. The model without lateral contraction exhibits higher values due to a stable, parallel frontal flow up to the crest, enhancing the discharge capacity. In contrast, lateral contraction creates a more complex flow pattern, combining frontal and lateral flows, which disrupts the flow and reduces the hydraulic efficiency of the weir.

3.6 *Effects of the number of cycles on labyrinth weir performance*

A labyrinth weir is typically composed of multiple cycles with identical geometry, assembled to form a continuous structure. The selection of the number of cycles depends on site constraints and the required discharge capacity. For a given width, the number of cycles can vary based on the desired configuration. To assess the impact of the number of cycles on discharge capacity, two weir models with identical geometry but different cycle numbers ($n=2$ and $n=3$) were tested.

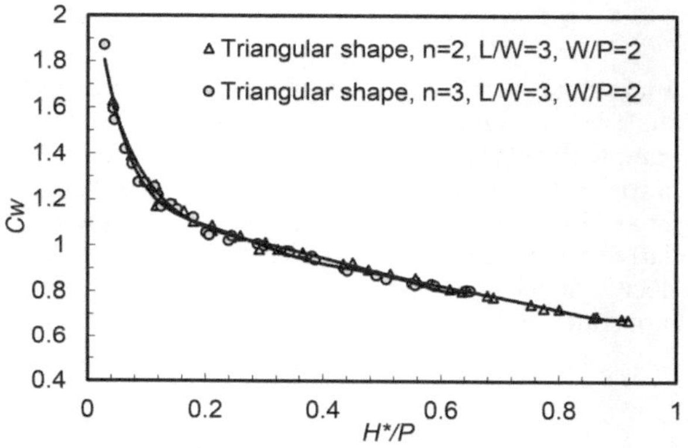

Figure 10. Discharge coefficient as a function of the number of cycles (Triangular shape, without lateral contraction).

Graphical representation (10) shows that the discharge coefficient curves as a function of the number of cycles are similar for both models, with no significant variation in trend. This indicates that the performance of the labyrinth weir is independent of the number of cycles. Thus, increasing the number of cycles for a given width is beneficial, as it reduces the transverse dimension of the weir and minimizes the volume of concrete required for the apron between the sidewalls.

3.7 *Labyrinth spillway with downstream overhang*

Following the creation of Hydrocoop-France in 1997, efforts were undertaken to optimize labyrinth weirs. In this context, an alternative configuration was proposed, incorporating a downstream overhang to facilitate their implementation on concrete dams with limited crest width (P. Blanc and F. Lempérière, 2001). This design aims to reduce the weir's footprint while maintaining high discharge capacity. This configuration was the subject of experimental studies conducted at the Laboratory of Hydraulic Planning and environment of the University of Biskra (Algeria), with the objective of evaluating its hydraulic performance.

Figure (12) shows that, for a moderate relative head, the discharge coefficient of the labyrinth weir with a downstream overhang is three times higher than that of a rectilinear weir. This type of weir achieves performance comparable to that of vertical-walled labyrinth weirs while benefiting from a reduced footprint due to its overhanging shape and improved downstream flow aeration.

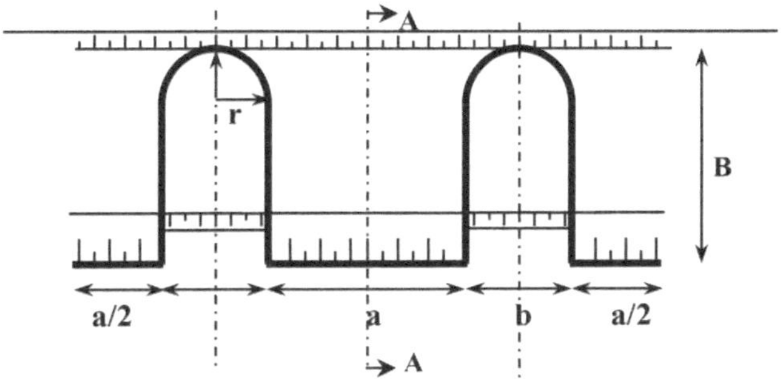

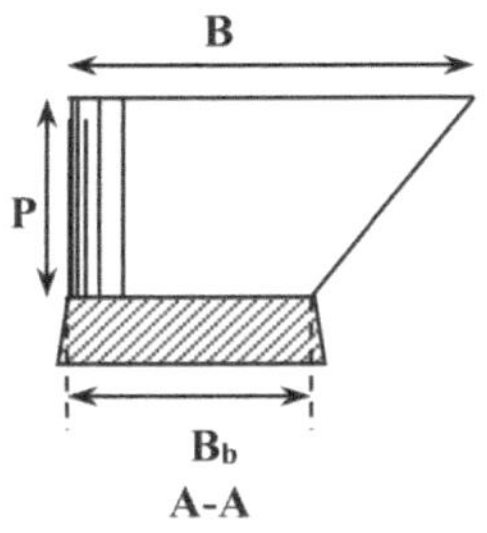

Figure 11. Diagram of the labyrinth weir model with downstream overhang.

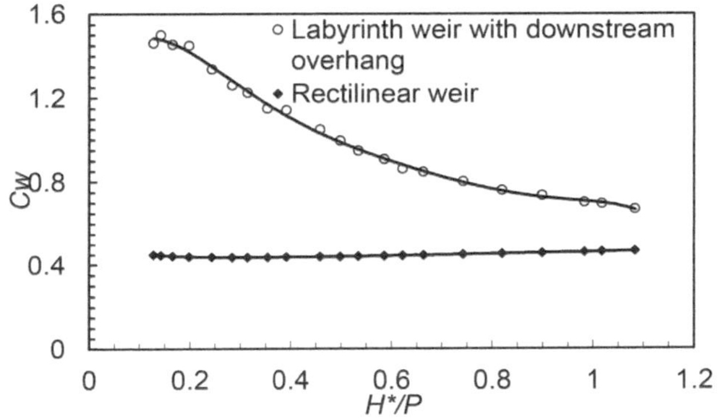

Figure 12. Flow over the labyrinth weir model with downstream overhang.

3.8 *Labyrinth weir with an inclined apron*

The use of an inclined apron in labyrinth weirs helps reduce wall height and minimize turbulence upstream and downstream while decreasing the amount of reinforcing steel required. However, this configuration increases the volume of concrete in the apron. To assess its impact on flow behavior, experiments were conducted on two models with identical dimensions: one with an inclined apron and the other with a horizontal apron.

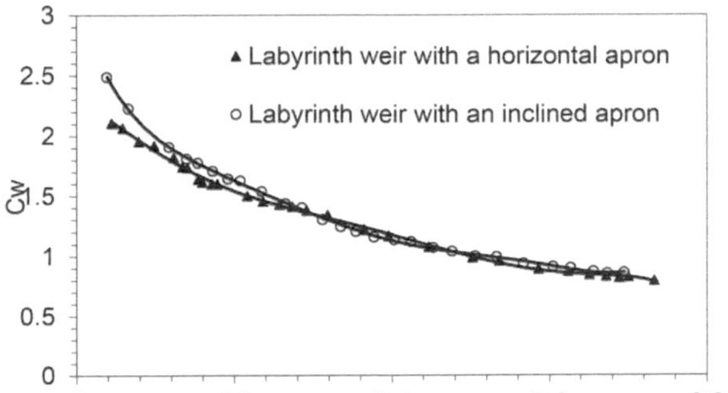

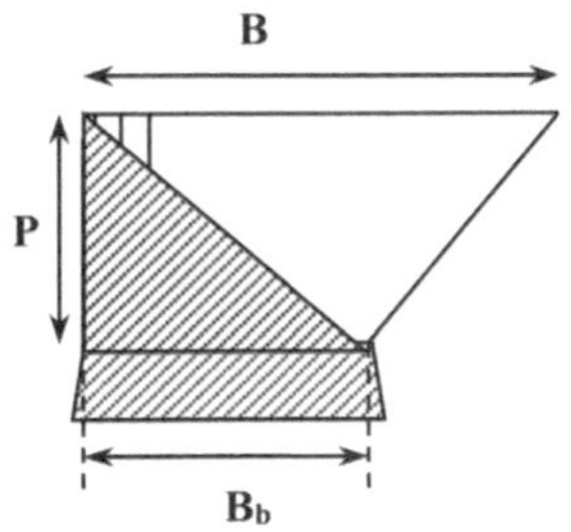

Figure 13. Discharge coefficient of the labyrinth weir as a function of the apron slope in the upstream and downstream cells, $L/W = 4$, $W/P = 1.5$.

Experimental tests indicate that the head on the weir is similar for both inclined and horizontal apron configurations. However, for relative heads $H^*/P < 0.3$, an inclined apron enhances the discharge coefficient, suggesting better hydraulic performance within this range of heads commonly used in practice (Figure 13). The choice between a horizontal or inclined apron should be based on structural assessment and site conditions.

The results of research conducted on labyrinth weirs at the University of Biskra in Algeria between 1995 and 2001 served as a basis for a new phase of study, initiated in 2001 in collaboration with Hydrocoop-France.

4 CONCLUSION

This paper presents only a portion of the research conducted between 1995 and 2001 on labyrinth weirs, which led to several significant improvements, including the adoption of a rectangular plan shape, the inclination of the apron, the use of overhangs, and the optimization of approach conditions. These advancements resulted in the design of the Piano Key Weir (PK-Weir), developed in 2003 by François Lempérière (Hydrocoop-France) and Ahmed Ouamane (University of Biskra-Algeria).

This innovative structure enhances flood discharge capacity while integrating efficiently with gravity dams of limited width.

REFERENCES

Afshar, A. 1988. The development of labyrinth spillway designs. *International Water Power & Dam Construction*, 40 (5): 36–39.

Blanc, P. & Lempérière, F. 2001. Labyrinth spillways have a promising future. *The International Journal on Hydropower and Dams, 8*, 129–131

Darvas, L.A. 1971. Performance and design of labyrinth weirs. *Journal of Hydraulic Division*, 08: 1246–1251

Falvey, H. T. 1995. Theory and design of labyrinth spillways. *Henry T. Falvey & Associates, Inc.*

Hay, N. & Taylor, G. 1970. Performance and design of labyrinth weirs. *Journal of the hydraulics division Proceedings of ASCE*, Vol 96, N° 11: 2337–2357.

Hinchliff, D. L. & Houston, K. L. 1984. Hydraulic Design and Application of Labyrinth Spillways. *Division of Research Engineering and Research Center, Bureau of Reclamation, Denver, Colorado.*

Houston, K. L. & DeAngelis C. S. 1982. A Site-Specific Study of a Labyrinth Spillway. *Proceeding of the conference Applying Research to Hydraulic Practice, Hydraulics Division of ASCE.* 86–95.

Indlekofer, H. & Rouvé, G. 1975. Discharge over Polygonal Weirs. *Journal of the hydraulics division Proceedings of ASCE.* 101. 3: 385–401

Lempérière, F.& Ouamane, A. 2003. The piano keys weir: a new cost-effective solution for spillways. *The International Journal on Hydropower and Dams.* 10.5

Ouamane A. & Lempérière F. 2006. Design of a new economic shape of weir. *International Symposium on Dams in the Societies of the XXI Century. Barcelona, Spain.* 463–470, CRC Press, Taylor & Francis

Lux III, F. & Hinchliff, D.L. 1985. Design and Construction of Labyrinth Spillways. *15th Congress on Large Dams, ICOLD*, Vol. IV, Q59, R 15. 249–274, Lausanne.

Magalhães, A. P. 1985. Labyrinth-Weir Spillways. *15th Congress on Large Dams, ICOLD*, Vol. IV, Q59, R 24. 395–407, Lausanne

Tullis, J.P.& Amanian, N. & Waldron, D. 1995. Design of Labyrinth Spillways. *Journal of hydraulic engineering*, 121, (3). 247–255

Yildiz, D.& Üzücek, E. 1996. Modelling the performance of labyrinth spillways. *The International Journal on Hydropower & Dams.* 3: 71–76

Zerrouk, N.& Marche, C. 1995. Les évacuateurs de crues en labyrinthe: dimensionnement, fonctionnement, exemples et particularités. *Canadian Journal of Civil Engineering*, 22, pp. 916–924

Labyrinth and Piano Key Weirs IV – Erpicum, Crookston & Paxson (Eds)
© 2026 The Author(s), ISBN 978-1-041-10527-5

Enhancing the hydraulic efficiency of Piano Key Weir – Recent research achievements

J. Yang, S.C. Li and A. Ansell
Civil & Architectural Engineering, KTH Royal Institute of Technology, Stockholm, Sweden

E. Skepparkrans
R&D Hydraulic Laboratory, Vattenfall AB, Älvkarleby, Sweden

C. Lin
Department of Civil Engineering, National Chung Hsing University, Taichung, Taiwan

ABSTRACT: As an effective flood control structure, the piano key weir (PKW) typically features a rectangular planform for its overhangs. However, this design offers room for improvement. To enhance hydraulic efficiency, two modified PKW layouts are proposed. In the first, the floor of each key was lowered with a semicircular cross-section, forming an elliptical planform and crest. In the second, the floor was lowered with an isosceles triangular prism, creating an equilateral triangular overhang. Both modified PKW models, manufactured by 3D printing, were tested in a large-scale experimental setup. Compared to the reference weir, the elliptical-overhang PKW increased the developed crest length by ~36% and enhanced discharge by 30–53% across the tested flow range. The triangular-overhang PKW extended the crest length by ~23%, resulting in a 16–20% increase in discharge. The modified inlet key crest, whether elliptical or triangular, extended further downstream, promoting jet break-up, air entrainment at low to medium-high flows, and improved energy dissipation. The elliptical or triangular overhang facade, combined with a lowered inlet key floor, reduced entrance energy losses, improving inflow conditions. The lowered floor also accommodates additional water volume, enhancing flow motion toward the crest. For the outlet key, the lower floor facilitates outflow and mitigates local submergence at high floods. These modifications offer more effective PKW designs, which is particularly beneficial when reservoir water level increases must be controlled during floods or when spillway construction space is limited—contributing to improved dam safety and cost efficiency.

1 INTRODUCTION

Based on laboratory model studies, the concept of the piano key weir (PKW) was first introduced in 2003 (Lempérière & Ouamane 2003; Ouamane & Lempérière 2006; Ouamane 2011). The weir features a rectangular plan, with tilted floors and overhanging apexes, making it an innovative and efficient hydraulic structure for flood discharge. The degree of discharge enhancement depends on the water head, as evidenced in numerous hydraulic studies, both physical and numerical (Laugier et al. 2011; Erpicum et al. 2013; Anderson & Tullis 2013; Crookston et al. 2019). With the construction of the first prototype at the Goulours Dam in 2006, Electricité de France (EDF) has since incorporated more than 10 such weirs in the upgrades of its existing dams. Due to its structural simplicity and unique hydraulic performance, over 35 PKWs have been implemented in various countries. More information can be found on the homepage: https://www.uee.uliege.be/cms/c_5026433/en/world-register-of-piano-key-weirs-prototypes.

DOI: 10.1201/9781003655541-06

The conventional PKW with a rectangular planform is undeniably an effective structure for flow release. Few other spillway types can be directly compared to it. However, its configuration is not without limitations. Notably, the rectangular shape results in non-negligible side contractions, which negatively impacts the inflow to the inlet keys. In the model study by Yang et al. (2023), the changes of crest water depth (h) along half a weir cycle are plotted at several water heads (H) (Figure 1), in which W, L = width and developed crest length of one weir cycle and x = curvilinear distance to the weir axis (measured along the crest centerline). H refers to the difference between the undisturbed upstream water stage and the weir crest elevation. At low heads, the water surface along the crest is roughly horizontal from up- to downstream. However, at high heads, the decline in crest depth becomes significant, indicating that the lower part of the crest is not as effective as its upper part. This trend is not an anomaly; it has been consistently observed across all weirs tested in our laboratory. If the impact of side contraction is reduced and the inlet key is enlarged, the adverse crest flow pattern should be counteracted.

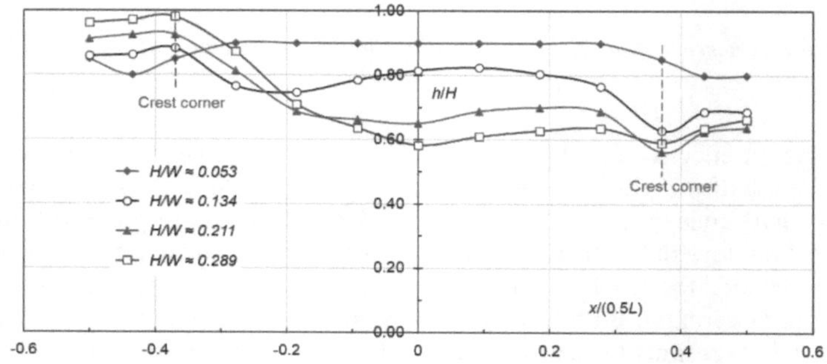

Figure 1. Changes of crest water depth (h) from up- to downstream across a conventional PKW.

Two modified PKW configurations have been worked out, one with an elliptical planform and the other with a triangular one. The inlet and outlet key floors of the rectangular weir are modified, resulting in either elliptical or triangular overhang apexes. High-quality models are manufactured by 3D printing; extensive laboratory tests are performed in a large-scale experimental setup. The study aims to devise weir modifications and evaluate their hydraulic performance. The goal is to recommend two improved PKWs with superior hydraulic efficiency for flood discharge.

2 ELLIPTICAL PKW

With a near-optimal PKW layout, a reference weir of rectangular planform is first defined. Modifications are made by replacing each of its key floors with a downward-facing semicircle in cross-section. In either key, its diameter is equal to the key width. This results in an elliptical overhang

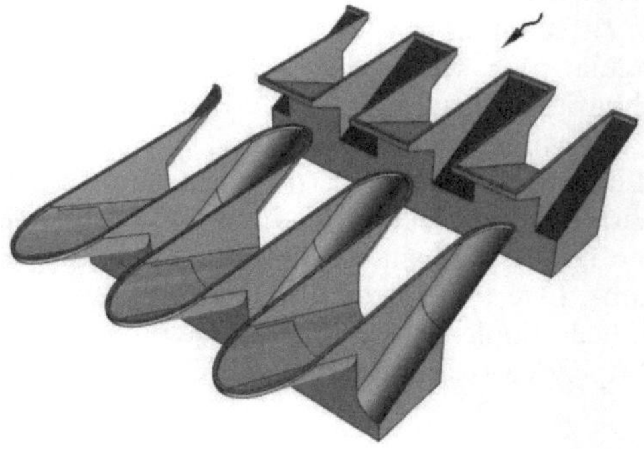

Figure 2. The elliptical PKW as compared to its reference weir (arrow indicating the flow direction).

and crest for both inlet and outlet keys. The exterior of each overhang is also elliptical. The elliptical crest's semi-major and semi-minor axes are dependent on wall thickness (T_s), key width (inlet W_i, outlet W_o), and floor angle with the horizontal plane (inlet α_i, outlet α_o). On each vertical surface of the weir base, the resulting floor shape is also elliptical. Details are given by Yang *et al.* (2023). Together with its reference weir, the modified layout is shown in Figure 2.

In the modifications, the geometrical parameters, W_i, W_o, α_i, α_o and T_s, of the reference weir remain unchanged. α_i and α_o may be equal or different. The dimensions of the elliptical PKW, along with those of its reference weir, are shown in Figures 3 and 4: $W_i = 20.0$ cm, $W_o = 15.0$ cm, $T_s = 1.5$ cm and $W = W_i + W_o + 2T_s = 38.0$ cm. The two floors are assigned the same angle, $\alpha_i = \alpha_o = 29.05°$.

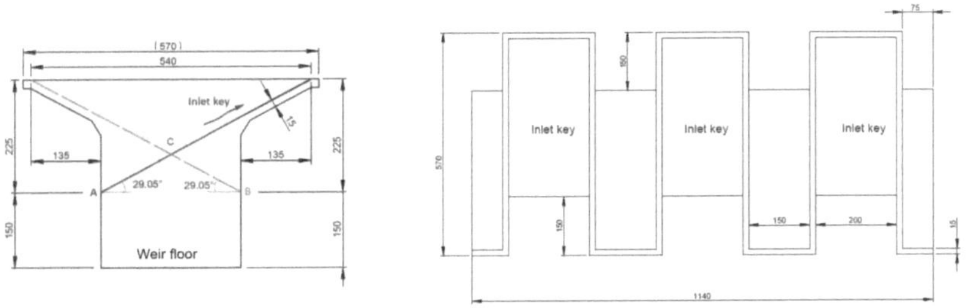

Figure 3. Dimensions of the reference weir for the elliptical-overhang weir (unit mm).

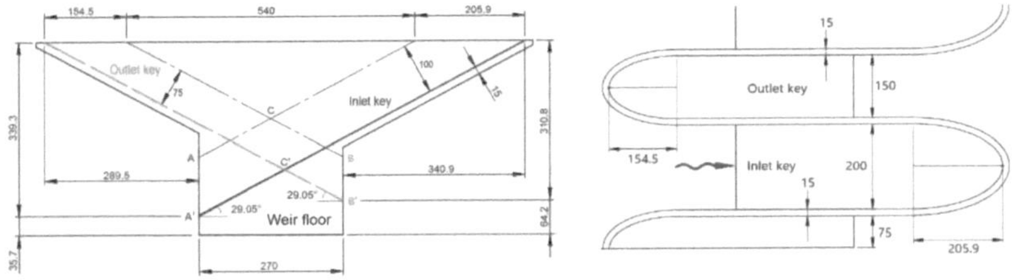

Figure 4. Dimensions of the elliptical-overhang weir (unit mm).

The unit width (W) of the modified weir remains the same. The two weirs also have the same total height (37.5 cm) and base width ($B_b = 27.0$ cm). The weir length, measured from edge to edge on the crest in the inflow direction, is $B = 57.0$ and 93.04 cm for the reference and the modified weir, respectively. The developed crest length changes from $L = 146.0$ to 199.20 cm, implying that the modified crest is $\Delta L = 53.20$ cm longer, a gain of $\beta = 36.4\%$.

3 TRIANGULAR PKW

The development of the triangular configuration is described in the study by Yang and Li (2024). By maintaining the weir length in the flow direction, an isosceles triangle of vertex angle θ is added to the rectangular crest of each key. At the (vertical) surface of the weir base, the key floor is lowered with an isosceles triangle of the same angle θ. By connecting the two triangular vertices at the crest and base, the key floor is lowered, forming the shape of an isosceles triangular prism. For each key, the floor sloping angle (α) differs somewhat from that in the reference weir. However, W_i, W_o and T_s of the reference weir remain unchanged.

Together with its reference weir, Figure 5 illustrates the equilateral triangular configuration ($\theta = 60°$). A parapet wall exists in the outlet keys. This triangular weir has a different reference weir than the elliptical one. Why not use the same reference for both? The reason is that the two studies were conducted a year apart.

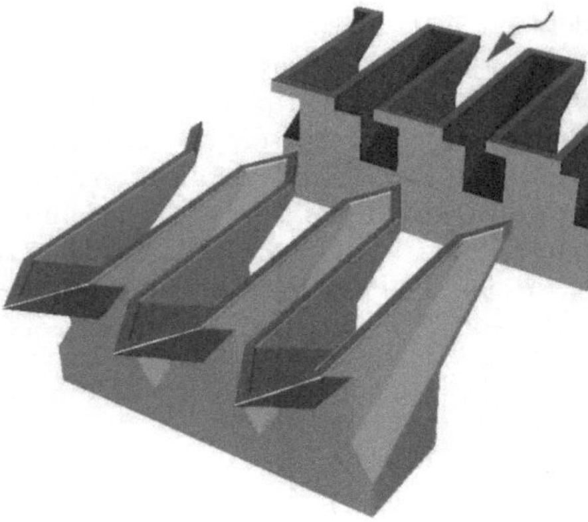

Figure 5. The triangular PKW configuration versus its reference layout (arrow showing the flow direction).

The dimensions examined in the experiments are given in Figures 6 and 7. They are assigned the same total height (51.25 cm) and base width (B_b = 31.25 cm). W_i = 15.0 cm, W_o = 11.67 cm, T_s = 2.92 cm. The α_i and α_o values are not the same in the reference weir; their values become somewhat larger after the modification. The weir length in the inflow direction is B = 69.58 and 97.69 cm, respectively. The developed crest length changes from L = 165.83 to 204.17 cm. This implies that the modified crest becomes ΔL = 38.33 cm longer, representing a β = 23.1% increase.

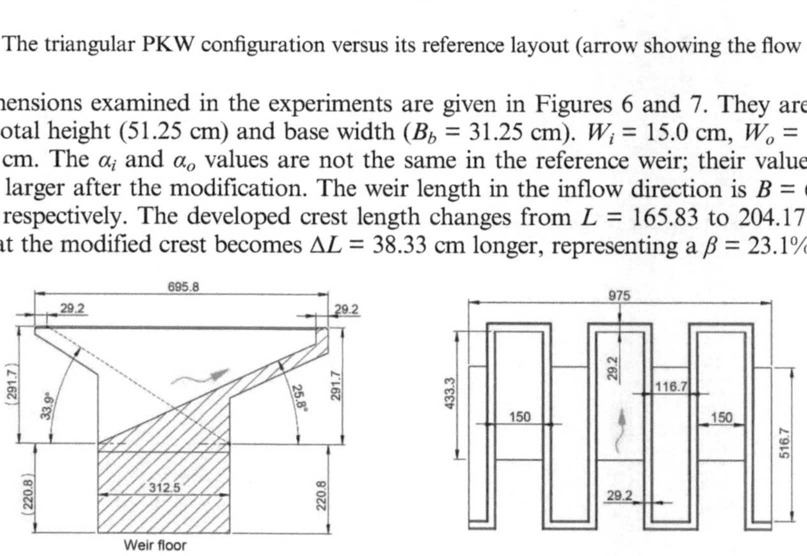

Figure 6. Dimensions of the reference weir for the triangular PKW (arrow indicating the flow direction).

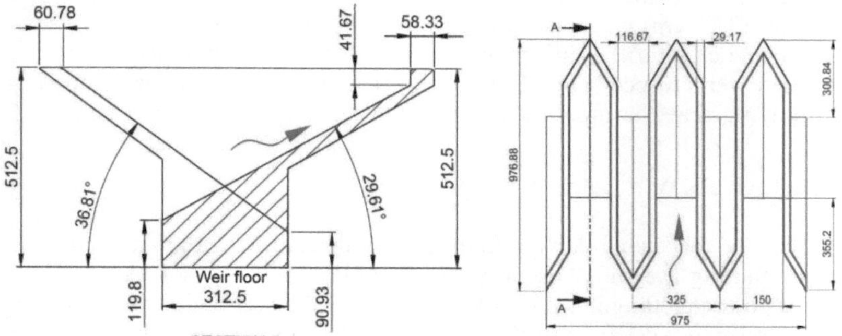

Figure 7. Dimensions of the triangular PKW (arrow indicating the flow direction).

4 EXPERIMENTAL LAYOUT

The PKWs are tested in a large flume measuring 11 m in length, 2 m in width, and 2 m in height (Figure 8). The lower part of the flume is adapted to form a channel, ensuring uniform approach flow conditions. For each weir, three units (width $3W$) are used, positioned at the downstream end of the flume for free outflow and symmetrically aligned about the flume centerline (Figure 9).

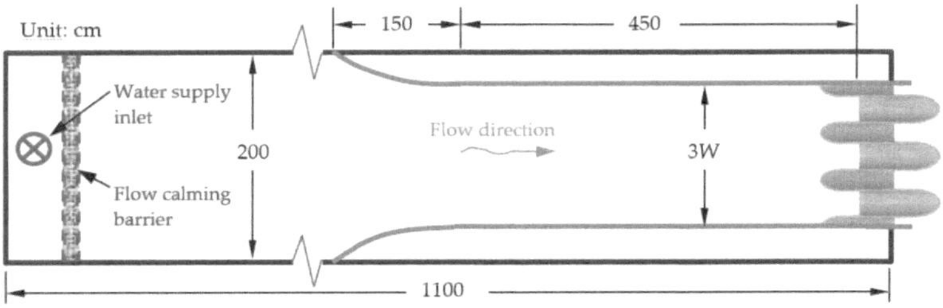

Figure 8. Layout of the test flume, with the PKWs placed at its downstream end for free outflow.

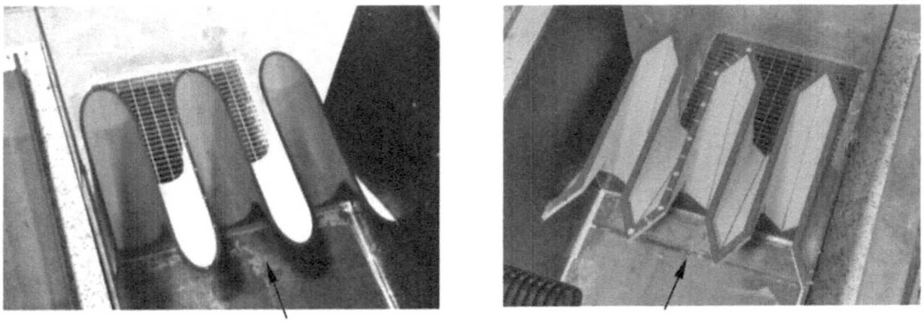

Figure 9. The elliptical and triangular weirs placed at the end of the channel of a $3W$ width.

A frequency-regulated pump supplies flow to the test rig, while a calibrated magnetic flow meter measures the flow rates with a relative error below ±1.5%. Water stages are monitored at several locations along the channel centerline. The falling jets do not enclose the space behind them, allowing the atmospheric pressure to act on both sides (free outflow). To ensure high and uniform quality, all models are produced using 3D printing. A 6 mm nozzle and a 1 mm layer height are used to achieve a relatively smooth surface finish.

5 RESULTS AND DISCUSSIONS

The improvement in discharge capacity and overall hydraulic behavior of the two modified weirs can be evaluated based on the following aspects.

5.1 *Discharge capacity*

The discharge capacity (Q) of a PKW can be expressed as

$$Q = \frac{2}{3} C L \sqrt{2 g H_T^3} \tag{1}$$

in which C = discharge coefficient, H_T = total water head [= $H + V_0^2/(2g)$], V_0 = average approach flow velocity in the channel, and g = gravitational acceleration. As compared with its reference, Figure 10 plots, as a function of H_T/W, the Q changes for both modified weirs. In addition, the relative discharge increase (ξ) and C are compared.

The results demonstrate that, at the same H_T/W ratio, either improved weir significantly enhances overflow discharge. Within the examined flow ranges, the capacity enhancement is

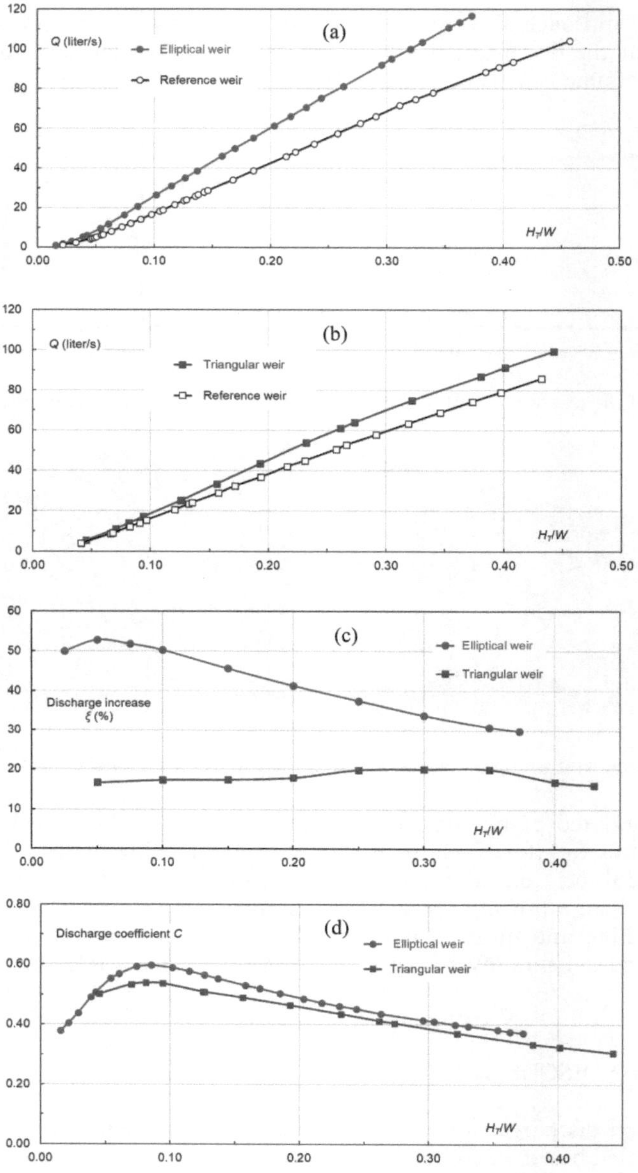

Figure 10. Changes in overflow behaviors: (a) discharge of the elliptical weir relative to its reference, (b) discharge of the triangular weir relative to its reference, (c) discharge capacity enhancement of both modified weirs, and (d) C changes of both modified weirs.

ξ = 30–53% in the elliptical weir and ξ = 16–20% in the triangular weir. For the elliptical, differences exist between low and high water heads, with a more distinct increment in the former. For the triangular, the discharge increase is almost independent of water head. Both modified weirs exhibit similar trend in C, and insignificant deviations from those of their reference weirs (Yang *et al.* 2023, Yang & Li 2024).

In the study by Shen and Ortel (2021), the crest length-to-width ratio (L/W) in their modified design is maintained identical to that of the reference design. This allows the flow efficiency improvements resulting from planform changes to be quantified independently. With all other parameters held constant, the discharge of a trapezoidal PKW can be further increased by up to 10% through optimizing the key width ratio. For the elliptical and triangular weirs, no such ratio optimization was performed on their reference designs. If optimization were applied, it is reasonable to expect that discharge efficiency could be further enhanced—particularly for the triangular weir, given the efficiency enhancement ξ = 16–20% at β = 23.1%.

5.2 *Crest flow parameters*

To further quantify the differences between the modified weirs and their reference designs, the crest water depth (h) and unit flow discharge (q) serve as key parameters. The water depth is measured along the crest centerline, extending from the outlet-key axis upstream to the inlet-key axis downstream, covering half of a weir cycle ($0.5L$ in length). Experimental results indicate that when H is small, the differences in the h variations are negligible. However, as H increases, these differences become pronounced. Figure 11 shows the h variations across the crest at $H \approx 8.0$ cm. The elliptical design gives rise to larger water depths along the weir, while the triangular weir leads to smoother water surface profiles. The trends at larger H values, not shown in here, are similar. This implies that, as compared to their reference layouts, the water depths of both modified weirs become larger and smoother along the crests.

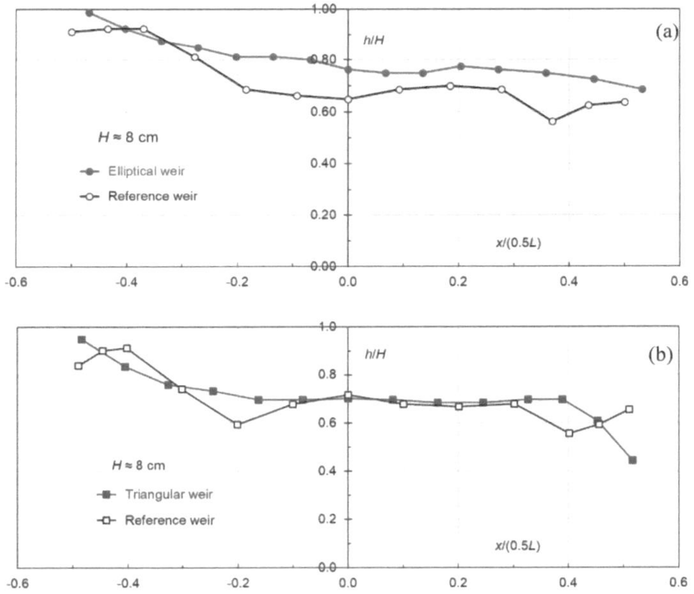

Figure 11. Variations of crest water depth (h) at $H \approx 8$ cm: (a) elliptical weir versus its reference, (b) triangular weir versus its reference. In the axis $x/(0.5L)$, L refers to the length of each individual weir.

For a given water head, the unit flow discharge is defined as $q = Q/L$, representing a lumped quantity without spatial differentiation along the crest. Figure 12 illustrates how q varies with H when comparing each modified weir to its reference. Per unit crest length, the elliptical weir demonstrates higher overflow efficiency than its reference. In contrast, the q values for the triangular weir are nearly identical with its reference, except for a minor downswing at high heads.

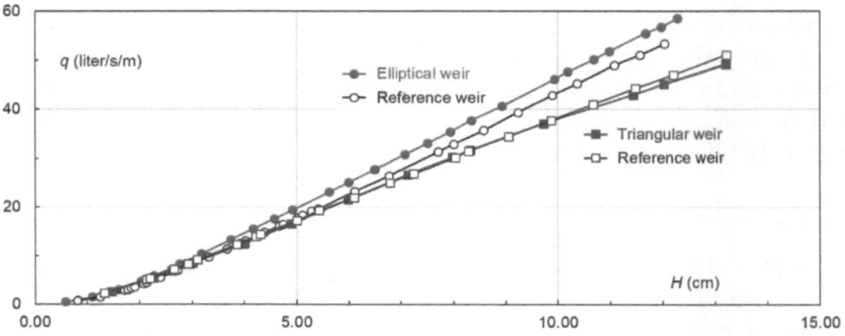

Figure 12. Changes of unit flow discharge (q) for the modified weirs relative to their reference ones.

5.3 *Flow patterns*

Overflow patterns are analyzed across a range of flow rates, from low to high. Figures 13 and 14 illustrate the flow behaviors at $H \approx 8.0$ cm as examples. This examination enhances the understanding of both similarities and differences between the weirs. Despite the increased flow capacity at the same water head, each modified weir behaves similarly to its reference counterpart. The submergence effect in the outlet key occurs at approximately the same water head (e.g. at $H = 8.0$–8.2 cm for the elliptical weir and $H = 7.8$–8.0 cm for its reference one).

(a) (b)

Figure 13. Flow over (a) the elliptical weir and (b) its reference weir. $H \approx 8$ cm.

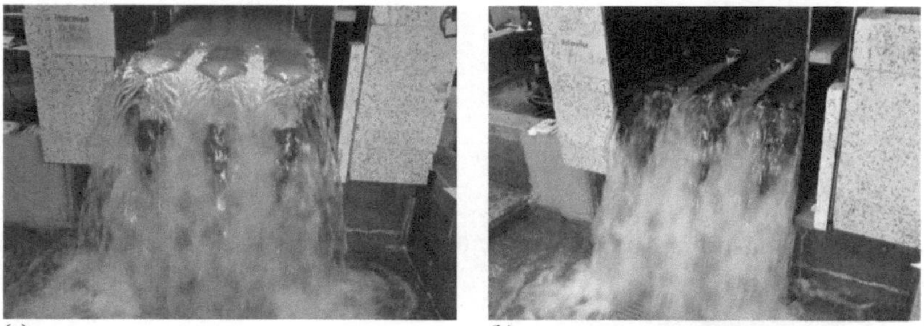

(a) (b)

Figure 14. Flow over (a) the triangular weir and (b) its reference weir. $H \approx 8$ cm.

Compared to its reference design, each modified layout extends further into the reservoir, creating more favorable approach flow conditions. Similarly, it projects an additional distance downstream. Experimental results indicate that the modified crest also spreads the jets more laterally. In other words, at the same head, the jets fan out more widely toward the impact location, which enhances air entrainment and energy dissipation.

5.4 *Key factors for flow improvement*

Compared to a conventional weir with a rectangular planform, improvements in flow discharge efficiency and overall flow behavior can be attributed to factors such as weir crest length, side contraction of inflow, and the volumes of both the inlet and outlet keys.

1. Each modified weir extends further both up- and downstream, increasing the overflow crest length. This extension is more pronounced in the elliptical weir (β = 36.4%) than in the triangular one (β = 23.1%), thus contributing to the discharge increase by ξ = 30–53% in the former and ξ = 16–20% in the latter.
2. The exterior of the upstream overhang is shaped in the vertical direction as either elliptical or triangular. The elliptical shape offers a more streamlined profile, while the triangular shape, though less streamlined, still mitigates side-contraction effects and improves flow into the inlet keys.
3. The lowered floor in the inlet key increases the inflow passage area (volume), facilitating smoother flow movement toward the crest.
4. The lowered floor of the outlet key enlarges the outflow cross-sectional area (volume) and reduces the submergence effect. However, its impact becomes significant only at higher flow rate.

6 CONCLUSIONS

The conventional PKW is highly effective for flood discharge. However, hydraulically, its upstream rectangular overhang presents a suboptimal geometry, causing significant side contraction of the inlet key flow. Furthermore, the crest corners function as hydraulic singularities, disrupting flow patterns. Research shows that under high-flow conditions, water depth along the crest decreases markedly in the flow direction, indicating that the lower crest contributes less to discharge than the upper portion. These findings point to opportunities for geometric optimization to improve hydraulic performance.

In each key of the PKW, reshaping the floor with a downward semicircular cross-section creates an overhang with an elliptical planform. The extent of crest extension primarily depends on the key width and the floor's slope angle. In the triangular layout, the key floor is shaped as a downward isosceles triangular prism, with crest lengthening influenced by both the key width and the vertex angle.

Among the examined models, the elliptical weir exhibits a greater increase in developed crest length compared to the triangular weir (β = 36.4% versus 23.1%). The former results in a more significant improvement in discharge efficiency (ξ = 30–53% versus 16–20% within the tested flow ranges). Both modified weirs contribute to a larger crest water depth and a smoother water surface across the crest. The extended crests reach further into the reservoir, enhancing approach conditions. Similarly, they project farther downstream, promoting jet spreading and improving energy dissipation, which are additional benefits of the improved layouts.

The elliptical or triangular extension into the reservoir may facilitate the passage of floating debris, though this has not been verified. The improved PKW configurations are particularly beneficial in scenarios where reservoir water level increases must be controlled during extreme floods or when spillway construction space is limited. These enhancements contribute to improved dam safety and cost efficiency.

ACKNOWLEDGMENTS

As part of research project "Innovative design & experimental-numerical studies of piano key spillway for enhanced discharge and hydraulic performance", the study is funded by the Swedish Centre for Sustainable Hydropower (SVC), contract no. VKU32022. As industry in-kind, Vattenfall R&D covers the costs for the test rig adaptation. The authors express their gratitude to Anna Helgesson, Anders Sjödin, Magnus Lagerkvist, Jennifer Persson, Isabelle Andersson, Hans

Lindström, and Patrik Lidberg for their valuable contributions to the project. James Yang prepared and revised the manuscript, and all the other authors consented to the submitted version.

REFERENCES

Anderson, R.M. & Tullis, B.P. 2013. Piano key weir hydraulics and labyrinth weir comparison. *Journal of Irrigation and Drainage Engineering* 139: 246–253.

Crookston, B.M.; Erpicum, S.; Tullis, B.P. & Laugier, F. 2019. Hydraulics of labyrinth and piano key weirs: 100 years of prototype structures, advancements, and future research needs. *Journal of Hydraulic Engineering* 145: 02519004.

Erpicum, S., Machiels, O., Dewals, B., Archambeau, P. & Pirotton, M. 2013. Considerations about the optimum design of PKW. In *Proc. the International Conference on Water Storage and Hydropower Development for Africa, Addis Ababa, Ethiopia*, 16–18 April 2013.

Laugier, F. 2007. Design and construction of the first piano key weir (PKW) spillway at the Goulours dam. *International Journal of Hydropower Dams* 14: 94–101.

Laugier, F., Pralong, J., Blancher, B. & Montarros, F. 2011. Development of a new concept of piano key weir spillway to increase low head hydraulic efficiency: fractal PKW. In *Labyrinth and Piano Key Weirs—PKW 2011*, London: CRC Press.

Laugier, F. & Vermeulen, J. 2017. Overview of design and construction of 11 piano key weirs spillways developed in France by EDF from 2003 to 2016. In *Labyrinth and Piano Key Weirs III—PKW 2017*, London: CRC Press.

Laugier, F., Vermeulen, J. & Lefebvre, V. 2013. Overview of piano key weirs experience developed at EDF during the past few years. In *Labyrinth and Piano Key Weirs II—PKW 2013*, London: CRC Press.

Lempérière, F. & Ouamane, A. 2003. The piano keys weir: a new cost-effective solution for spillways. *International Journal on Hydropower & Dams* 10: 144–149.

Ouamane, A. 2011. Nine years of study of the piano key weir in the University Laboratory of Biskra "lessons and reflections". In *Labyrinth and Piano Key Weirs—PKW 2011*, London: CRC Press.

Ouamane, A. & Lempérière, F. 2006. Design of a new economic shape of weir. In *Proc. the International Symposium on Dams in the Societies of the 21st Century, ICOLD, Paris, France*, 18 June 2006.

Shen, X. & Oertel, M. 2021. Comparative study of nonsymmetrical trapezoidal and rectangular piano key weirs with varying key width ratios. *Journal of Hydraulic Engineering* 147(11): 04021045–1.

Yang, J., Li, S.C., Helgesson, A., Skepparkrans, E. & Ansell, A. 2023. Geometric modification of piano key weirs to enhance hydraulic performance and discharge capacity. *Water* 15(23): 4148.

Yang, J. & Li, S.C. 2024. Piano key weir (PKW) – improvement in conventional geometry for augmented discharge capacity. *Water* 16(23): 3375.

Labyrinth and Piano Key Weirs IV – Erpicum, Crookston & Paxson (Eds)

An innovative association of Fusegate® and Piano Key Weir

F. Lempérière
HydroCoop, Paris, France

A. Chapuis
ISL Ingénierie, Paris, France

F. Del Rey
Hydroplus, Rueil-Malmaison, France

C. Bessière
Ingérop, Rueil-Malmaison, France

ABSTRACT: The combination on the same dam of Piano Key weir (PK-Weir) and fusible structures offers a highly effective passive solution for managing flood discharges. This concept first introduced by Hydrocoop in 2015 as « Innovative Associated Spillways » can leverage the strength of each system. More than 30 years after its first application, the returns of experience of the Hydroplus Fusegate® system applications on more than 80 dams in 25 different countries have confirmed their reliability in terms of sequentially overturning on the downstream side for the correct design flood. With their high discharge capacity at low head, PK-Weirs optimize the use of the flood storage volume by decreasing the peak of the downstream flood. As shown in the presented study case, the combination of these two innovative passive dispositions provides robust solutions for ensuring reliable safety during common and extreme, even unpredicted, floods.

1 A SHORT HISTORY OF THE FUSEGATE® SYSTEM

1.1 *The genesis of an idea*

In the 90s, the city of Lussas France was looking for solutions to increase the storage capacity of its water supply reservoir without having to change the maximum water level. François Lempèriére was involved in the discussion at the time and, he started thinking about some innovative solutions to solve this problem. The key to the solution lied in the optimization of the dam spillway. Indeed, the normal reservoir level was managed by the spillway sill elevation and, its discharge capacity was driven by its shape and by the difference between the maximum water level and the normal reservoir level. With these hypotheses, he made an analysis of how spillways are designed. Basically, spillways can be classified in two types: gated or ungated.

Ungated spillways are the safest type of spillways. They involve no mechanical components and, thus, require little maintenance. However, ungated spillways are generally more expensive than the gated spillways and, they require large difference in elevation between the normal reservoir level (top of spillway sill) and the maximum water levels. This difference causes a significant loss in the storage capacity of the reservoir.

On the other hand, they allow controlling the downstream flooding and minimize the difference in elevation between normal reservoir level and the maximum water level. However, their biggest disadvantages are the operation and maintenance requirements, and the need for human

DOI: 10.1201/9781003655541-7

intervention. In addition, the probability of a gate failure or a miscalculated operation during a flood is not negligeable.

The concept of the Fusegate® system was born after searching for a solution which would combine the advantages of both type of spillways while eliminating the disadvantages. To find a system without any electrical and mechanical components, requiring no human intervention to operate and with minimal maintenance. At the end, a third type of spillway was invented.

The idea behind the new system is to install "elements" on the spillway sill control structure in order to keep a sufficient difference in elevation between the reservoir level and the maximum water level to pass common floods. In the event of extreme flooding, these elements will sequentially overturn towards the downstream side.

1.2 *Functioning of the Fusegate® system*

The operating principles of these elements created the Fusegate® system. Fusegate® are stable standalone structures retrofitted on the spillway sill by gravity and are secured by toe abutments anchored behind them. Fusegate® consist of three parts: a water retaining structure, a pressure chamber on the underside, and a water inlet well above their crest.

Installed atop the spillway sill, they form a watertight barrier and raise the overspilling crest. For most common floods, water is discharged over the Fusegate®, such as a free spillway.

In case of major flooding, a Fusegate®'s pressure chamber starts to fill when the water level exceeds the crest of the inlet well. As soon as the drain holes reach the saturation point, pressure under the Fusegate® lifts the unit, tips it in the downstream side and washes it away in the spillway channel. It is important to note that Fusegate® begin to overturn only in the event of very low-probability floods. The tipping of the first Fusegate® will generally occur somewhere between the 100-year to 1 000-year range flood events.

Figure 1. Fusegate® functioning.

1.3 *Implementation of the Fusegate® system*

The Fusegate® system was first implemented on the Lusas dam located in Ardeche France in 1991. The storage capacity was increased by 30%. The spillway was retrofitted with ten labyrinth type Fusegate® designed for moderate overspilling. Site works commenced in April 1991 and were completed by June the same year. During the large flood recorded in the region in January 1994, one Fusegate® had overturned as designed. It was quickly immediately replaced to restore the normal reservoir level.

In 1992, a second project was completed with Electricite de France (EDF). The Saint Herbot dam France was equipped with 5 Fusegates® 1.5-meter high. This pilot project with EDF was subject to a full-scale testing by releasing water from an upstream reservoir.

Figure 2. Lussas dam (18m high) and Saint Herbot dam (11m high) with the Fusegate® system.

In both projects the Fusegate® had a labyrinth shape to improve the discharge before the first tipping.

In 1994, a major step was taken for the Fusegate® System in India with the installation of 33 straight crested type units, 20-meter wide and 2-meter high at Wanakbori dam. The increase in storage capacity has been significant with an additional 25.53 Mm^3 which represents 70% of the initial capacity.

Figure 3. Wanakbori dam and its overspilling with the Fusegate® system.

As of 2025, more than 80 dams have been equipped with the Fusegate® system in 25 countries worldwide.

2 SHORT HISTORY OF THE PIANO KEY WEIRS

2.1 *Preliminary studies*

As mentioned in the paper (Lempérière & Ouamane, 2003), a very different design for free-overflow spillway shape has been studied since 1999 by Hydrocoop, with the support of preliminary model tests conducted first at the EDF laboratory in Chatou in 1999, and then in 2002 at Roorke University in India and Biskra University in Algeria.

Detailed tests on preselected shapes were also conducted in 2003, at the Biskra laboratory and some other tests using a very large flume at the EDF laboratory in Chatou.

The conclusions of these preliminary studies showed that, compared to Creager type weir profiles, PK-Weirs may divide by a factor 3 the spillway length, or by a factor 2 the upstream head.

2.2 *First applications*

The first PK-Weir was designed and build in 2006 in France, at the EDF's Goulours dam successfully (Laugier, 2007).

Additional research and development studies were performed in parallel in France where EDF collaborated with the Ecole Polytechnique Fédérale de Lausanne laboratory, and with the Liège laboratory in Belgium, other studies include ones in Algeria, and another one in Vietnam with the contribution of M. Ho Ta Khanh for the theoretical and practical developments of PK-Weirs.

International workshops were then organized between the main actors, the first being held in Liège Laboratory in 2011, the second in Paris in 2013 and, the third in Vietnam in 2017. These workshops also inspired other engineering companies and laboratories for launching research and studies (for example in Pretoria - South Africa, and at the Utah University - USA).

Initiated by the hydraulic engineering department of EDF, the Vietnam National Committee on Large Dams and Water Resources Development, the Liège University – Hydraulics in Environmental and Civil Engineering, and the EPFL– Laboratory of Hydraulic Constructions, a tentative "World Register of PK-Weirs prototypes" has been established and will be updated during this 2025 congress.

By the end of June 2020 more than 35 PK weirs have been listed in the world. (>15 in France, >6 in Vietnam and at least further in Sri Lanka, Canada, UK, Australia, India, Switzerland, Algeria, USA and South Africa).

Even though they are much older than the Piano Key weir concept proposed for the first time in 2003, two Algerian dam spillways were added to this list. Indeed, their labyrinth design with overhangs is close to the Piano Key weir concept. They can then be considered as "ancestors" of modern Piano Key weirs and probably deserve to be part of this world list.

2.3 *Performances*

Each of these prototypes are specific cases adapted to their site conditions. Hazelmere PK-Weir with its 10 meters height is the highest in the world. With a total width of 103 meters, it is quite standard application Type A, compared to the preliminary studies. It was designed to pass the 50-year flood, with an inflow of 1 090 m^3/s, under a nappe depth of 0.95 m, and the 100-year flood, with an inflow of 1 360 m^3/s under a nappe depth of 1.16 m.

The table here below gives the PK-Weir discharges issued from the tests performed at the DWS laboratory in Pretoria, with a comparison with the theoretical discharge of a classical Creager weir for the same head. The ratio Q_1/Q_0, with Q_1 the PK-Weir discharge and Q_0 the Creager sill discharge, decreases significantly when the water head continues to rise.

Table 1. Hazelmere PK-Weir' spillway compared to a creager spillway.

Upstream water depth	0.95 m	1.16 m	1.40 m	2.62m	3.20 m
Q_1 PK-Weir discharge m^3/s	1080	1347	1645	3175	4288
Q_0 Creager sill discharge m^3/s	210	283	375	961	1297
Ratio Q_1/Q_0	5,14	4,76	4,38	3,30	3,30

Figure 4. Hazelmere PK-Weir section and its top view.

3 ASSOCIATED INNOVATIVE SPILLWAYS (AIS) CONCEPT: PK-WEIRS AND FUSEGATE® SYSTEM ASSOCIATION

3.1 *Associated innovative spillways (AIS)*

More than 30 years after the first application of the Hydroplus Fusegate® system at the Lussas dam, and about 20 years after the construction of the first PK-Weir at Goulours dam, the various returns of experience for these two solutions have confirmed the reliability of the Fusegate® in terms of tipping for the correct design flood and the large discharge capacity potential of PK-Weirs particularly under low heads.

The "Associated Innovative Spillways" has been first introduced by Hydrocoop in 2015 (Lempérière & Vigny, 2015) under the French concept "Déversoirs Innovatifs Associés". It was later further developed and presented in a French communication at a congress of the French Committee on Dams and Reservoirs (CFBR) in 2018 (Chapuis, 2018) and highlighted at ASIA 2018 congress (Deroo, 2018).

3.2 *Opportunity and advantages*

The synergy between PK-Weirs and Fusegate® system, both passive solutions without mechanical components, delivers a robust approach, addressing moderate as extreme flood scenarios. In both cases no external power source and no human intervention are required for the spillway operation. In addition, maintenance operations are minimized. PK-Weirs maximize spillway capacities at low water head within limited space, efficiently managing frequent and moderate floods, while maintaining the increase of the upstream water levels.

With their high discharge capacity at low head, PK-Weirs optimize the use of the flood storage volume by decreasing the peak of downstream flood. Fusegate® before their overturning can also, if required, manage the frequent and moderate flood in the same way as the PK-Weir.

In combining both alternatives, the first Fusegate® overturning could trigger for a greater flood than the one without the PK-Weir, serving as a lever to reassure operators. The first tilt of a Fusegate® may be delayed up to a flood of 500 years return period or more. This means that most owners will never witness a Fusegate® tilt – except in very rare situations that could be labelled as force majeure. In such cases:

- Fusegate® will guarantee the flood will be evacuated correctly through the spillway.
- Temporary loss of storage capacity will not be the primary concern.
- Exceptional operational costs for dam owners will be an acceptable trade-off.

The Fusegate® system ultimately provides additional safety by absorbing the logarithmic discharge gap between the frequent and moderate floods managed by free spillways, and extreme floods.

This innovative pairing is so particularly suited for adaptation to climate-induced flood variability, and compliance with enhanced safety requirements.

This combination of these two passive structures eliminates operational risks, requires no operating instructions and thus maximizes dams' protection against floods. Thanks to the Fusegate® system, they allow to maintain a comfortable discharge capacity beyond their design limits, unlike tulip spillways, lateral spillways, or gated spillways converted into orifice spillways, for example. It represents a forward-looking strategy for sustainable and resilient dam infrastructure.

3.3 *Limitation*

However, this combination cannot always be applied, particularly when:

- Dam's operating constraints require reservoir level regulation and/or control of released flows, necessitating an active regulation mechanism (gated spillway) if the bottom outlet capacities are insufficient.
- Required footprint is not available to implement the solution.
- No difference is allowed by environmental and social surveys between the normal reservoir level and the maximum reservoir level.

4 APPLICATION TEST AT SHONGWENI DAM

4.1 *Introduction*

Safety inspections in 1988 revealed critical issues at the 70-year-old Shongweni Dam northwest of Durban on the Mlazi River, including inadequate spillway capacity (only 25% of the required Safety Evaluation Flood -SEF- of 5,000 m³/s) and structural instability at a level just lower than the non-overspill crest level. A dam failure could have released a 13-meter-high flood wave, endangering thousands of lives and causing extensive damage to industrial development and to the overall environment.

In 1994, the dam operator, Umgeni Water, decided to increase the discharge capacity and the storage capacity. Traditional gate solutions were costly and unreliable, leading to the selection of the Hydroplus Fusegate® system as the most economical and environmentally viable option. Ten Fusegate® (6.50m high, 9.75m wide) were installed with a first Fusegate® tip-off occurring only when the 240-year flood is exceeded.

Figure 5. Shongweni Fusegates®.

In April 2022, a major storm (300 mm of rainfall in 24 hours) caused severe flooding and over 400 deaths in the Durban area. The Shongweni Dam withstood the event, with two Fusegate® tipping as designed, confirming their reliability in extreme conditions.

Figure 6. Shongweni after 2 Fusegates® triggering (left) and Durban after the flood (right).

4.2 *Mix spillway with Fusegate® and PK-Weir on the Shongweni dam*

4.2.1 *Theoretical analysis*

The purpose of the following theoretical analysis is to get the better of the two systems in combining PK-Weirs with Fusegate® to get the first triggering of the Fusegate® system for the 500-year flood and for the 1,000-year flood instead of the 240-year flood. The fixed input data are:

- Full Supply Level=298.10 m,
- Maximum Water Level=301.00 m,
- SEF=5,000 m^3/s.

The design of the PK-Weir is as suggested by bulletin ICOLD B172 and by the parameter P, corresponding to the total height of the PK-Weir. Regarding the Fusegate® system, the design has been performed by Hydroplus. Four alternatives were analyzed and developed.

4.2.2 *Alternative 1: PK-Weir = 97.50 m with P=6,50m*

The alternative 1 corresponds to a PK-Weir having the same height as the Fusegate® and the same total width to determine the new MWL. For the SEF, the calculated maximum water level is 303.29m which is 2.29m above the allowed maximum water level. Therefore, this solution is not relevant.

4.3 *Alternative 2: PK-Weir with MWL=301m and P=6.50m*

The alternative 2 corresponds to a PK-Weir having the same height as the Fusegate® and the width allowing to stay below the MWL. For the SEF, the width of the PK-Weir is 298.10 m which must be compared to the 97.5m of the Fusegate® solution. In Shongweni dam, this footprint will not have been available. Therefore, this solution is not appropriate.

4.4 *Alternatives 3a & 3b: Fusegate® height 6,5 m and PK-Weir with MWL=301m and P=6,50m*

The alternative 3a corresponds to a combination of the PK-Weir having the same height as the Fusegate® and the Fusegate® element allowing to stay below the MWL with a first triggering for the 500-year flood, as opposed to the alternative 3b which corresponds to a 1000-year flood triggering.

According to (Chapuis, 2018), it can be introduced the rate of the AIS system $R_{AIS} = W_{PKW}/W$; the linear ratio of PK-Weir crest length to the total crest length W, with $W = W_{PKW} + W_{HF}$.

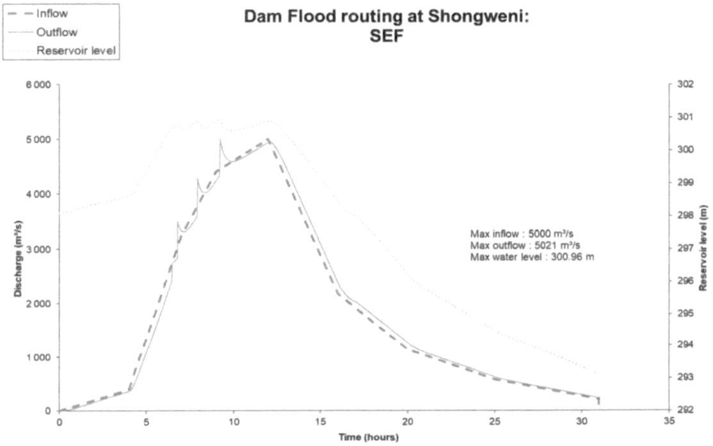

Figure 7. Dam Flood routing at Shongweni, with a triggering for the 500-y flood.

For the alternative 3a, the Fusegate® width part is 78 meters, and the PK-Weir Width part is 45 meters. The R_{AIS} of this pairing solution allowing to postpone the first Fusegate® tip-off above the 500-year flood is about 36%. The total footprint of spilling section is 123 m, which is 26% larger than the current configuration with only 10 Fusegate® with a first tilting around 240-year flood.

For the alternative 3b, the R_{AIS} allowing to postpone the first Fusegate® tip-off above the 1000-year flood is about 60%. The total footprint of spilling section is 143 m, which is 46% larger than the current configuration. In Shongweni dam, this footprint would have been available for both alternatives.

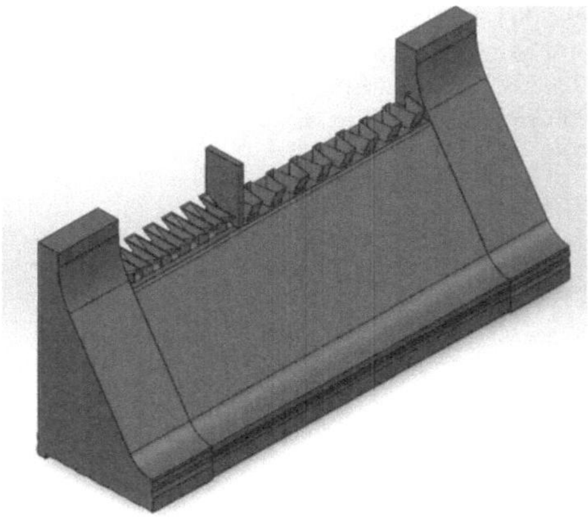

Figure 8. Shongweni spillway– alternative 3a.

4.5 *Synthesis*

The Table 2 shows the summary of the four alternatives.

Table 2. Summary of the alternatives.

Study	Current conf.	1	2	3a	3b
PK Weir height (m)	-	6,50	6,5	6,50	6,50
PK Weir length (m)	-	97,50	190,00	45,00	85,00
Fusegate® height (m)	6,50	-	-	6,50	6,50
Fusegate® length (m)	97,50	-	-	78	58,5
MWL for PMF (El. M)	300,99	303,29	300,98	300,97	300,97
1st Tipping Return period	>200 y	-	-	>500y	>1000y
ΔQ_1 after 1st Tipping	21%	-	-	14%	6%
Q_{out} Max (m^3/s)	4 932	4 928	4 966	5 008	5 008

5 CONCLUSION

Climate change has important impacts on the hydrology with causing floods to be more frequent and much greater than in the past. To increase the resilience of dams, 21st-century engineering improvements offer of two innovative passive solutions with flood discharge potential: Fusegate® system and PK-Weirs. Together these Associated Innovative Spillways (AIS) provide robust solutions for ensuring reliable safety during common and extreme floods while minimizing construction footprint and investment costs. The combination of these innovative spillways should become a standard approach in the study of design alternatives, provided they are compatible with the operational constraints of the reservoir.

REFERENCES

Bessière, C. 1992. Hausses submersibles et fusibles Hydroplus, La Houille Blanche n°2/3

Booyse, D. 2017, "Raising of Hazelmere dam by means of PKW", Labyrinth and Piano Key weirs III – PKW.CRC Press

Chapuis, A. 2018. Winning scheme of "Associated Innovative Spillways" for dam passive rehabilitation dealing with flood issues – Theory and case study, Colloque CFBR 2018: « Méthodes et techniques innovantes dans la maintenance et la réhabilitation des barrages et des digues » 27–28 novembre 2018

Deroo, L. & Chapuis, A. & Lapeyre, O. 2018. Innovative solutions to increase flow capacity of spillways and maximize dam safety, ISL Ingénierie, ASIA 2018

Laugier, F. & Vermeulen, J. 2018, "Réhabilitation de barrages existant avec des évacuateurs de crue labyrinthe de type PKW" First International Conférence on Dams ICDBiskra

Lempérière, F. & Ouamane, A. 2003. "The Piano Keys Weir: a new cost-effective solution for spillways The International Journal on Hydropower & Dams, Issue Four

Lempérière, F. & Vigny, J.P. 2015. Déversoirs Innovatifs Associés: en associant des Blocs Fusibles à des PK Weirs on augmente de 60 % la capacité du déversoir Piano Keys Weir et de 300 % le débit d'un seuil Creager. HydroCoop

Labyrinth and Piano Key Weirs IV – Erpicum, Crookston & Paxson (Eds)

An innovative use of Piano Key Weirs or labyrinth weirs for long dams in plains

M. Ho Ta Khanh
VNCOLD-CFBR Independent Consultant, France

ABSTRACT: Piano Key Weirs (PKW) are labyrinth type weirs but with upstream and/or downstream overhangs which allow them to have a reduced footprint at their base. The new solution of a combination of gates and PKW (or labyrinth weir) for a long dams has many advantages of costs for investment, operation and safety during high floods. If a new dam is built on a broad river where a large footprint is possible, the choice between a PKW and a labyrinth weir can arise as the difference of specific discharges of the two alternatives are relatively small and the construction of the labyrinth weir with only vertical formworks is simpler. This papier presents the Van Phong, Xuân Minh and My Son dams with PKW and the Phu Phong dam with labyrinth weir built in Vietnam since 2015. The proposed hybrid solution, compared with a traditional barrage, shows that its implementation, even on a broader stretch of the river, can also facilitate the diversion works.

1 INTRODUCTION

Piano Key Weirs (PKW) are labyrinth type weirs but with upstream and/or downstream overhangs which allow them to have a reduced footprint at their base and can be installed at the top of most concrete gravity or arch dams (Lempérière & Ouamane, 2013).

As for the labyrinth weirs (Tullis et al. 1995), the PKW discharge capacity rises fast according to the hydraulic head and this feature can be used to address the issue of hydrological increase due to climate change for existing dam (Laugier e al. 2017). As free flow spillway, PKW operate without any mechanical equipment, human operator or energy source. They are intrinsically safe and significantly improve general dam safety. The PKWs require little maintenance: no joints and moving parts, no risk of corrosion for concrete structures. The PKWs are simple to construct (rectangular and triangular formworks) and can be used for more or less complete factory prefabrication for concrete or steel works.

As disadvantages, compared to the labyrinth weirs, the construction of PKW with overhang(s) requires scaffolding for the formwork and maybe a risk of jamming with large floating debris (logs). Recently in Vietnam, the PKW and labyrinth weirs have replaced a part of the traditional barrages in plains (Ho Ta Khanh et al. 2013). This combination has revealed many advantages by optimizing cost, reliability and other technical issues such as reservoir level control or sediment passage. This papier presents 4 examples of PKW and labyrinth weir used for long dams in Vietnam since 2015.

2 THE VAN PHONG DAM

2.1 *General*

The Van Phong barrage [5] is about 70km from Quy Nhon. The dam derives the water through a main canal for irrigation by gravity of 10,815 ha in the plain near the ocean. The

DOI: 10.1201/9781003655541-8

reservoir FSL is determined by the highest possible entrance canal level and the MWL must avoid large inundated zones upstream with probable huge resettlement issues.

Among the 3 alternatives examined (Figure 1) in order to maximize the FSL and to make this scheme possible:

- Alternative 1: a barrage with 28 slide gates.
- Alternative 2: a barrage with 28 radial gates.
- Alternative 3: a dam with a combination of 60 PKW units on the Left Bank and Right Bank, and 10 radial gates (H= 3.5m, L= 15m) in the central part.

Alternative 3 was found as the most appropriated for the cost and the dam safety. A minimum of gates is necessary to lower the FSL if necessary and to evacuate the sediments and the gates are only opened for the major floods.

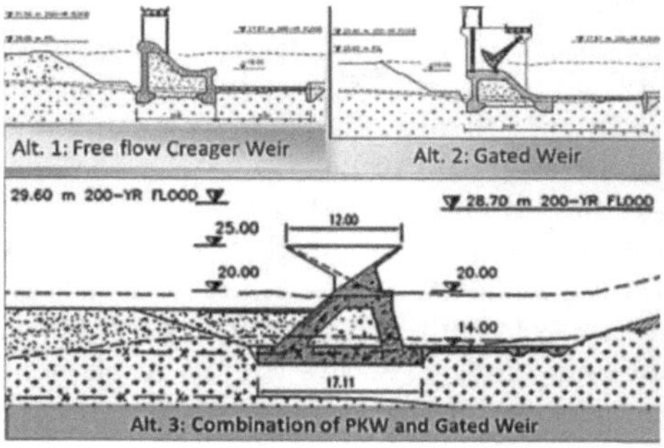

Figure 1. Comparison of 3 Van Phong alternatives at the feasibility stage.

The investment cost (2014) for the alternative with PKW is USD 26.68 x106 and the lowest cost for the alternatives totally gated is USD 37.52 x106, but this advantage is definitely greater if the operating and maintenance costs and the safety of functioning are taken into account in this comparison.

2.2 *Main barrage characteristics*

The PKW were selected as part of the project design as they can spill a significant discharge for a low nappe depth, which fits the site constraint of low allowable upstream water level. This combination is also safer and cheaper than solutions with either all surface gates or flaps gates.

The main characteristics of the Van Phong dam (Figures 2 to 5) are:

- Total length of the dam: 474m, with a length of 181m for the PKW on the R.B and a length of 121m for the PKW on the L.B. The length of the gated spillway in the central part is 172m.
- Crest level of the PKW= 25.00 masl.
- Max height of the dam on foundation: 12m to 16m with 7m on the riverbed.
- The dam, with 60 PKW units and 10 radial gates, was tested with a large-scale hydraulic model. The 1000-year flood event is 15,350 m3/s and the maximum discharge of the PKW is 8700 m3/s.

For floods higher than the 100-year flood, the barrage is completely submerged. During the peak flood (3465 m3/s) on the 16th of December 2016 (Figures 6 and 7), the max PKW discharge was 1773 m3/s and the discharge by the gates was 1692 m3/s. The max upstream water level was 25.72 masl and the max downstream water level was 25.42 masl, very near the values measured on the hydraulic model and conform to the design. The small 0.30m head loss through the dam – as for a totally gated barrage - shows the good efficiency of the PKW for evacuating the floods with a small raising of the upstream water level.

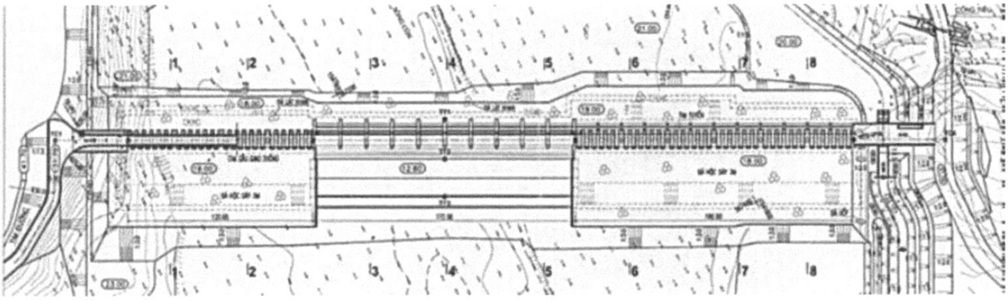

Figure 2. Plan view of the Van Phong dam.

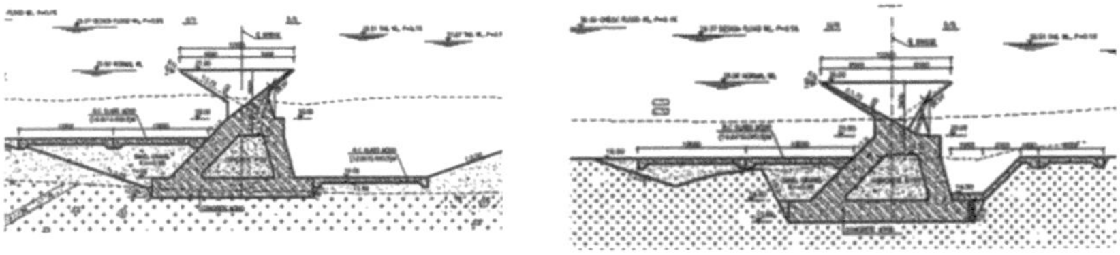

Figure 3. Cross sections (inlet and outlet keys) of the Van Phong PKW type A.

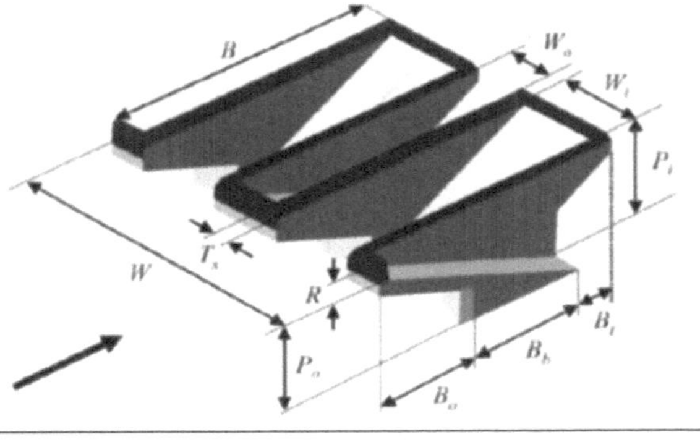

P (m)	5
W (m)	301.75
L (m)	1704
W_i (m)	2.4
Number of Outlets	60
Wo (m)	2.0
Ts (m)	0.20

Figure 4. Main features of the Van Phong PKW keys.

Figure 5. The Van Phong dam (left) and during a low flow event (right).

Figure 6. Photo of the dam (looking downstream) during the 16th of December 2016 flood.

Figure 7. Photo of the dam (looking upstream) during the 16th of December 2016 flood.

3 THE XUAN MINH DAM

3.1 *General description*

The Xuân Minh dam (Tables 1 and 2; Figures 8 to 12) is in the province of Thanh Hoa. It is located downstream the Cua Dat hydropower and about 7 kilometres upstream the Bai Thuong dam: the 3 HPPs are in cascade on the Chu river. The main purpose of the Xuân Minh dam is the hydroelectricity and the development of the regional economy.

The Xuân Minh spillway includes a 2 surface gates spillway on the left part (for each gate H=10.5m, L=10.0m, the spillway crest is at El. 17.5 masl) and the PKW type B on the right part. The maximum discharge of Xuân Minh is 7260 m3/s for the check flood at the site, but it can also evacuate 11,830 m3/s (the 10,000 year-flood of the Cua Dat reservoir). At the selected site, a small power station (15MW) was built on the left bank and has been fully operated since 2019.

3.2 *The design*

Two alternatives of dam were contemplated: (1) In the initial design, a traditional barrage with 10 gates (H=10.5m, L=10m) installed on the central part of dam. The dam crest level is at El. 17.5 masl to evacuate the design flood. Taking into account the investment and maintenance costs and also considering the possible incidents during the operation of the 10 gates, this alternative presents several disadvantages. (2) The Owner decided to replace the barrage by a PKW and a 2-slide

Table 1. Water levels and discharge of the PKW and the gates for different flood frequencies.

	Floods	Total Q. m^3/s	U/S WL m	D/S WL m	PKW Q. m^3/s	Gated Weir Q. m^3/s
1	10%	3 353	28.98	25.24	2064	1289
2	5%	3 394	29.00	25.29	2099	1295
3	1.5%	3 466	29.04	25.36	2147	1319
4	0.1%	8 312	33.38	30.12	6559	1753
5	0.01%	11 928	36.12	32.64	9721	2207

Table 2. Main features of the Xuân Minh dam.

Design parameter	m	Geometrical parameter	m	Geometrical parameter	m	Ratio	Value
El. crest	27.5	W_i	3.9	B	18.0	W_i/P_p	0.7
H_{DF}^*	1.5	W_o	3.3	B_i	0	W_i/W_o	1.2
H_{CF^*}	4.0	T_s	0.3	B_o	6.0	P/W_u	1.0
W	140.0	P_O	7.5	S_i	1.0	B/P_P	3.0
		P_i	6.0	S_o	0.5	N	6.0

*Note *: H_{DF}: Head over the Crest on PKW for design flood. H_{CF}: Head over the Crest on PKW for check flood.*

gated spillway. The PKW with an upstream overhang, 140m long, is installed on the right bank. The PKW crest is at El. 27.50 masl. This FSL is calculated to maintain the required normal operation of the Cua Dat hydropower. The bottom of the river varied from El. 12.00 masl to El. 17.5 masl. The maximum dam heights vary from 12.0m to 15.5m with a deep foundation.

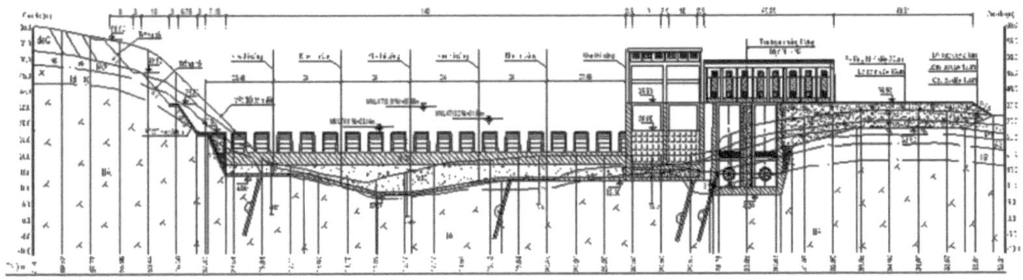

Figure 8. The Xuân Minh dam. Longitudinal section.

4 THE MY SON DAM

Location: 11° 43′ 40.47"N; 108° 48′ 44.64"E. The My Son Hydropower Project was built in 2023, located in Ninh Thuan province, Vietnam. The project's spillway is a Piano Key Weir type A with a total length of 350.0m. This is the longest Piano Key Weir spillway so far constructed in Vietnam.

5 PKW AND LABYRINTH WEIR

If a new dam is built on a large river where a large footprint is possible, the choice between a PKW and a labyrinth weir can arise as the difference of specific discharges of the two alternatives are relatively small(Truong Chi hien et al. 2006) and the construction of the labyrinth

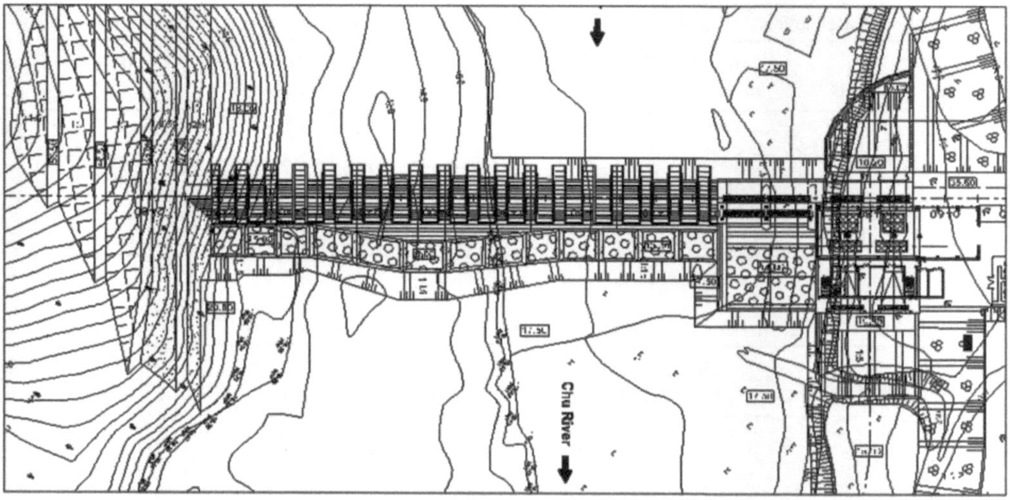

Figure 9. The Xuân Minh dam. Plan view.

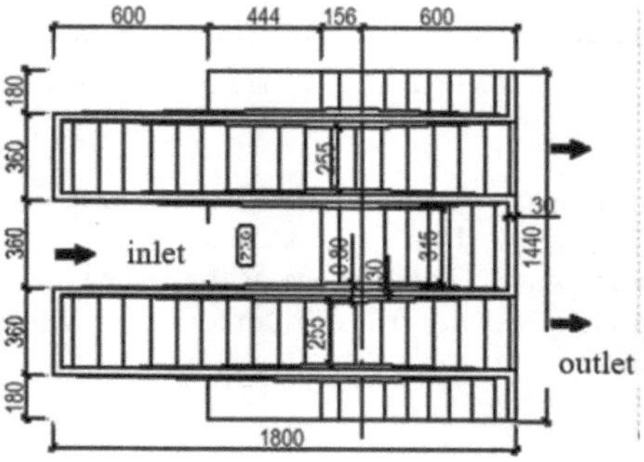

Figure 10. Plan view of an inlet and an outlet keys.

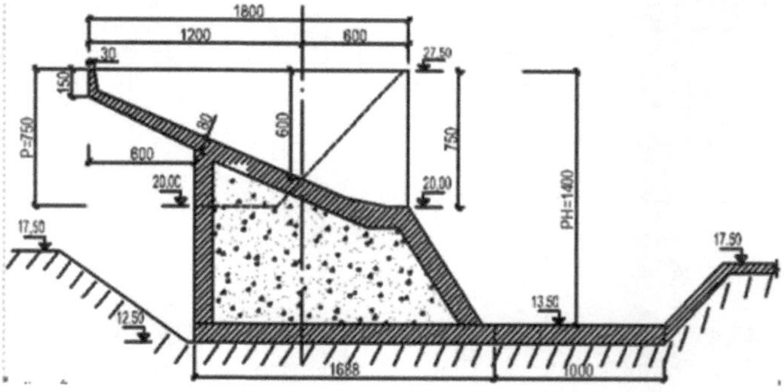

Figure 11. Cross section of an outlet key.

with only vertical formworks is simpler. Note: The PKWs have generally rectangular keys with overhangs (upstream and/or downstream), but different other alternatives of PKW exist. Figure 17 below shows that a PKW type D, with a flat bottom and without cantilever, is a labyrinth weir with a rectangular shape in plan view.

Figure 12. The Xuân Minh dam during a flood.

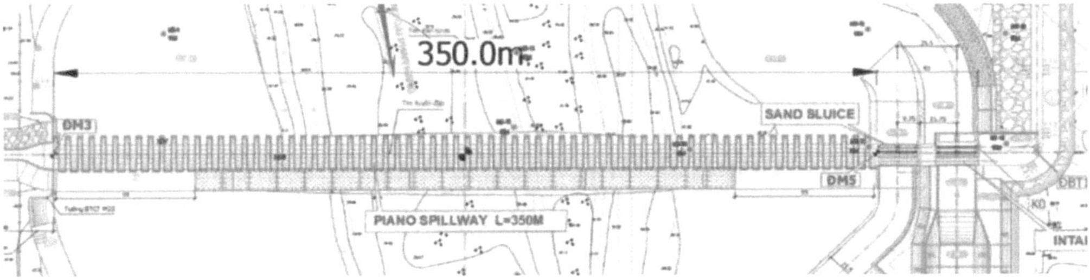

Figure 13. Plan view of the My Son dam.

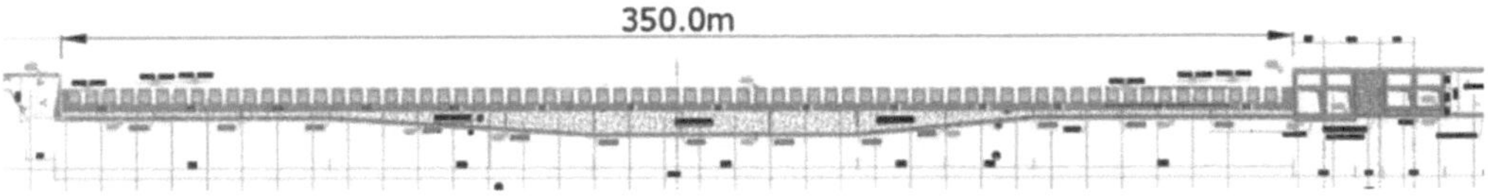

Figure 14. Longitudinal cross section of the My Son dam.

Figure 15. D/S view and outlet cross section of the My Son dam.

The Figure 18, with in abscissa the nappe H and in ordinate the specific discharge q, shows that a PKW (type A with U/S and D/S overhangs-upper line) has only a bit higher capacity than a trapezoidal labyrinth weir (medium line) and a PKW type D (lower line). Other laboratory tests have shown a bit larger differences (about 15%) for the low discharges. In any cases, the corresponding variation of the upstream MWL is not very important for the most frequent floods.

Figure 16. Overflow on the PKW and inlet cross section of the My Son dam.

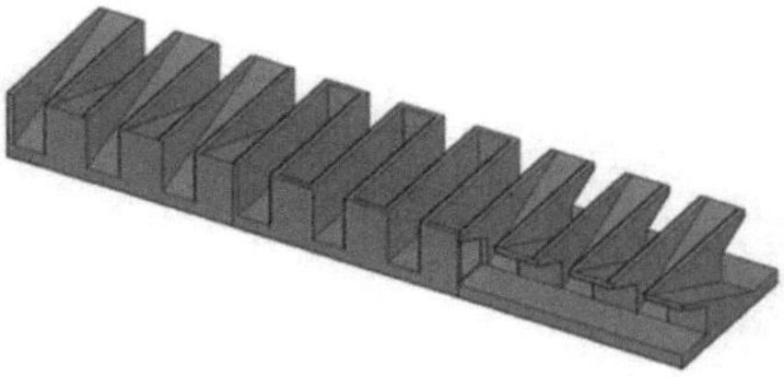

Figure 17. A Type D PKW.

The choice between a PKW and a labyrinth weir could be based consequently on the dam cost that depends on several factors such as the foundation, the characteristics of the structure and the mode of construction.

The Phu Phong dam (Figures 19 to 21), just downstream the Van Phong dam, is over 589m long. The centre part, 350m long, includes 20 slide-gated bays 15m wide each. On both sides of the dam there are 239m wide trapezoidal labyrinth weirs. Except during the major floods, the gates are closed.

The optimization of the Van Phong dam design (combination of radial gates and PKW) and the optimization of the Phu Phong dam design (combination of slide gates and labyrinth), with different characteristics, resulted in a comparable total cost for the two structures of about $30x10^6$ USD (2022), lower than a traditional barrage on these two sites. A comparison between only the Van Phong PKW and the Phu Phong labyrinth would be irrelevant as those two spillways have different characteristics.

6 CONCLUSIONS

In Vietnam, there are many infrastructures (road/building) or rice fields along rivers in plain. It is therefore necessary to minimize the rise of the river level during the floods when dams are built on these rivers. Hybrid structures combining PKWs or labyrinths with some gates — such as Van Phong, Xuân Minh, My Son and Phu Phong dams — appear to be the optimal solution with advantages of costs for investment and operation and safety during high floods.

The choice between a PKW and a labyrinth should be based on the dam cost which depends on many factors (depth and nature of the foundation, reinforced concrete volume, method of construction, etc.), as the difference of the discharge capacity is not very important between these two alternatives. The proposed hybrid dam solution (PKW or labyrinth + gates), compared with a traditional barrage, shows that its implementation, even on a broader stretch of the river, can also facilitate the diversion works.

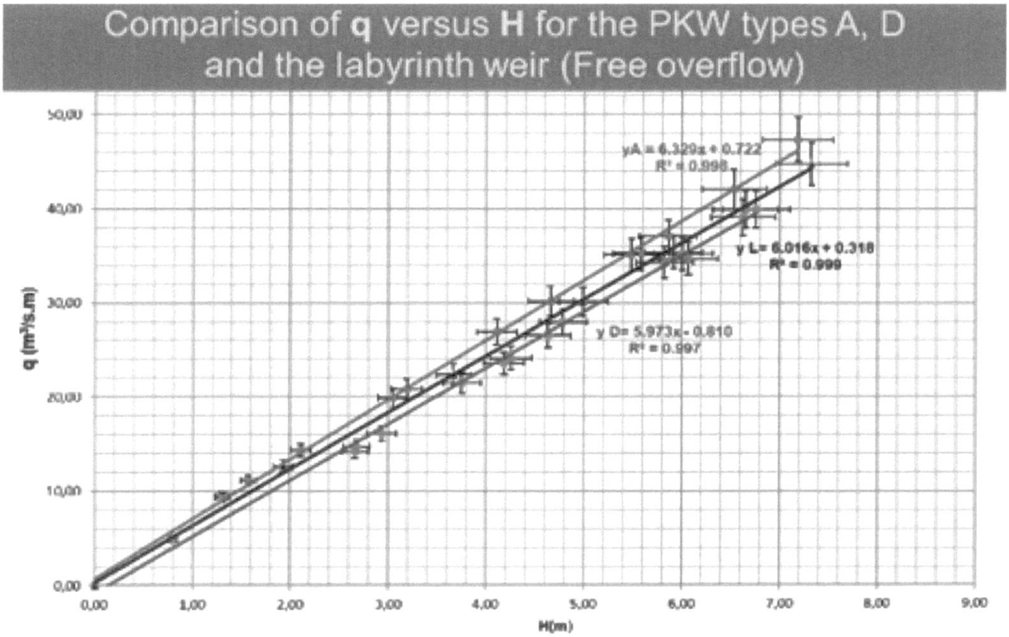

Figure 18. Comparison of the specific discharges vs. the nappe depths for 3 types of labyrinth weirs.

Figure 19. The trapezoidal labyrinth weir on the Phu Phong dam under construction (L.B).

Figure 20. The trapezoidal labyrinth weir on the Phu Phong dam during a low overflow (R.B).

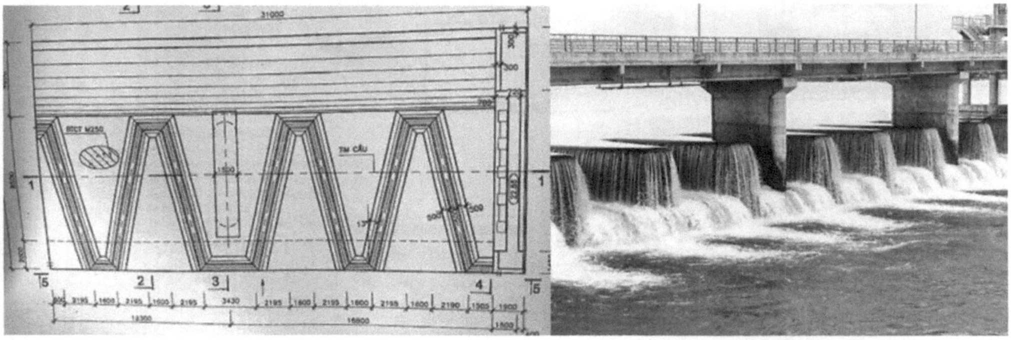

Figure 21. Drawing and detail view of the trapezoidal labyrinth weir on the Phu Phong dam.

REFERENCES

Lempérière, F. & Ouamane, A., *The Piano Keys weir: a new cost-effective solution for spillways*, Hydropower & Dams, Issue Five, 2003.

Tullis, J.P., Amanian, N. & Waldron, D., *Design of Labyrinth Spillways*, Journal of Hydraulics Engineering, Vol. 121, N° 3, March 1995.

Laugier, F., Vermeulen, J. & Blancher, B. "*Overview of design and construction of 11 Piano Key Weirs spillways developed in France by EDF from 2003 to 2016*" - Labyrinth and Piano Key Weirs III – PKW 2017 – Erpicum et al. (Eds) © 2017 Taylor & Francis Group, London, ISBN 978-1-138-05010-5".

Ho Ta Khanh, M., Truong Chi Hien, Pinchard,T. & Pralong, J., 2013. "*Utilization of Piano Key weirs for low barrages*". (2010) International Workshop on Labyrinth and Piano Key weirs, Paris, France.

Dinh Sy Quat, Nguyen Luong Am & Nguyen Manh Hung. "*Study, Design and Construction of the Van Phong Piano Key Weir*". Labyrinth and Piano Key Weirs III – PKW 2017 – Erpicum et al. (Eds) © 2017 Taylor & Francis Group, London, ISBN 978-1-138-05010-5".

Truong Chi Hien, Huynh Thanh Son & Ho Ta Khanh. M. "*Results of some Piano Key weir hydraulic model tests in Vietnam*". (2006) Q 87, R 39, 22d ICOLD Congress in Barcelona.

I.3 Floating objects and sedimentation

Labyrinth and Piano Key Weirs IV – Erpicum, Crookston & Paxson (Eds)
© 2026 The Author(s), ISBN 978-1-041-10527-5

Behavior of floating trees at coexisting piano key weir and gated spillway

J. Yang
R&D Hydraulic Laboratory, Vattenfall AB, Älvkarleby, Sweden
Civil & Architectural Engineering, KTH Royal Institute of Technology, Stockholm, Sweden

A. Helgesson
R&D Hydraulic Laboratory, Vattenfall AB, Älvkarleby, Sweden

L. Johnsson
Sydkraft Hydropower AB, Sundsvall, Sweden

S.C. Li
Civil & Architectural Engineering, KTH Royal Institute of Technology, Stockholm, Sweden

ABSTRACT: During extreme flood events, floating trees (driftwood) can pose serious risks to the safety and operation of dams. In upgrading an existing dam, a piano key weir (PKW) was introduced to replace the aging embankment dam and to increase the discharge capacity of its gated spillway. This upgrade necessitated a detailed understanding of driftwood behavior at both discharge structures, which differ significantly in design and operating conditions. To investigate this, hydraulic model tests were conducted using small, freshly cut trees—some with branches and roots—to simulate natural trees moving towards the dam. The results indicated that tree accumulation at the gated spillway or PKW depended on flow discharge, cross-sectional position in the reservoir and water levels. Blockage of the gates—or simultaneous blockage of both the gates and the PKW—caused a greater rise in reservoir water level than blockage of the PKW alone. However, the rise was still limited. This study highlights the combined use of the PKW and the gated spillway as cost-effective, as they complement each other in managing driftwood and mitigating water-level rise.

1 BACKGROUND

Hydropower dams must operate safely in cold climates. In Sweden, as in many other regions, another issue of safety concern is the production of floating debris (trees) following a high flood. The trees comprise typically spruce and pine trees, with trunk lengths often up to 20–25 m. They are transported down the river and accumulate gradually upstream of a dam if no boom is installed to intercept them. No driftwood incident has been reported to date in the regulated Swedish rivers. However, available documents substantiate that a heavy rain once eroded ~10 000 m^3 trees that became driftwood in a small river basin. In addition, the majority of Swedish dams have only two or three surface gates. This implies that every opening is crucial for flood discharge, and driftwood clogging would lead to an undesirable reduction in discharge capacity.

In light of the dam-safety guidelines, a large number of Swedish dams have been upgraded so as to increase their safety standard (Yang et al. 2002, Yang & Cederström

DOI: 10.1201/9781003655541-9

2002). Safety upgrading is still ongoing for many existing dams. Studies show that production of floating trees is likely to occur in many reservoirs due to heavy rainfall (Åstrand & Persson 2017). For a chosen upgrade solution, it is desirable that the driftwood behaviors at the spillway be well understood and the residual capacity estimated (Crookston et al. 2015, Pfister et al. 2013).

The piano key weir (PKW) is an innovative structure for effective flood release (Lempérière & Ouamane 2003, Ouamane 2011). Similar to its use in France, the PKW has emerged as a prominent option in dam-safety projects in Sweden, owing to its reliability and cost-effectiveness (Yang et al. 2024, 2025). In the dam upgrade in question, it is chosen to replace the existing embankment dam and to enhance the discharge capacity. In combination with the existing spillway gates, the design flood is discharged by allowing a temporary rise above the full reservoir water level (FRWL). Hydraulic model tests are performed to help finalize the solution and to understand the performance of floating trees and evaluate hydraulic aspects of the safety concern.

2 REBUILDING ALTERNATIVE

Situated in Mid Sweden, the dam was commissioned in 1965. Figure 1 shows its downstream view. The facility is composed of a 42 m spillway section, a 30 m gravity dam to its left, an embankment dam to its right and a power station. The embankment has a max. structural height of ~10 m, and its crest is 5 m wide and 110 m long. The spillway has three openings equipped with upward-going tainter gates, with a total net width of 34 m. The leftmost and middle openings (A and B), each 12 m wide, are of the same make, with their thresholds 5.35 m under the FRWL. Due to its poor condition, the rightmost opening (C), 10 m wide, is to be modified into an overflow weir with a sill elevation at the FRWL. The reservoir is shallow at the spillway, with a river bed elevation ~1.0 m below the thresholds of the A and B openings. They discharge ~500 m^3/s at the FRWL, which can be compared with the design flood 1 000–1 200 m^3/s.

Figure 1. The existing dam with spillway gates, looking upstream (courtesy AFRY).

In light of the safety evaluations, the upgrading solution is finalized. The studies show that the most favorable and cost-effective option is the construction of a PKW downstream of the embankment dam (Figure 2). Thus, the existing spillway is on the left side of the river, and the PKW on the right side. The PKW is assigned a L-shaped planform, connecting to the existing spillway to its left and extending to the right riverbank. Placed on a low concrete dam, it is ~100 m long, with a crest elevation set the same as the FRWL. This means that the PKW starts to operate once the reservoir level exceeds the FRWL. The right PKW wing, ~67 m long, is aligned at a right angle with the river flow, while the left wing, ~33 m long, is at a 20° angle.

Figure 2. Finalized rebuilding alternative of a piano key weir (PKW) downstream of the existing embankment dam that is to be partially removed.

Upon completion of the PKW construction, the embankment dam is excavated to an elevation 3.0 m below the FRWL, which is determined in hydraulic model tests. More excavation would undoubtedly further improve the approach flow conditions, but the costs would be higher. At this elevation, the remaining embankment dam does not encroach much on the discharge capacity. The existing cofferdam in the reservoir is un-touched.

3 PKW CONFIGURATION

The parameter definition of a typical PKW with both up- and downstream overhangs is shown in Figure 3, in which W_i, W_o = inlet and outlet key width, P_i, P_o = inlet and outlet key height, B_i, B_o = crest length of inlet and outlet key overhang (excluding T_s), B_b = weir base length, S_i, S_o = parapet wall height of inlet and outlet key, α_i, α_o = angle of the inlet and outlet key floor with the horizontal plane, T_s = wall crest thickness, P = total weir height, and subscripts i and o denote the inlet and the outlet key, respectively.

The width of a single weir unit and its streamwise crest length (inclusive of the crest walls) are

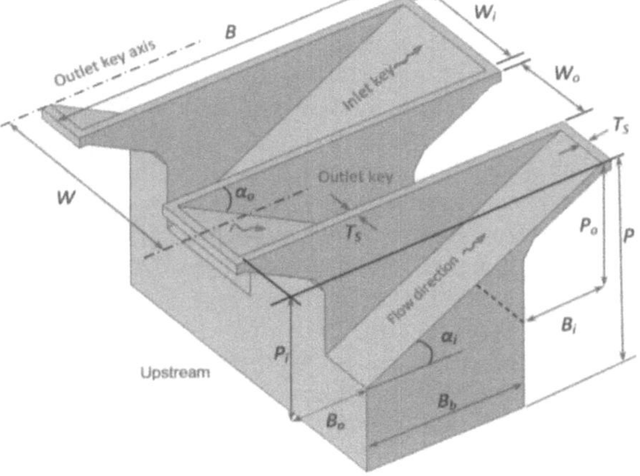

Figure 3. Parameter definition of PKW with both up- and downstream overhangs and parapet walls.

$$W = W_i + W_o + 2T_s \tag{1}$$

$$B = B_i + B_o + B_b + 2T_S \tag{2}$$

The angle of each sloping key floor with the horizontal plane reads

$$\tan(a_i) = (p_i - S_i)(B_b + B_i) \tag{3}$$

$$\tan(a_o) = (p_o - S_o)(B_b + B_o) \tag{4}$$

Along the crest centreline, the developed crest length (referring to one single weir cycle) is

$$L = 2B + W_i + W_o \tag{5}$$

Based on separate flume studies of two layouts at two scales (1:12 and 1:35), a relatively favorable configuration is finalized and approved by the dam owner (Yang and Helgesson 2024, Yang et al. 2025). The chosen dimensions are listed in Table 1, resulting in W = 3.90 m, P = 3.75 m, B = 8.35 m and L = 19.9 m. The PKW design involves 23 regular (identical) units, a half unit at each end, and an irregular unit in the corner. The upstream angle between the two straight wings is ~110°. Counted along the wall crest centreline, their total developed crest length is L_T = ~488 m, which is equivalent to $L_T/L \approx$ 24.5 regular units. The PKW structure is placed on a concrete foundation of varying heights adapted to the river bottom topography.

Table 1. PKW parameters chosen in the dam upgrade design (unit m).

W_i	W_o	W	T_s	B_i	B_b	B_o	B
1.80	1.40	3.90	0.35	1.45	3.75	2.45	8.35
S_i	S_o	a_i	a_o	P_i	P_o	P	L
0	0.50	33.9	25.8	3.50	3.50	3.75	19.90

4 HYDRAULIC MODEL OVER THE FACILITY

To evaluate the PKW design and analyze the behavior of floating debris, hydraulic model tests were conducted at a scale of 1:35. The model simulated a reservoir length of 150 m and a downstream river valley of 100 m, encompassing the full river width. The physical model measured 10 m in length, 6 m in width, and 0.60 m in height.

For high accuracy, the reservoir bathymetry, downstream river terrain, the PKW, and the existing spillway were all fabricated using precision machine milling. This technique ensured a smooth surface finish, which was essential given the relatively small dimensions of the PKW model. Figure 4 presents a view of the model. To replicate natural topography and concrete structures, Styrofoam (density 300 kg/m^3) and Ureol (density 700 kg/m^3) were employed. Flow rates were measured using a calibrated magnetic flow meter with a diameter of 200 mm.

Initial tests were conducted to determine the individual discharge capacities of the PKW and the existing spillway, as illustrated in Figure 5. At the FRWL, the existing spillway releases 500 m^3/s. When the water level is raised by 1.0 meter above the FRWL, the discharge capacities of the existing spillway and the PKW increase to 670 m^3/s and 530 m^3/s, respectively. The PKW's discharge refers to its total capacity, without differentiating between the left and right wings.

To better understand the flow patterns, surface flow velocities were measured at a cross-section located 100 m upstream of the existing spillway, at several discharges

Figure 4. Hydraulic model of the dam with the PKW and part of the existing spillway.

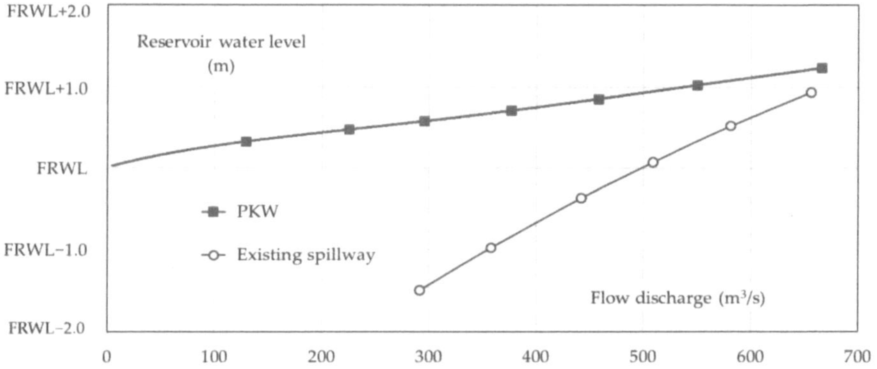

Figure 5. Individual discharge capacity of the PKW and the existing spillway.

(Figure 6). The origin, X = 0, is defined along the centerline of the right pier of spillway opening C (Figure 2). When looking downstream, positive X-values extend to the right of this reference line. As expected, flow velocity in the reservoir increases with discharge. However, higher velocities consistently occur on the left side of the reservoir compared to the right, regardless of the flow rate. This indicates that floating trees, approach the PKW at lower velocities on the reservoir's right side.

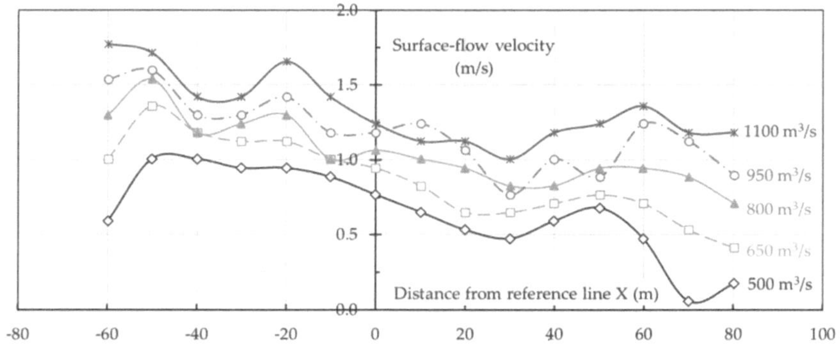

Figure 6. Surface velocity along a cross-section 100 m from the existing dam axis (looking downstream).

Site studies indicate that floating trees may originate from both the upper and lower catchment areas of the dam, with approximately 80% originating from the former (see Table 2). Tree lengths can reach up to 25–30 m and are categorized into five size classes. Trees from the upper catchment typically consist of bare trunks, as limbs and branches are lost during the long transport distance. In contrast, trees from the lower catchment may retain limbs and branches, with ~20% of these trees still bearing branches. From both catchment areas, around 50% of the trees retain their root systems, with root diameters reaching up to 10% of the tree length.

Table 2. Potential source and lengths of floating debris from the dam site.

Category	I	II	II	IV	V	Total (%)
Prototype tree length	< 10 m	11–15 m	16–20 m	21–25 m	26–30 m	
From upper catchment (%)*	21	25	23	10	1	80
From lower catchment (%)**	7	5	5	3	0	20

*trunk without limbs/branches, **trunk with limbs and limited branches.

Model trees were prepared based on the field data, following the five length categories listed in Table 2. Their dimensions were scaled to match the hydraulic model at a 1:35 ratio. In total, 150 small, freshly cut trees were trimmed for use in the experiments (Figure 7). The trees retained their natural densities, ranging from 890 to 970 kg/m³. In terms of buoyancy and the likelihood of passing over the PKW, this should lead to conservative estimates. Each tree trunk tapered from root to top, reflecting natural geometry. For example, trunks in Category I had diameters ranging from approximately 10 mm at the base to 5 mm at the top, while those in Categories IV and V ranged from about 25 mm down to 10 mm. During testing, trees of varying lengths were introduced into the model randomly—one or two at a time—to simulate natural debris movement.

Figure 7. 150 small trees trimmed in accord with the field data.

5 TEST RESULTS AND DISCUSSIONS

The tests include the following aspects: (a) contribution from the PKW in limiting reservoir water level as compared to the case without the PKW, (b) patterns of tree accumulation at the dam, and (c) resulting water-level rise in three debris approach scenarios.

5.1 *PKW contributions*

At the FRWL, the capacity of the dam is 500 m^3/s, with no water flowing over the PKW and weir C. The tests are made at a discharge kept constant at 550 m^3/s. If a tree follows the flow through the spillway, it is fetched and released again into the water. First, 20 trees of mixed lengths are released on the left side of the reservoir, one tree or two at a time. When the flow becomes stable, another 20 trees are dropped into the water and so on. This means that the tests are run with 20, 40, 60, 80, 100 and 150 trees afloat to the spillway, and the accumulation pattern and water level rise are monitored at each step. For comparison, the tests are repeated for the situation without the PKW (the flow to the weir is blocked).

Either with or without the PKW, all the trees accumulate at the gated openings. Figure 8 shows the water level rises (averaged) as a function of the number of trees. With the PKW, the water level rise is 0.15–0.30 m. If the gated spillway is debris blocked, more flow is shifted to the PKW. Owing to the long PKW crest, the rise is limited, almost independent of the number of trees at the openings. Without the PKW, the water level increases by 0.60–1.30 m, which is significant. It is observed in the tests that the first 20–40 trees account for the major increase in water level, the effect from the trees that follow become less obvious.

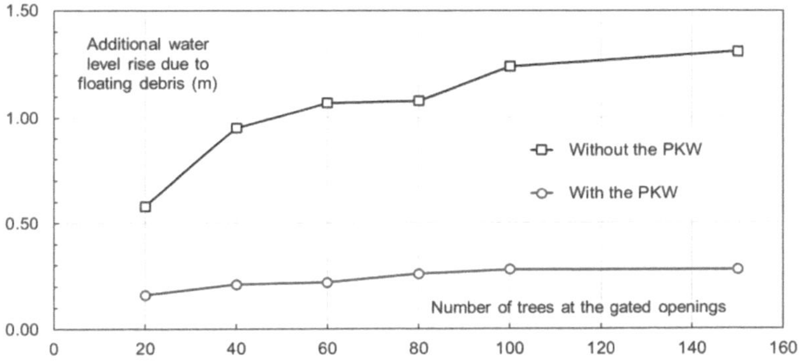

Figure 8. Water level rise in the reservoir due to clogging of the gate openings, with and without the PKW, at constant flow discharge 550 m^3/s.

5.2 *Debris accumulation patterns*

Repeated tests are run at discharges varying from 500 to 1100 m^3/s in a few steps at 50–100 m^3/s intervals. The debris behaviors at the dam are recorded, with the following results.

(1) Being lighter than water, the trees remain partially visible above the surface. The accumulation at either the gates or the PKW takes place at a random manner, depending on factors such as tree length, shape (root, branch), trunk orientation relative to the flow, density etc.

(2) Some trees pass the gated spillway in the beginning. However, if one tree gets stuck, trees that follow start accumulating immediately upstream of the gates. The clogging and the resulting in capacity reduction are mainly governed by the first incoming 20–40 trees.

(3) Depending on the trunk orientation in relation to the flow, trees become wedged in the gate openings or pushed against their pier noses. Due to limited water depth at the openings, the trees are pulled down, covering a major part of the water depth.

(4) Some trees may enter the inlet keys of the PKW. Trees are pushed against the upstream overhangs, many of which are dragged beneath the overhangs at high flows. However, due to limited rises in reservoir water stage, no trees are pushed over the PKW and further downstream.

(5) The trees, some of which have roots and branches, become always interdigitated at the discharge structures. The interlocking pattern forms a relatively stable network that is somewhat compacted by the flowing water and hardly changes with time.

(6) If the gated openings are first blocked by debris, this leads to a rise in reservoir water level and more flow and trees are shifted to the PKW and vice versa.

In addition to certain scale effects, the tests are plagued by randomness. Each test leads to a somewhat different accumulation pattern and water level rise, which is however expected. To simulate potential prototype events, three major scenarios are identified and examined below.

5.3 *Debris approaching scenarios*

It is observed in the tests that the trajectory of movement of the trees and the rise in reservoir water level are dependent on flow discharge as well as on its cross-sectional position in the reservoir. Three tree approaching scenarios are therefore examined: DS1 – roughly equal distribution in reservoir cross-section; DS2 – transported only along the left half of the reservoir; and DS3 – transported only along its right half.

In scenario DS1, the situation of tree accumulation is shown in Figure 9. At flow discharges 500–600 m^3/s, all the trees move, in the beginning, to the existing openings A–C. A partial clogging reduces its capacity, which results in a rise in reservoir water level, thus pushing more flow to the PKW. At discharge 500 m^3/s, ~90% of the trees finally accumulate upstream of the spillway and only ~10% at the PKW. With an increase in flow discharge, more and more trees move to the right wing of the PKW. At discharge ~800 m^3/s, the number of trees at the gated openings is approximately the same as at the PKW. As the flow increases gradually up to 1 100 m^3/s, 60–65% are aggregated at the PKW, many of which are pushed under the overhangs.

In Figure 2, the straight line in the reservoir, aligned with the centerline of the right pier at spillway opening C, marks approximately the middle of the main river course, which is 120 meters wide (a significant port of the right riverbank is shallow). As shown in Figure 10 for scenario DS2, due to the higher velocity on the left river side, the trees are transported to the

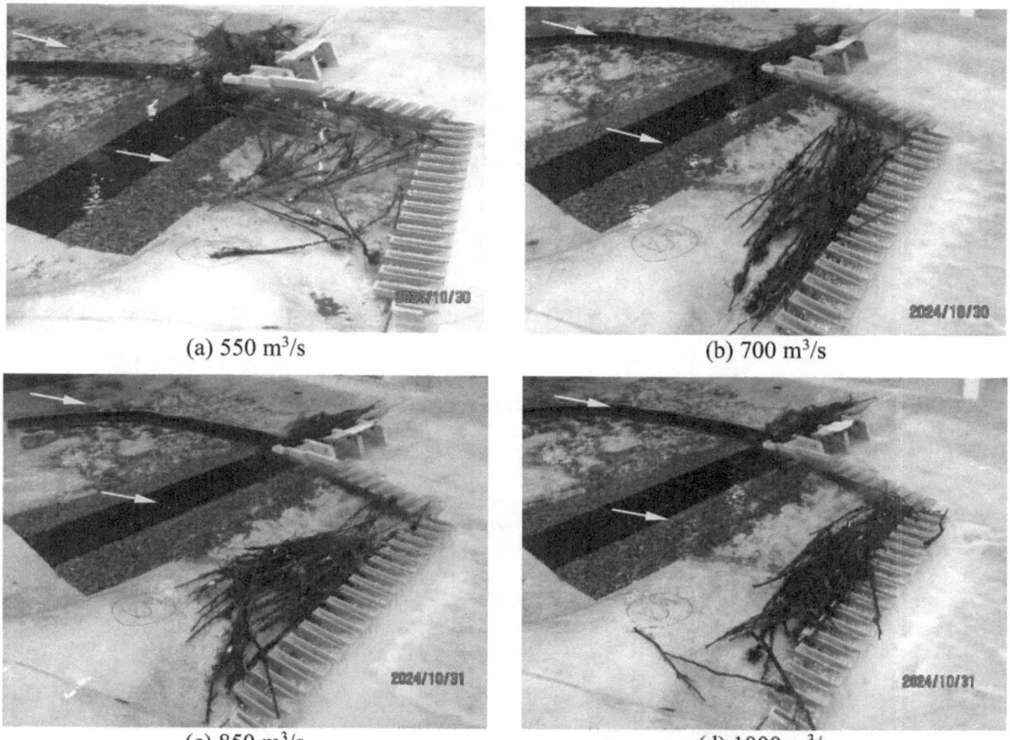

(a) 550 m^3/s (b) 700 m^3/s

(c) 850 m^3/s (d) 1000 m^3/s

Figure 9. DS1 – approximately equal distribution of floating trees along the river cross-section. The arrows indicate the flow direction in the reservoir. (a) 550 m^3/s, (b) 700 m^3/s, (c) 850 m^3/s, (d) 1000 m^3/s.

(a) 550 m^3/s (b) 700 m^3/s

(c) 850 m^3/s (d) 1 000 m^3/s

Figure 10. DS2 – floating trees along the left half of the river course. The arrows indicate the flow direction in the reservoir. (a) 550 m^3/s, (b) 700 m^3/s, (c) 850 m^3/s, (d) 1 000 m^3/s.

existing gate openings, which is almost independent of the discharges. The only exception is that a few trees move to the PKW at flow discharges above 900 m^3/s.

Figure 11 shows the test results in scenario DS3. At flow discharges <550 m^3/s, the surface flow on the right side of the reservoir is too weak to transport the trees to the PKW. The first trees move therefore to the spillway openings. A partial clogging of openings A–C leads to a reduced discharge capacity and a higher reservoir level, which in turn results in more flow to the PKW and higher velocity on the right river side. At flow discharge 550 m^3/s, the trees are relatively even distributed across the dam. At 650 m^3/s, ~90% of the tree move finally to the PKW and the remaining upstream of openings B and C. With a further increase in discharge, almost all the trees follow the flow to the PKW; only a few are wedged at opening C.

Figure 12 plots the effects of the debris clogging on reservoir water level. Each test is repeated 3–4 times; each resulting reservoir water level represents the average of the repeated test results. Due to the clogging, the reservoir water-level rise is discharge dependent in either scenario. Obviously, a higher discharge gives rise to a larger rise in reservoir water level. As compared with the situation with floating trees, scenarios DS1 and DS2 lead to similar effects, corresponding to a 0.30–0.65 m rise within the examined flow range. In scenario DS3, the rise is within 0.15–0.25 m. This implies that clogging of only the PKW, not the gated openings, is the most favorable scenario.

6 CONCLUSIONS

In an existing dam equipped with gated openings, a PKW is proposed to replace the embankment section and enhance the overall discharge capacity. Model studies are conducted to

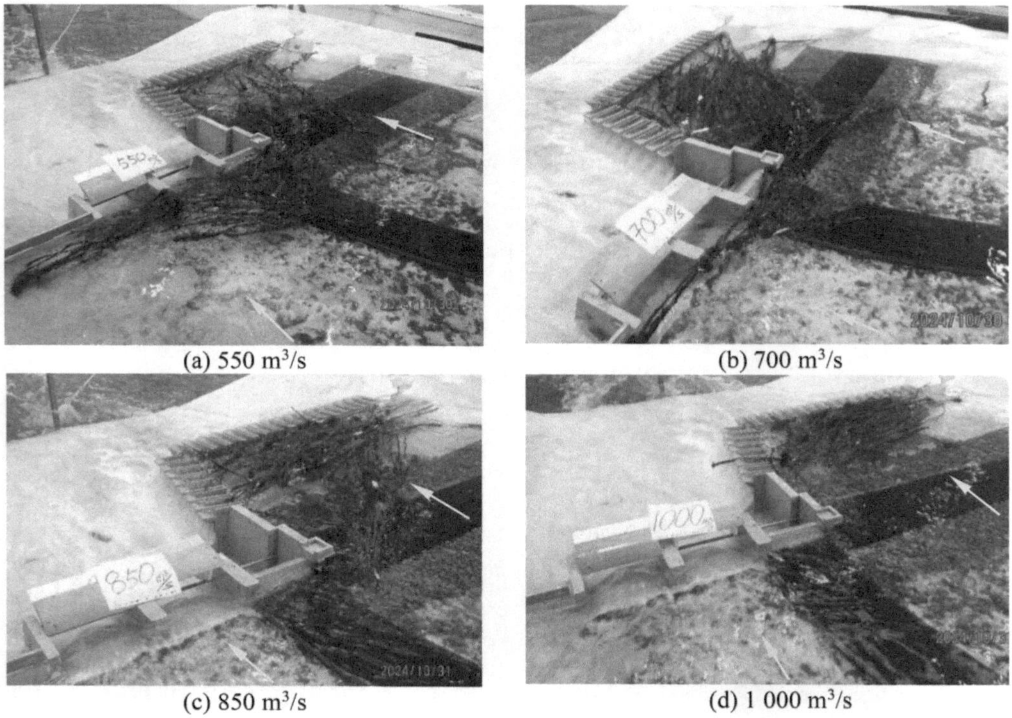

Figure 11. DS3 – floating trees along the right half of the river course. The arrows indicate the flow direction in the reservoir. (a) 550 m^3/s, (b) 700 m^3/s, (c) 850 m^3/s, (d) 1 000 m^3/s.

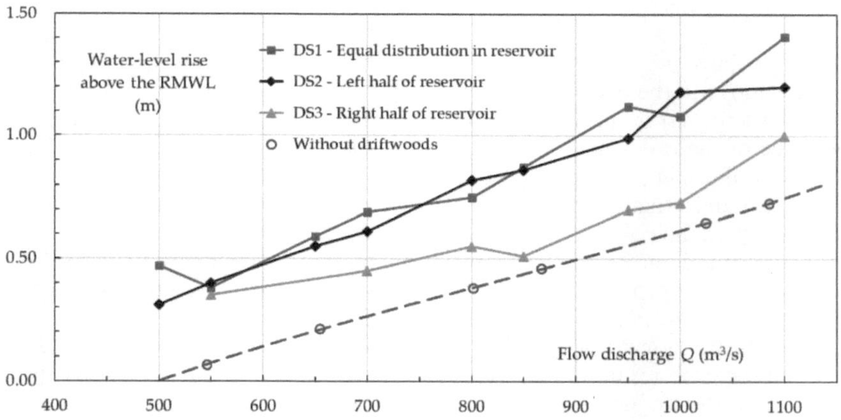

Figure 12. Water-level rise due to floating trees in three scenarios. For each flow discharge of a scenario, the plotted value represents the average of repeated test results.

examine the potential clogging effects caused by floating trees. Using a total of 150 small trees of varying lengths—representative of prototype conditions—the following conclusions are drawn.

The pattern of tree accumulation at the gate openings or the PKW depends on several factors, including tree length, shape, and orientation relative to the flow direction. The resulting water level rise due to clogging is primarily influenced by the flow discharge and the cross-sectional position of the approaching trees within the reservoir.

Under a roughly uniform distribution of floating debris in the reservoir (DS1), approximately 90% of the debris initially accumulates at the gated openings at flow rates of 500–550 m^3/s. As the flow increases, more trees are redirected toward the PKW. At flows between 1,000 and 1,100 m^3/s, 60–65% of the trees accumulate at the PKW. When trees approach the dam from the left half of the reservoir (DS2), nearly all debris is directed toward the gated spillway, regardless of the discharge rate. In contrast, when trees approach from the right half (DS3), accumulation occurs at both the gated spillway and the PKW for flows below 600 m^3/s. However, at flows above 850 m^3/s, nearly all debris is conveyed to the PKW, with only a few pieces becoming trapped at the gate openings.

When floating debris accumulates at the gates or at both structures, the resulting rise in reservoir water level is comparable and greater than when only the PKW is clogged. This indicates that clogging limited to the PKW represents the most favorable scenario. Nevertheless, regardless of the specific clogging pattern, the gated spillway and the PKW function complementarily, creating a synergistic effect that helps mitigate the overall rise in reservoir water level.

It should be noted that laboratory tests involving floating trees are subject to both scale effects and inherent randomness. While the latter may reflect the natural variability seen in prototype conditions, these limitations should be considered when interpreting the results. As such, the qualitative observations may offer more reliable insights than the quantitative data. Nevertheless, the studies enhance understanding of debris behavior at the dam and offer insights for designing potential debris management measures, even for other dams with comparable layouts.

ACKNOWLEDGEMENTS

The study, conducted for Uniper Sydkraft Hydropower AB, was carried out at Vattenfall's Hydraulic Laboratory in Älvkarleby. Project contributors included Mats Billstein, Pär Nilsen, Erik Skepparkrans, Anders Sjödin, Magnus Lagerkvist, Jennifer Persson, and Patrik Lidberg. The manuscript was prepared and revised by James Yang, with all co-authors having reviewed and approved the submitted version.

REFERENCES

Crookston, B., Mortensen, D., Stanard, T., Tullis, B.P. & Vasquez, V.M. 2015. Debris and maintenance of labyrinth spillways. *USSD 2015 Annual Conference, Louisville KY, USA. April 13–17, 2015.*

Lempérière, F. & Ouamane, A. 2003. The piano keys weir: a new cost-effective solution for spillways. *International Journal on Hydropower & Dams* 10: 144–149.

Ouamane, A. 2011. Nine Years of Study of the Piano Key Weir in the University Laboratory of Biskra "Lessons and Reflections". *Proc. Labyrinth and Piano Key Weirs—PKW 2011, Liège, Belgium, 9–11 September 2011.*

Pfister, M., Capobianco, D., Tullis, B. & Schleiss, A.J. 2013. Debris-blocking sensitivity of piano key weirs under reservoir-type approach flow. *Journal of Hydraulic Engineering* 139(11): 1134–1141.

Yang, J., Johansson, N. & Cederström, M. 2002. Towards safer dams – refurbishment examples in Vattenfall's dam-safety program. *Proc. HydroVision 2002, Portland, USA, July 29–August 2, 2002.*

Yang, J. & Cederström, M. 2002. Floods, new design-flood criteria and dam refurbishment in Sweden. *Proc. 2nd International Symposium on Flood Defense, Beijing, China, September 10–13, 2002.*

Yang, J. & Helgesson, A. 2024. *Lillflyforsen discharge safety – hydraulic model studies 2024* (in Swedish). Report no. VRD-R51-2024, Älvkarleby, Sweden: Vattenfall R&D.

Yang, J., Nordström, E., Li, S.C. & Kocahan, H. 2024. Labyrinth and piano key weirs in cold climate – experiences from prototype facilities and laboratories. *Proc, 92nd ICOLD Annual Meeting & International Symposium, New Delhi, India, 29 September 3 October, 2024*

Yang, J., Frisk, A., Billstein, M., Mattiasson, P. & Johnsson, L. 2025. Dam upgrade with piano key weir for discharge of extreme floods. *Proc. 28th ICOLD Congress, Chengdu, China, 16–23 May 2025*

Åstrand, S. & Persson, F. 2017. *Method for analysis and management of floating debris* (in Swedish). Energiforsk report no. 2017:446, Stockholm, Sweden.

Labyrinth and Piano Key Weirs IV – Erpicum, Crookston & Paxson (Eds)
© 2026 The Author(s), ISBN 978-1-041-10527-5

Effect of driftwood on the hydraulic performance of A-type Piano Key weirs situated in-stream or at reservoirs

P. Venetz
Geoformer igp AG, Brig-Glis. Formerly: Laboratory of Hydraulic Constructions (LCH), Ecole Poly-technique Fédérale de Lausanne (EPFL), Switzerland

M. Pfister
Haute école d'Ingénierie et d'Architecture de Fribourg (HEIA-FR, HES-SO), Switzerland

ABSTRACT: Driftwood is a natural part of most river ecosystems. The knowledge of its interaction with hydraulic structures, such as PKWs, is essential for two reasons: (1) to avoid uncontrolled inundations because of the level rise initiated by trapped driftwood (reducing the weir discharge coefficient), and (2) to initiate the passage of the wood over the PKW, so that it remains in the river to further fulfill its environmental services. Knowledge of the conditions motivating driftwood to block at a PKW is essential. The present study accordingly focuses on the blocking probability of driftwood at PKWs and the resulting head increase – providing a basis to limit both. Extensive physical model tests were conducted with broad boundary conditions to achieve a certain significance of results. This is a challenging goal, since driftwood-related phenomena are known to show a partially random behavior. The observations suggest that driftwood typically blocks up to "mid-level" heads, and that a jam is likely removed during floods (large heads). An acceleration of the flow upstream of the PKWs and wide inlet keys showed less blockage and a reduced level increase. These findings are partially in conflict with the maximum hydraulic PKW efficiency: The reduction of the flow head to an absolute minimum by optimizing the PKW geometry might support blockage of driftwood.

1 INTRODUCTION

In-stream driftwood (floating debris) is a natural element of rivers draining forested basins. It provides shelters for juvenile fishes and other species. While it offers multiple ecological benefits during periods with frequent discharges, it might become a threat during floods. Then, driftwood, often combined with urban waste, can be trapped at structures such as weirs (Bezzola and Hegg 2007), bridge decks (Bezzola et al. 2002; Schmocker and Hager 2011) or bridge piers (Melville and Dongol 1992; Pagliara and Carnacina 2010). Those represent confined hydraulic sections, across which the driftwood – particularly if entire trees are transported – can possibly not pass. If the wood blocks, then the water must alter its flow.

Weirs within rivers represent often such "obstacles", prone to driftwood blockage. Two key questions thus arise:

- How likely is the blockage of driftwood at a certain weir type?
- Which level rise does the latter provoke upstream of the blocked weir?

The nonlinear nature of Piano Key weirs (PKW) allows for significantly longer weir crests than could be installed using a traditional linear weir for a given river width. An efficient rating curve with small flow heads results, typically generating a small impulse to transport wood from the leading edge over the weir. Further, narrow inlet keys provide low currants along the weir top. One could thus suggest that PKWs are prone to driftwood blockage. Numerous studies have been published on the effect of debris at PKWs so far, few are mentioned hereafter.

DOI: 10.1201/9781003655541-10

Ouamane and Lempérière (2006) conducted early physical model tests with driftwood. They observed that driftwood got stuck in the inlet key for small discharges, resulting in a slight reduction in discharge capacity of about 10%. Laugier (2007) reported of model tests with driftwood at the PKW spillway of Goulours Dam (France), concluding that an increasing head tends to remove previously blocked debris and that most debris pass the PKW for heads above H=1 m.

Pfister et al. (2013) investigated individual trunks as well as debris batches approaching three different A-type PKW installed on a dam configuration. The approach flow was thus of "reservoir" type. They concluded that the dimensionless trunk diameter D/H is the dominant parameter indicating whether an individual trunk will be trapped by or pass over a PKW. Note that D= trunk diameter and H=PKW flow head. A 50% blocking probability was observed if the driftwood diameter equaled the PKW critical flow depth (i.e., $D=(2/3)H$). The blocking probability was defined as the number of blocked trunks divided by the number of the totally supplied trunks. As for the tests with approaching batches, the relative weir head increase caused by a blockage was only significant when the reference (index r, before debris occurrence) H_r was small, i.e., for smaller discharges. For large heads, typically used as PKW design values, the relative increase was negligeable, since the wood was pushed over the PKW by the flow. Note that the test procedure of Pfister et al. (2013) was close to the one applied herein, so that their data could be integrated in the present analysis as boundary condition.

Vaughn et al. (2021) investigated the blocking sensitivity of Labyrinth weirs in channel and reservoir applications. Heterogeneous driftwood approaching the weir as a batch or as single trunks were tested in a physical model. It was found that the blocking probability depended on D/H_r, as for PKWs, but with a generally lower likelihood. The relative head increase for wood accumulation at the weir was shown in function of the channel width.

Benet et al. (2021) studied driftwood blockage at gated ogee crests in a physical model. Large quantities of heterogenous driftwood were supplied. The head increase was addressed via a reduced discharge coefficient. To reduce or eliminate a level rise, racks, overhanging piers or the removal of the piers were studied.

Panthi and Crookston (2025) presented an extensive literature review on PKWs. More importantly, they conducted model tests on the combined effect of driftwood and sediments on PKWs under river-type approach flow. The relative head increase under blockage was determined in function of the wood characteristics (size, blocked volume, and obstructed area). Their trunks disposed of branches, what led to a blockage concentrated at the upper PKW edge (outlet key crest), whereas Pfister et al. (2023) tested trunks without branches. Their wood blocked on the entire PKW surface. Panthi and Crookston's driftwood also blocked partially below the overhangs, a phenomenon not observed for the "plain" trunks of Pfister et al.

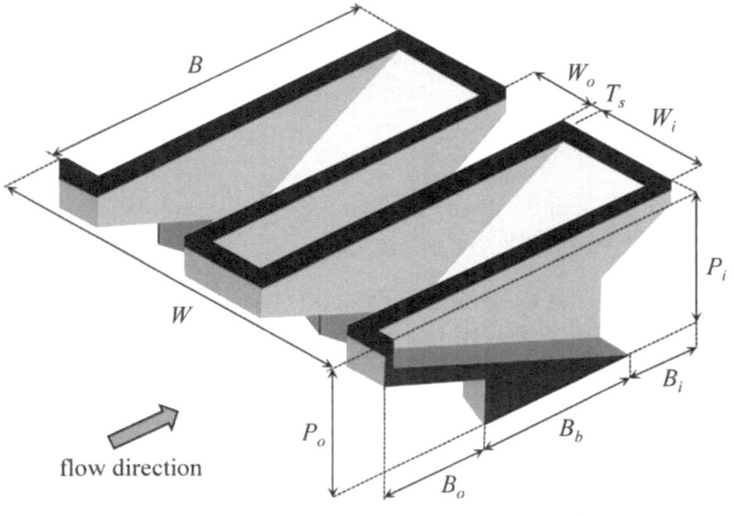

Figure 1. Common PKW notation according Pralong et al. (2011, sketch from Pfister et al. 2013).

Combining the results of these two studies indicates that branches might affect the behavior of the wood via interconnection and batch permeability. Panthi and Crookston concluded that for higher discharges with consequent high upstream heads, the chances of driftwood jams are lower. To end, Pralong et al. (2011) defined a PKW nomenclature, which is used herein (Figure 1).

2 PHYSICAL MODEL

Systematical physical model tests were conducted at the Laboratory of Hydraulic Constructions (LCH) of EPFL Lausanne (Venetz 2014).

2.1 *Channel and PKW*

Three A-type PKWs I, II and III were installed at the end of a 5 m long channel. The PKWs were different in terms of their number of cycles and geometrical dimensions, to broaden the outcomes. Figure 2 shows the PKWs and Table 1 lists their geometrical dimensions. The width W of the channel corresponded approximately to the width W of the tested PKW.

To modify the approach flow depth (and thus the approach flow conditions influencing the driftwood behavior), the PKWs were mounted on sills of different height, being S=0.07 and 0.19 m for PKW I, S=0, 0.05 and 0.25 m for PKW II, and S=0.05 m for PKW III.

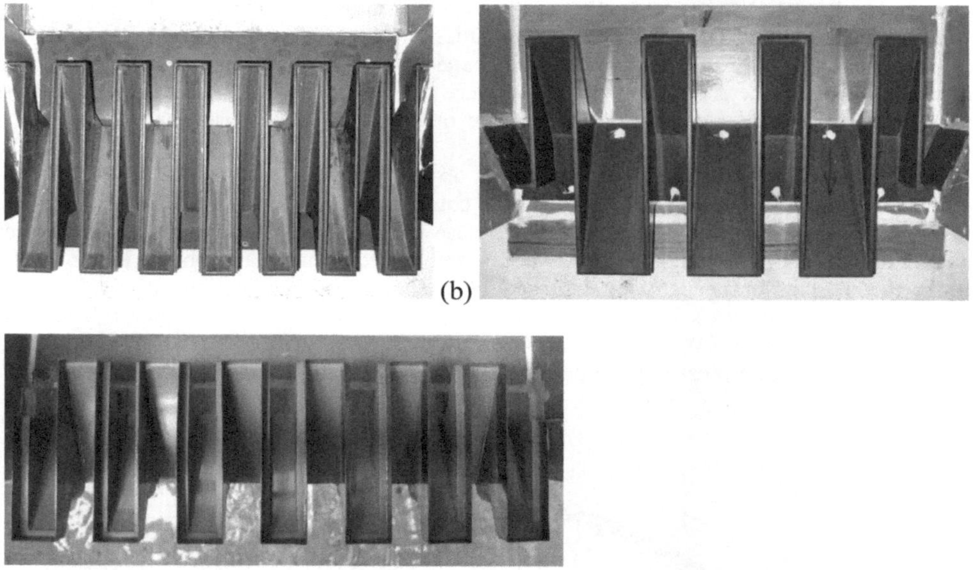

(a) (b) (c)

Figure 2. Photos (plan views) of the tested PKWs (a) I, (b) II and (c) III mounted at the channel end, with a flow direction from top to bottom.

The discharge Q was measured with the in-house MID up to ±0.5% full span. The overall flow depth h_1 (Figure 3) was measured by a point gauge (±0.5 mm) at about 1.0 m upstream of the PKWs, to be outside of their drawdown curve.

Minor scale effects related to the rating curve occur if the weir head H<0.03 m (Erpicum et al. 2013, Pfister et al. 2013b), given that viscous and surface tension forces are not considered in the Froude similarity (Pfister and Hager 2014). However, a relative comparison is presented herein pointing at the PKW head *increase* due to the presence of driftwood, whereas the determination of the absolute head may follow Leite Ribeiro et al. (2012). All tests were free of tailwater submergence, given that the model channel ended with the PKW.

As for the approach flow along the channel, the discharge Q, channel width W and flow depth h_1 allowed for determination of the relevant hydraulic parameters, being the flow velocity $V=Q/(h_1 W)$, the velocity head $h_v=V^2/2g$, the specific energy head (energy line) as sum $E=h_1+h_v$, and finally the PKW head $H=E-(P+S)$.

The upstream channel aimed to represent quasi river-type approach flow conditions. The effect of the upstream configuration, affecting the velocity and thus the momentum of the arriving driftwood, is hydraulically often described with the approach flow Froude number F (Schmocker and Hager 2010). The latter results, for a rectangular channel as considered herein, as $\mathsf{F}=Q/(Wg^{0.5}h_1^{1.5})$.

Table 1. Geometrical dimensions of the tested PKWs.

		PKW I	PKW II	PKW III
W	[m]	0.665	0.545	0.555
W_i	[m]	0.040	0.082	0.044
W_o	[m]	0.033	0.050	0.039
W_u	[m]	0.093	0.148	0.085
T_s	[m]	0.010	0.008	0.001
B	[m]	0.325	0.325	0.190
B_o	[m]	0.100	0.118	0.055
B_i	[m]	0.075	0.090	0.050
$P_i=P_o$	[m]	0.150	0.096	0.084
R	[m]	0.020	0.000	0.017
L/W	[-]	7.80	5.20	5.50
W_i/W_o	[-]	1.21	1.64	1.13
B/P	[-]	2.17	3.39	2.26
B_i/B	[-]	0.23	0.28	0.26
B_o/B	[-]	0.31	0.36	0.29

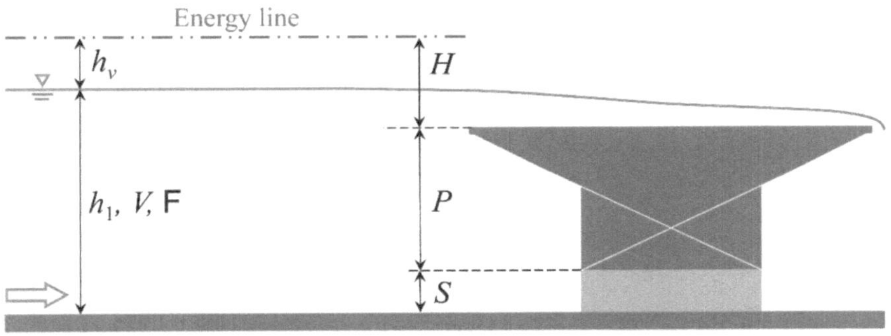

Figure 3. Definition sketch of hydraulic parameters upstream of the PKW.

2.2 *Driftwood scenarios and test procedure*

The choice of the model driftwood was based on the following aspects:

- "Natural" trunks and rootstocks without branches (but rootstocks with roots),
- a wide spectrum of trunk lengths T to achieve a general application of the results,
- a ratio of $T/D \approx 20$, derived from the survey of Bezzola and Hegg (2007), and
- watered during 2 hours before the tests, with a density of around 700 to 900 kg/m^3.

Nine driftwood trunk classes (TC) plus standard rootstocks (RS) were defined, as given in Table 2 and shown in Figure 4a. The mentioned lengths T and diameters D are average values per TC. The relative trunk length T compared to

- a PKW cycle $W_u = W_i + W_o + 2T_s$ included $0.36 \leq T/W_u \leq 4.65$ for PKW I, $0.23 \leq T/W_u \leq 2.92$ for PKW II, and $0.39 \leq T/W_u \leq 5.01$ for PKW III, and
- the channel width (quasi-equal to the PKW width W) included $0.05 \leq T/W \leq 0.65$ for PKW I, $0.06 \leq T/W \leq 0.79$ for PKW II, and $0.06 \leq T/W \leq 0.78$ for PKW III.

The driftwood was supplied some 3 m upstream of the PKW to allow its acceleration up to V before reaching the drawdown curve generated by the latter.

In situ, a wide variety of driftwood characteristics and volumes may arrive at a PKW. The prediction of the latter is challenging. The incertitude linked to the arrival of driftwood is probably larger than scale effects linked to the model trunks. These mainly focus on the trunk stiffness and the blockage permeability, both being overestimated in the model.

Table 2. Driftwood trunk classes (TC), with index M as maximum.

TC	1	2	3	4	5	6	7	8	9
T [m]	0.033	0.067	0.100	0.133	0.167	0.233	0.300	0.367	0.433 (T_M)
D [m]	0.002	0.003	0.005	0.007	0.008	0.012	0.015	0.018	0.022 (D_M)

Figure 4. Photos of (a) trunk classes TC (Table 2), and (b) one of the heterogeneous accumulative batches (TC 1 to 9 from left to right, two trunks per class).

The following three driftwood occurrence scenarios were tested:

1. *Individual trunk tests*. For a constant discharge, 50 trunks of each TC (Table 2) were individually supplied one by one at arbitrary locations across the channel and with random orientation. A blocked trunk was removed before the arrival of the subsequent trunk, intentionally avoiding driftwood accumulation effects. It was noted if a trunk blocked or passed (to derive the blocking probability Π). Every PKW was tested with different sill heights S and under various discharges, so that finally 205 data points resulted. If including the data of Pfister et al. (2013), collected for the same three PKW but with reservoir conditions, 98 additional data points join our experiments.
2. *Accumulative tests with homogenous batches*. Four homogenous batches including each 20 similar trunks (of the same TC, Table 2) were added subsequently to the model, so that finally 80 similar trunks were inserted. Before driftwood supply and after the addition of every individual batch, the resulting flow depth (h_1 in Figure 3) was measured, indicating the upstream head H (influenced by the trapped driftwood). The flow depth measurement

prior to the first batch addition severed as reference (index r) head H_r without driftwood. Further, the number of blocked trunks was counted after the addition of every batch, resulting in the blocking probability. No trunks were removed, so that prior batches influenced the behavior of subsequent batches. The discharge was kept constant during a test. All three PKW configurations were tested with different sill heights, as well as for all TC, so that over 180 accumulative tests were conducted (with 926 data points).

3. *Accumulative tests with heterogeneous batches*. Heterogeneous batches were supplied sequentially. A batch consisted of two trunks per TC plus two rootstocks (20 elements, Table 2, Figure 4b). All 10 batches (with finally 200 elements) were added subsequently under constant discharge, without removing blocked trunks or rootstocks. Before the addition of the first batch, and after every following batch, the upstream flow depth h_1 and the number of trapped elements (trunks plus rootstocks) were recorded. We conducted 41 tests and added the 12 tests of Pfister et al. (2013) to our data. The total of 52 tests included 582 data points.

2.3 *Test limits*

According to the three driftwood scenarios investigated, the hydraulic parameters as shown in Table 3 were tested.

It might seem that only "small" approach flow Froude numbers F were tested herein (up to 0.22). Note that, for typical PKWs, $P>H$ and H correspond roughly to the critical weir head (at least for the reservoir type). Thus, low in-stream F emerge even for high discharges.

Table 3. Test limits.

	Discharge Q [m³/s]	Reference Head H_r [m]	Velocity V [m/s]	Froude number F[-]	H_r/P [-]
Individual trunk tests (including data of Pfister et al., 2013)	0.002 to 0.077	0.002 to 0.090	0.00 to 0.31	0.00 to 0.20	0.013 to 0.684
Accumulative tests with homogenous batches	0.002 to 0.063	0.004 to 0.071	0.01 to 0.26	0.005 to 0.22	0.027 to 0.611
Accumulative tests with heterogeneous batches (including data of Pfister et al., 2013)	0.005 to 0.055	0.009 to 0.063	0.00 to 0.24	0.00 to 0.17	0.074 to 0.746

3 INDIVIDUAL TRUNK TESTS

As stated in previous studies, the blocking probability Π of a particular trunk – entered individually at a PKW – depends mainly on its diameter D and on the PKW head H_r (prior to blockage). Figure 6a shows the link between D/H_r as given by Pfister et al. (2013), who conducted quasi-identical tests in terms of driftwood, PKW geometry and procedures as we did.

However, Pfister et al. (2013) studied reservoir-type conditions without in-stream approach flow velocity. Their driftwood was nevertheless slightly accelerated in the vicinity of the PKW because of the drawdown curve. The driftwood of the present study enterd the PKW drawdown zone with an initial velocity V due to the channel flow. It has thus more momentum to pass the PKW. The latter is visible in Figure 6a, which shows the present data as well as those of Pfister et al. (2013), together with their blocking probability function $\Pi=1.5(D/H_r)-0.5$ (valid for individual trunks and within $0<\Pi<1$). Their function lies on the left of the present data points, indicating that a certain blocking probably (e.g. $\Pi=0.5$) is achieved for small D/H_r at reservoirs, but for larger D/H_r within rivers. Thus, as for reservoirs, "smaller" trunks block than within rivers, under otherwise similar conditions. Vice versa, a certain trunk may

Figure 5. Impression of trapped driftwood at a PKW in the model (accumulative tests with homogenous batches, PKW II, TC 8, S=0 m, Q=0.015 m^3/s).

probably block on a PKW installed at a dam, but not if facing an in-stream PKW. It seems thus evident to add the effect of the approach flow Froude number F to the function proposed by Pfister et al. (2013). By formulating the latter as (1+F) allows the inclusion of both data sets, since (1+F)=1 for reservoir conditions with F≈0.

Further, the data analysis showed a effect of W_i/W_o on Π. Relatively seen "wide" inlet keys W_i let pass more flow along the horizontal weir plane, keeping the trunks moving in streamwise direction to the end of the PKW.

Considering all the phenomena mentioned, the blocking probability Π of individually arriving trunks can be estimated as (R^2=0.76, Figure 6b):

$$\Pi \cong 1.5\left(\frac{\mathrm{D}}{H_r}\right)(1+\mathrm{F})^{-1}\left(\frac{W_i}{W_o}\right)^{-1} - 0.4 \qquad (1)$$

Equation (1) is similar to that proposed by Pfister et al. (2013) but includes now the effect of the approach flow Froude number and of the relative inlet key width. Equation (1) applies within $0 \leq \Pi \leq 1$ as well as for the limits mentioned in chapter 2.3.

Knowledge of the blocking probability of a single trunk is of minor importance in practice. More interesting might be the question about the "limit" trunk diameter, which certainly does not block – or certainly blocks – under specific conditions. Consequently, the extremes of blocking are specified, i.e., $\Pi \geq 0.9$ for "certain" blockage and $\Pi \leq 0.1$ for "certain" passage. These limit values of Π were inserted in Equation (1), showing that a trunk probably

- gets trapped if roughly $(D/H_r)/[(1+\mathrm{F})(W_i/W_o)] \geq 0.87$, and
- passes if roughly $(D/H_r)/[(1+\mathrm{F})(W_i/W_o)] \leq 0.33$.

The *head increase* due to driftwood blockage is negligible for individual trunks.

4 ACCUMULATIVE TESTS WITH HOMOGENOUS BATCHES

All accumulative tests indicated a similar trend: The blockage asymptotically reached an equilibrium (usually a maximum) if the number of arriving trunks was "large" enough. With the herein 80 supplied trunks (4 batches with 20 trunks each), the latter equilibrium was reached.

The analysis of the *blocking probability* focused on the test-end with its "final" (synonym for "stable") blockage. Thus, only the 180 data points giving the final blockage stage (out of

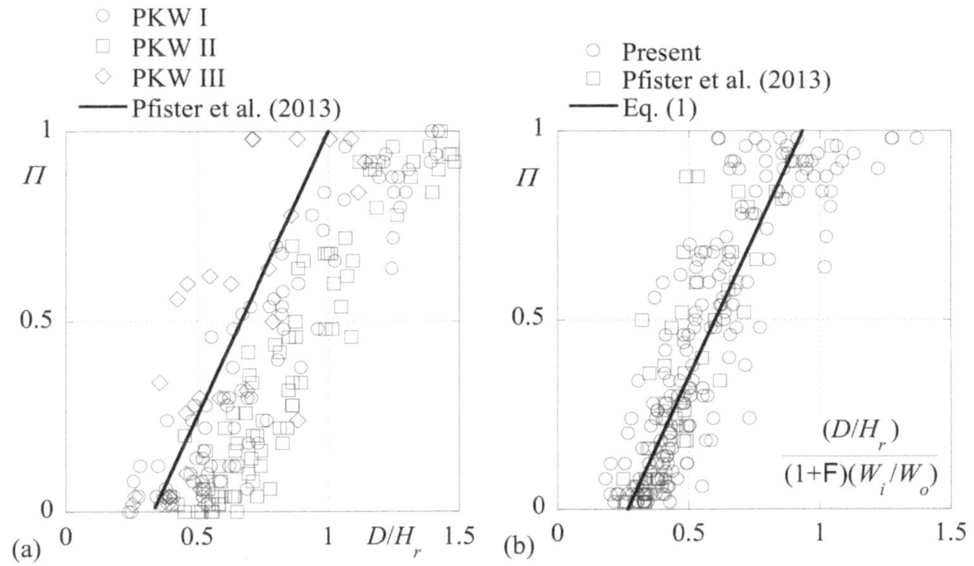

Figure 6. Blocking probability Π versus (a) D/H_r, for the present data (channel approach), and (b) $(D/H_r)/[(1+F)(W_i/W_o)]$, for the present data and those of Pfister et al. (2013, reservoir approach), all for individual trunk tests.

the 926 points, chapter 2.2) were studied. The other points represented the build-up of blocking, being ignored in the present analysis.

The final blockage probability Π for homogenous batches was similar to that of individual trunks. The determinant parameters were the same, namely, the trunk diameter relative to the reference head D/H_r, the channel flow velocity expressed via F, and the "flow intensity" on the PKW top plain given by W_i/W_o. Figure 7a shows the batch data, normalized as before in Figure 6b. The data points again basically collapse, however with the typical scatter for driftwood blockage. Equation (1) was included in Figure 7a, indicating that the batch points lay left of the latter. Batches are thus more prone to block than single trunks, what seems evident.

Considering all the phenomena mentioned, the final blocking probability Π of trunks arriving in batches can be estimated as (R^2=0.56, Figure 7a):

$$\Pi \cong 3\left(\frac{D}{H_r}\right)(1+F)^{-1}\left(\frac{W_i}{W_o}\right)^{-1} - 0.8 \qquad (2)$$

Equation (2) is similar to Equation (1). The comparison of the single trunk arrival with approaching batches indicates that the blocking starts (Π=0) under quasi-identical conditions, but batches then block almost instantly up to Π=1, whereas individual trunks tend to smaller probabilities for larger abscissa values. The interference between the many trunks (among each other and with the channel boundary) of a batch allows for higher blocking probabilities, as Figure 5 illustrates for a specific test. Equation (2) applies within $0 \leq \Pi \leq 1$ as well as for the limits mentioned in chapter 2.3.

Next, the extremes of blocking are specified, i.e., $\Pi \geq 0.9$ for "certain" blockage and $\Pi \leq 0.1$ for "certain" passage. These limit values of Π were inserted in Equation (2), showing that a batch of similar trunks probably

- gets trapped if roughly $(D/H_r)/[(1+F)(W_i/W_o)] \geq 0.57$, and
- passes if roughly $(D/H_r)/[(1+F)(W_i/W_o)] \leq 0.30$.

The two limits are close (closer that for single trunk arrival), indicating that the blocking probability approaches a binary behavior: Either all trunks of the batch block or none. Such a limit would correspond to Π=0.5, being $(D/H_r)/[(1+F)(W_i/W_o)]$=0.43. The above given limits are smaller than those of a single trunk arrival. As stated before, batches are likelier to get trapped at PKWs than single trunks. A full blockage (Π>0.9) of batches is observed at only 2/3 of the normalized condition $(D/H_r)/[(1+F)(W_i/W_o)]$, as compared to an individual trunk.

The analysis of the *relative level rise* included also the build-up of the blockage. All 926 data points (chapter 2.2) were considered. The maximum observed level rise upstream of all tested PKWs and hydraulic settings was H/H_r=1.48. It occurred for the following conditions: PKW II, batch no. 4 (80 trunks supplied), TC 8, D/H_r=0.98, H_r/P=0.20, F=0.09, and Π=0.91. These conditions include thus "long" trunks, "large" supplied volumes, an "average" discharge with a relatively "small" approach flow velocity, and a "full" blockage.

The relative head increase H/H_r is often identified as key parameter for PKW design. We hence focused on the latter and tested it against numerous parameters, being systematically varied in the model. Literature mentions, as parameters with a determinant effect, wood volumes and flow heads. Despite an extensive analysis, no satisfying correlation could be derived. This was not surprising, as other authors faced similar problems – or derived relations for comparable small data sets with a restricted variation of the boundary conditions. We thus focused on an extreme value analysis.

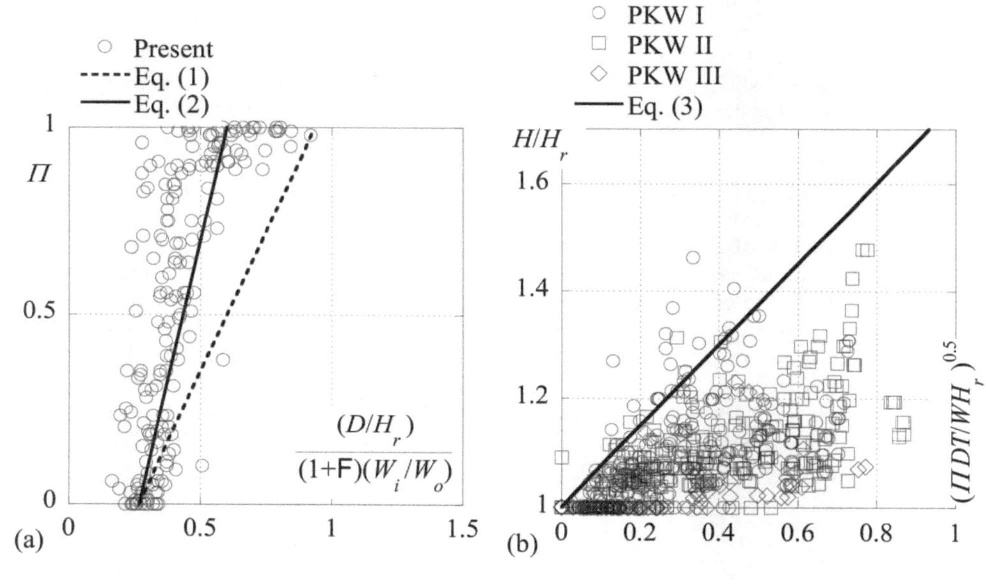

Figure 7. (a) Blocking probability Π versus $(D/H_r)/[(1+F)(W_i/W_o)]$ for the accumulative tests with homogenous batches, and (b) relative head increase H/H_r versus $(\Pi \cdot TD/WH_r)^{0.5}$.

The hydraulic head is dominated by the condition directly at the PKW. Theoretically, it is close to critical flow occurring on the crest. If the latter is disturbed by driftwood, then the necessary head to spill the flow increases. We thus focused on the wood at the PKW crest and ignored depositions upstream in the channel. For that perspective, two aspects become decisive:

- Which part of the hydraulic section at the PKW is *potentially* blocked by trapped driftwood? The maximum blocked section depends on the trunk size and its position. For a trunk trapped parallel to the PKW the obstructed section is TD. The hydraulic open section (including the head instead of the critical flow depth) can be approximated as WH_r. The potentially blocked part of the hydraulic section is consequently a multiple of TD/WH_r.
- Which part of the hydraulic section at the PKW is *effectively* blocked by trapped driftwood? Here, the blocking probability Π indicates if wood gets trapped or not.

Testing all 926 data points of the accumulative tests with homogenous batches for these two aspects gave an envelope function (i.e., the maximally observed H/H_r):

$$\frac{H}{H_r} \leq 1 + \frac{3}{4}\sqrt{\Pi \cdot \frac{DT}{WH_r}} \tag{3}$$

When illustrating H/H_r versus $(\Pi \cdot TD/WH_r)^{0.5}$, then the data is arranged as shown in Figure 7b. Almost all points are covered by Equation (3). The latter applies within $0 \leq H/H_r \leq 1.6$ (see chapter 4) as well as for the limits mentioned in chapter 2.3.

5 ACCUMULATIVE TESTS WITH HETEROGENEOUS BATCHES

Once more, the accumulative tests with heterogenous batches all showed a blockage up to an equilibrium (maximum) if the number of arriving trunks was "large enough". With the herein 10 supplied batches (each with 18 trunks und 2 rootstocks, resulting in 200 supplied elements per test), the latter equilibrium was reached.

The analysis of the *blocking probability* focused on the final blockage. Only tests with 200 (or more for Pfister et al. 2013) supplied elements (trunks and rootstocks) were considered. In total, 65 data points respected this criterium. The other points represented the build-up of blocking, being ignored in the present analysis.

The same normalization as before was tested and produced – at least for driftwood-related questions – satisfying results. If expressing the blocking probability with Equating (2) ($0 \leq \Pi \leq 1$), then the data arranged as shown in Figure 8a. One assumption had to be made: The trunk diameter D was set to $0.9D_M$, i.e., to 90% of the largest arriving trunk. This was necessary since various trunk sizes (all TC, Table 2) arrived at the tested PKWs. The statements made in the context of the homogenous batches (chapter 4) similarly apply to heterogenous batches.

As for the *relative level rise*, the complete set of 582 data points (all blockage stages) could be included in the analysis, comprising those of Pfister et al. (2013, reservoir approach). Tests with "small" trapped volumes, while the blockage was building up, were also involved.

The maximum relative level rise upstream of all tested PKWs and hydraulic settings was H/H_r=1.67 for heterogenous batches (at $(\Pi \cdot TD/WH_r)^{0.5}$=1.52, not visible in Figure 8b). It occurred for PKW III, with 180 trunks supplied, at H_r/P=0.07, F=0, and Π=0.99. These conditions include thus a "large" supplied volume, a "small" discharge with a reservoir inflow type, and a "full" blockage.

The data of the accumulative tests with heterogenous batches were normalized according to Equation (3). The determinant trunk length T and diameter D were again set as $0.9T_M$ and $0.9D_M$ (Table 2), corresponding approximately to the maximum driftwood dimensions occurring within the batches. Like the blocking probability, the "largest" trunks were once more decisive. All 582 data points are shown in Figure 8b, including a wide parameter range. Again, most points are limited by the maximum relative head increase indicated by Equation (3).

6 CONCLUSIONS

If a catchment is forested, then the potential effects of driftwood on the hydraulic performance of PKWs should be considered as part of the design. An analysis of the wood in the vicinity of the water course or potentially sliding into the latter provides maximum dimensions. These, together with the PKW geometry and the river approach flow characteristic can provide information about the:

- Blocking probability of driftwood at the PKW. By trend, large trunk diameters are probe to block, under small discharges (heads), low approach flow velocities (Froude numbers)

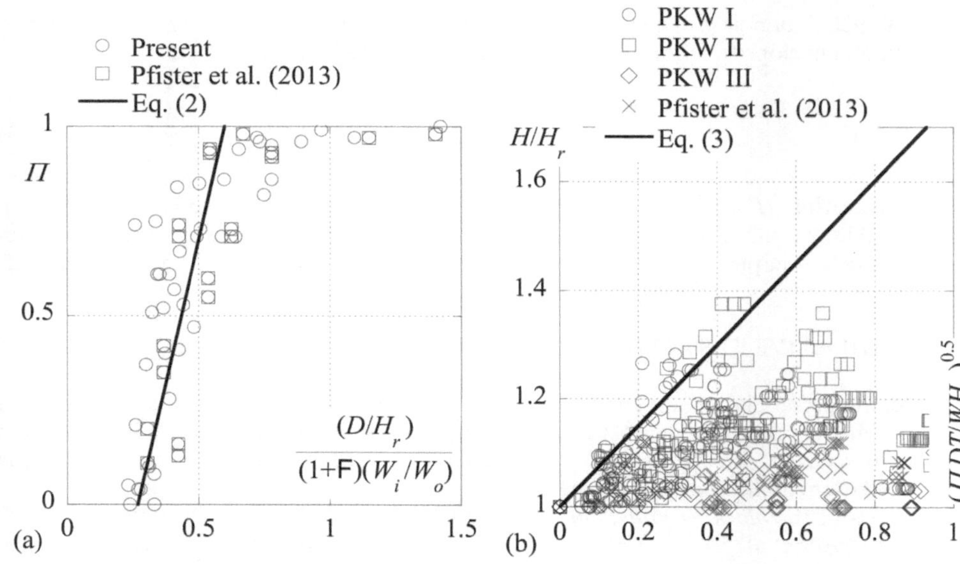

Figure 8. (a) Blocking probability Π versus $(D/H_r)/[(1+\mathrm{F})(W_i/W_o)]$ for the accumulative tests with heterogenous batches, and (b) relative head increase H/H_r versus $(\Pi \cdot TD/WH_r)^{0.5}$.

and narrow inlet keys. This behavior is favorable! Often, floods mobilize large wood volumes. But floods also generate high weir heads and approach flow velocities, supporting the passage of driftwood. The same effect have relatively wide inlet keys – what is hydraulically efficient. PKW might block under smaller discharges but should be cleaned once the flow (head) increases.

- Upstream level rise initiated by a jammed PKW. The maximum head increase reported in literature and observed herein is typically below 1.5 to $1.8H_r$, what is in fact small since H_r is generally low at PKWs. Further, these maxima occur typically for small discharges, and not for flood conditions. The aspects mentioned above to lower the blocking probability also reduce the risk of a significant level rise.

It seems that the meticulous optimization of the PKW rating curve exclusively centering on hydraulics aspects might neglect natural boundaries, like driftwood (large heads remove jams) and potential ice blockage, hydrology, etc.

Knowledge of driftwood blockage probability and subsequent level rise is important to provide safety against inundations. In parallel, one could try to motivate the wood to pass over the PKW. Various studies indicate that wood is mainly trapped at the upstream appendix (first contact with the PKW, particularly if branches are present, Panthi and Crookston 2025) and on the horizontal weir surface. These measures could support the passage of driftwood:

- Adding elements (e.g., ramps) at the upstream outlet key crests potentially lifting trunks: Inertia and drag could then push the trunks on the horizontal weir face. Devices against ice at bridge piers might serve as inspiration, and
- increasing the current on the top horizontal weir surface by installing wide inlet keys.

ACKNOWLEDGMENT

Mr. Ole-Christian Herrmann conducted the model tests of PKW III. Mr. Cédric Bron, head technician of LCH, was responsible for the model set-up including the measurement technique. The authors thank both for their excellent support.

REFERENCES

Bénet, L., De Cesare, G., & Pfister, M. (2021). Reservoir level rise under extreme driftwood blockage at ogee crest. *J. Hydraul. Eng.*, 147(1), 04020086.

Bezzola G.R., Abegg J., & Jäggi M. (2002). Verklausung von Brückenquerschnitten. VAW-*Mitteilung*, Nr. 175, 87–97, ETH Zürich (in German).

Bezzola, G.R., & Hegg, C., eds. (2007). *Ereignisanalyse Hochwasser* 2005, Bundesamt für Umwelt & Eidgenössische Forschungsanstalt für Wald, Schnee und Landschaft, Bern, Switzerland (in German).

Erpicum, S., Silvestri, A., Dewals, B., Archambeau, P., Pirotton, M. (2013). Escoulourbe Piano Wey weir: Prototype versus scale model. Intl. Workshop *Labyrinth and Piano Key Weirs* II (PKW 2013), 65–72, CRC Press, Boca Raton.

Laugier F. (2007). Design and construction of the first Piano Key Weir spillway at Goulours Dam. *Intl. J. Hydropower Dams*, 13(5), 94–100.

Leite Ribeiro M., Pfister, M., Schleiss A.J. & Boillat J.-L. (2012). Hydraulic design of A-type Piano Key Weirs. *J. Hydraulic Res.*, 50(4), 400–408.

Melville, B.W., & Dongol, D.M. (1992). Bridge pier scour with debris accumulation. *J. Hydraul. Eng.*, 118(9), 1306–1310.

Ouamane A. & Lempière F. (2006). Design of a new economic shape of weir. Proc. Intl. Symp. *Dams in the Societies of the* 21st *Century*, Barcelona, 463–470, Taylor & Francis.

Pagliara, S., & Carnacina, I. (2010). Temporal scour evolution at bridge piers: effect of wood debris roughness and porosity. *J. Hydraulic Res.*, 48(1), 3–13.

Panthi, M., & Crookston, B. (2025). Self-Cleaning Ability of the Piano Key Weir with Driftwood Blockages. *J. Hydraul. Eng.*, 105(4), 04024022.

Pfister M., Capobianco, D., Tullis B. & Schleiss A.J. (2013). Debris-Blocking Sensitivity of Piano Key Weirs under Reservoir-Type Approach Flow. *J. Hydraul. Eng.*, 139(11), 1134–1141.

Pfister, M., & Hager, W.H. (2014). History and significance of the Morton number in hydraulic engineering. *J. Hydraul. Eng.*, 140(5), 02514001.

Pfister, M., Battisacco, E., De Cesare, G., & Schleiss, A.J. (2013b). Scale effects related to the rating curve of cylindrically crested Piano Key weirs. Intl. Workshop *Labyrinth and Piano Key Weirs* II (PKW 2013), 73–82, CRC Press, Boca Raton.

Pralong J., Vermeulen, J., Blancher B., Laugier F., Erpicum S., Machiels O., Pirotton M., Boillat J-L., Leite Ribeiro M. & Schleiss A.J. (2011). A naming convention for the piano key weirs geometrical parameters. Proc. Int. Conf. *Labyrinth and Piano Key Weirs*, Liège, 271–278.

Schmocker L. & Hager W.H. (2011). Probability of drift blockage at bridge decks. *J. Hydraul. Eng.*, 137 (4), 470–479.

Vaughn, T., Crookston, B., & Pfister, M. (2021). Floating woody debris: Blocking sensitivity of labyrinth weirs in channel and reservoir applications. *J. Hydraul. Eng.*, 147(11), 06021016.

Venetz, P. (2014). Sensitivity of piano key weirs regarding driftwood blockage under channel-type approach flow. M.Sc. *thesis*, Laboratory of Hydraulic Constructions, EPFL Lausanne.

Labyrinth and Piano Key Weirs IV – Erpicum, Crookston & Paxson (Eds)

Impact of sedimentation on the discharge capacity of labyrinth weirs

F. Belzner and M. Gebhardt
Federal Waterways Engineering and Research Institute, Germany

ABSTRACT: This study investigates the hydraulic performance of a trapezoidal labyrinth weir under different sedimentation conditions with numerical modeling. A physical model with a half-round crest was tested in a flume, while numerical simulations were conducted using OpenFOAM®. The numerical model was validated against experimental data, showing strong agreement in upstream water levels and discharge coefficients. Further numerical simulations were conducted to investigate the influence of sedimentation. Results indicate that for sediment heights below 50% of the weir height, the impact on the hydraulic capacity is minimal. However, at full sedimentation, the overflow height may increase by up to 15 %, which must be considered in the design process.

1 INTRODUCTION

1.1 *Fixed weirs at german waterways*

With a total length of 7,350 km, more than 300 weirs and 450 lock chambers Germany has the largest inland waterway network in Western Europe. Every year up to 240 million tons of cargo are transported on these waterways.

Currently, there are considerations to replace movable weirs outside of the core network by fixed weirs to reduce operation and maintenance costs. Therefore, established weir types like side weirs or newer developments like labyrinth weirs and piano key weirs are in consideration. Results were presented in Gebhardt *et al.* (2017) and Belzner *et al.* (2017).

The first three labyrinth weirs are currently under construction at the Upper Havel-Waterway in Eastern Germany.

1.2 *Sediment transport*

Unlike movable weirs, labyrinth weirs are passive structures, meaning that sediments cannot be actively flushed downstream. Investigations on the influence of over- and underflow gates on sediment transport show that above a distance of three times the upstream water depth there is no significant difference between the over- and underflow gates with respect to the vertical-velocity distribution upstream of the weir (Gebhardt *et al.*, 2013). This indicates that the impact of the weir type on sediment remobilization is minimal when the headwater level remains constant. Thus, only sediments in close proximity to the weir can be actively flushed downstream.

Wilmore (2004) conducted four model tests for a labyrinth weir to investigate the effect of sedimentation on the upstream head of the weir for a given discharge. Sedimentation was simulated with a horizontal double bottom upstream of the weir, which corresponds to the effect of a completely sedimented impoundment. Different heights were tested and the weir height was reduced by up to 50%, which resulted in a 25% increase in the upstream head H_u.

Nevertheless, research confirm the self-cleaning ability of labyrinth weirs (Gebhardt *et al.*, 2018) and piano key weirs (Noseda *et al.*, 2019). The densimetric Froude number is shown to have a significant influence on the mobilization of sediment deposited in a labyrinth weir.

DOI: 10.1201/9781003655541-11

However, these studies focused on loose, uncompacted sediment, which is primarily deposited during bedload transport. In contrast, during suspended sediment transport fine particles can clog the pore spaces in the riverbed during water infiltration, contributing to the solidification of the substrate and increasing the erosion resistance. It is therefore assumed in this study that existing sediment deposits in the inlet keys are not erodible.

In discussions with stakeholders of planned labyrinth weirs the impact of sedimentation is a frequently asked question. Thus, the purpose of this paper is to show the effect of partly silted inlet keys of a trapezoidal labyrinth weir on the hydraulic capacity.

2 EXPERIMENTAL SETUP

2.1 *General setup*

A half round crowned trapezoidal labyrinth weir (P = 0.46 m, B = 0.72 m, W = 0.60 m) with two outlet keys and one and two half inlet keys (Figure 3) was positioned in a flume with a width of W = 0.60 m. The side wall angle of the labyrinth weir was α = 8°, the thickness of the walls was T_s = 0.03 m. Due to the limited knowledge regarding the exact shape of potential sediment deposition, the velocity fields from existing numerical simulations were first analyzed. Based on this, the possible shape of the sediment deposit was estimated and modeled as a rigid body with a triangular shape, consisting of several elements (Figure 1). The angle of the sediment wedge (α_{sed} = 31.8°) was determined based on the length of the labyrinth weir B and the maximum sedimentation height $P_{sed,\ max}$ = 0.43 m. The height of the sediment body was varied by removing the lowest sediment element and shifting the overlying elements downwards by 0.07 m, so that the gradient remained the same. Systematic experiments with discharges Q between 0.025 $m^3 \cdot s^{-1}$ and 0.300 $m^3 \cdot s^1$ in steps of 0.025 $m^3 \cdot s^{-1}$ were conducted. Only free flow conditions were considered. As a reference, tests without sediment were carried out.

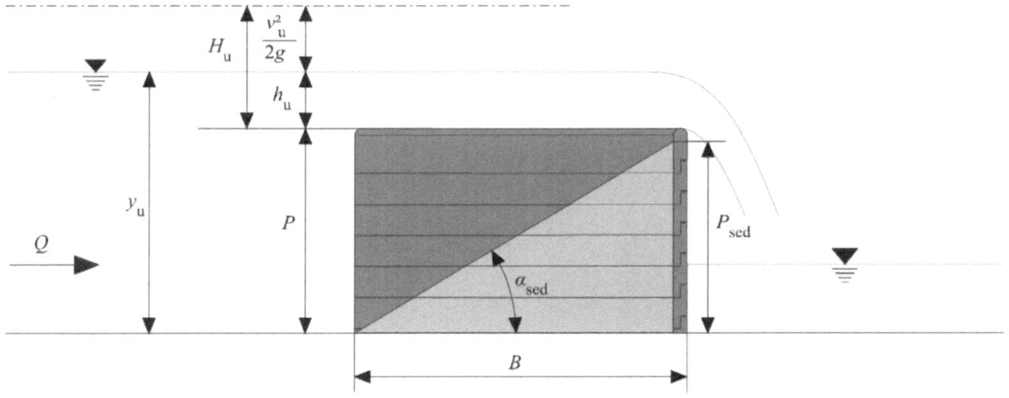

Figure 1. Definition sketch (drawing is not to scale).

2.2 *Physical model*

The physical model was used to generate validation data for the numerical model. For the physical tests a flume with a length of 18 m, a height of 1.20 m and a width of W = 0.60 m was used (Figure 2). The water level measurements were carried out with floating water level probes on measuring containers, which were connected to the flume by the principle of communicating pipes. The discharge was measured by magnetic-inductive flowmeters (MID) and the downstream water level was controlled by a flap gate. The upstream water level was measured approximately 7 meters from the weir. A full description of the physical setup and the measurement technique can be found in Gebhardt and Belzner (2025). In the physical model no artificial aeration war provided.

2.3 *Numerical model*

The numerical simulations were performed with the open source CFD toolbox OpenFOAM® in version v2212 with the two-phase transient solver interFoam. A detailed description of the numerical algorithm can be found in Rusche (2002). A 3D model with identical dimensions like the flume was created and discretized with a hexahedral dominant mesh with a base mesh size of 0.1 m. Areas where small-scale phenomena like flow detachment occur (e.g. the crest) were discretized up to a mesh size of 0.00625 m, which corresponds to about five cells across the width, resulting in an overall mesh size of about 350,000 cells. Further grid refinement in the region of the crest showed no further change in overflow height. For turbulence modeling a k-Omega-SST model was used. Three dimensional transient calculations with stationary boundary conditions were conducted until a stable upstream water level (quasi-stationary state) was achieved. A fixed discharge (Dirichlet boundary) and a variable water level (Neumann boundary) were specified as boundary conditions for the inflow. At the outlet, a fixed downstream water level (Dirichlet boundary) and variable discharge (Neumann boundary) were specified to ensure free flow. This set of non-standard boundary conditions is described in Thorenz (2024). In order to achieve a nappy behavior comparable to that of the laboratory model, at the upstream and downstream walls patches were installed to provide optional artificial ventilation.

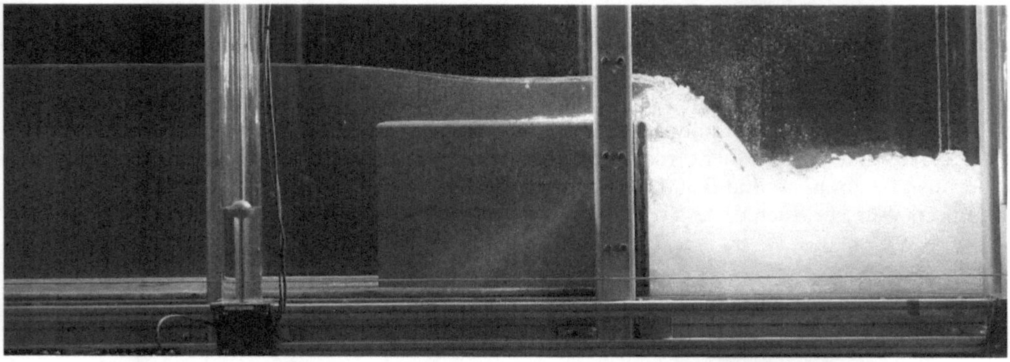

Figure 2. Side view of laboratory model.

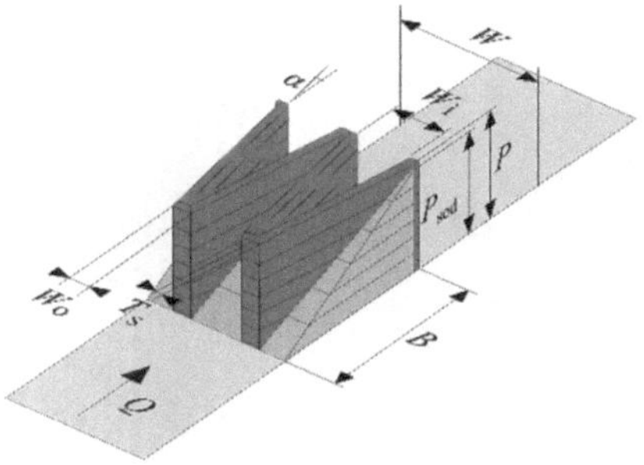

Figure 3. Isometric representation of the experimental setup.

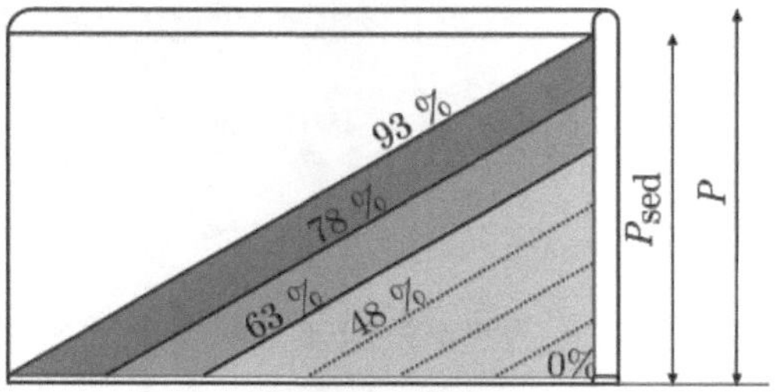

Figure 4. Labyrinth weir with sediment. Side view (drawing is not to scale).

The calculations were performed on a high-performance cluster with AMD Epyc 9654@2.4 GHz CPUs and 192 Cores per node. To save computational time, initially a simulation with a coarser mesh was carried out to generate an initial state. A fixed courant number was set ($CFL = 2$) so that the length of a timestep and the simulation time were depending on the flow velocity and the cell size. Thus, an overall simulation time of 3.5 h was needed for the highest discharge (Q = 0.300 $m^3 \cdot s^{-1}$) and 1 h for the smallest discharge (Q = 0.025 $m^3 \cdot s^{-1}$) for a simulated time of 100 s.

3 NUMERICAL RESULTS

3.1 *Comparison to physical model*

The discharge coefficient C_d was determined with Poleni's Formula from the results of the numerical simulations:

$$C_d = \frac{Q}{2/3 \cdot L \cdot \sqrt{2g} \cdot H_u^{1.5}} \tag{1}$$

where Q = discharge; L = developed crest length; g = 9.81 m·s^{-1} and H_u = upstream energy head.

For validation of the numerical model the overflow heights (Figure 5) and the discharge coefficients C_d (Eq. 1, Figure 6) estimated with the laboratory model were compared with the numerical results for conditions with (+) and without (x) artificial aeration. The overflow heights indicate an almost exact correlation for q < 0.15 m²·s^{-1}. For larger discharges, there is a difference between the aerated and non-aerated condition with the aerated condition being closer to the results measured in the laboratory. The consideration of the discharge coefficients C_d in Figure 6 confirms this observation. For $H_u/P < 0.13$, however, the laboratory results show higher performance than the numerical model. In this region the flow conditions described by Crookston (2010) as *clinging* and the transition between *clinging* and *aerated* were observed. The results indicate that this effect is not correctly reproduced by the numerical model. For $H_u/P > 0.13$ a strong agreement between the numerical and the physical model results for the aerated condition was observed. No artificial aeration was provided in the physical model and no clear statement could be made about the aeration status based on the observation of the physical model. However, the comparison with the numerical model shows that the nappe must be numerically aerated for $H_u/P > 0.13$. Thus, aerated conditions were assumed for the simulations with sediment.

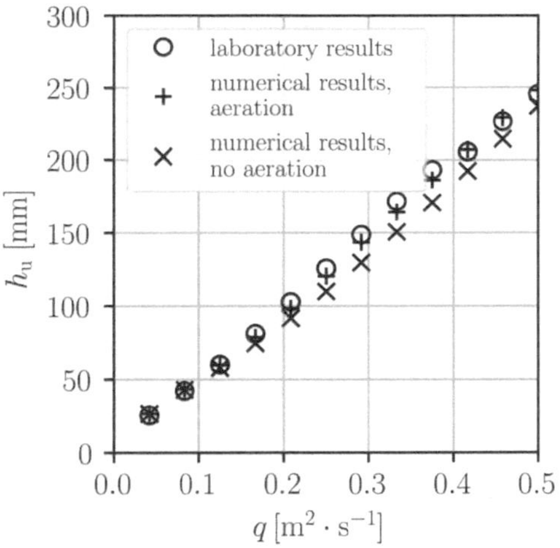

Figure 5. Comparison of the overflow height h_u estimated with the laboratory (o) and the numerical model with aeration (+) and without aeration (x).

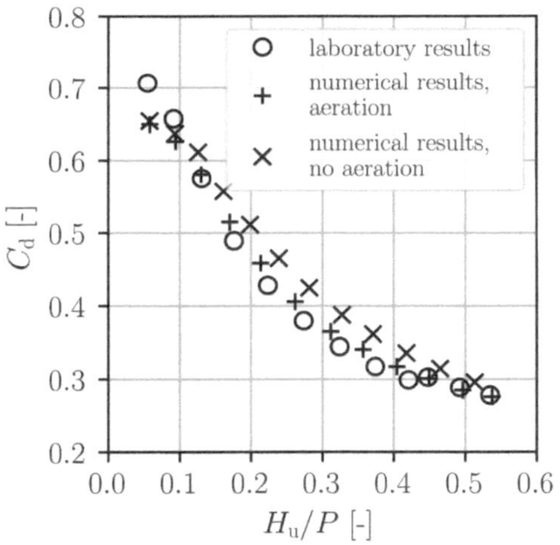

Figure 6. Comparison of the discharge coefficient C_d estimated with the laboratory (o) and the numerical model with aeration (+) and without aeration (x).

3.2 *Rating curve*

The numerically determined rating curve, the results from Belzner *et al.* (2017) as well as the C_d values presented in Crookston and Tullis (2013) are shown in Figure 7. A comparison shows an underestimation of the discharge capacity in the present study compared to Crookston and Tullis (2013) for the same H_u/P. Even the measurements presented in Belzner *et al.* (2017) slightly differ from the present study.

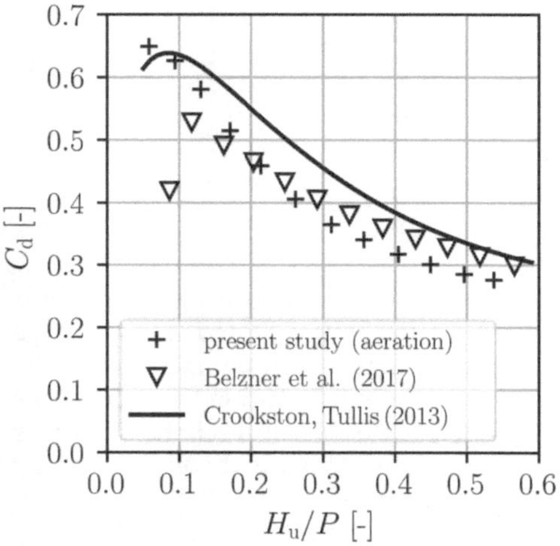

Figure 7. Comparison of the rating curves from different studies.

Table 1. Comparison of experimental setups.

	n [-]	P [m]	L/W [-]	W/P [-]	orientation
present study	2	0.46	5	1.3	inverse
Belzner et al. (2017)	5	0.25	5	4.9	normal
Crookston and Tullis (2013)	2	0.31	5.8	2	normal

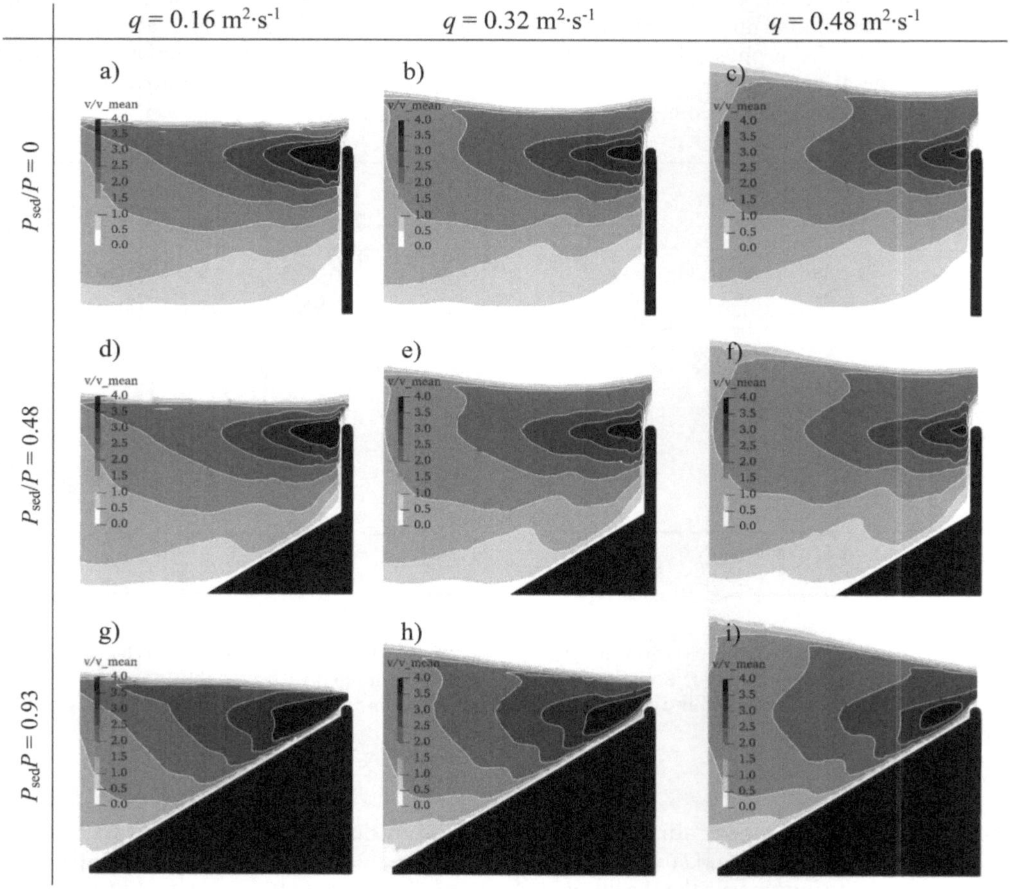

Figure 8. Relative velocity v/v_{mean} in the inlet key.

The reason for the different discharge coefficient curves is not exactly known, but some possible explanations can be discussed. Investigations of Houston (1983) indicate that the discharge capacity of inverse orientated labyrinth weirs (edge keys are inlet keys), like used in the present study, is lower than for the normal configuration. Furthermore, the model scale and number of keys are different between Crookston and Tullis (2013), Belzner *et al.* (2017) and the present study. The comparison of different experimental setups (Table 1) and the resulting rating curves indicate that there might be a dependency of scale, orientation and the influence of the lateral boundaries that needs to be further investigated.

3.3 *Influence of sedimentation*

Figure 8 illustrates the relative velocity field in the inlet key, derived from numerical simulations for sedimentation levels P_{sed}/P = [0, 0.48, 0.93] and flow rates q = [0.16, 0.32, 0.48] m²/s. The relative velocity v/v_{mean} was calculated as the ratio of the inlet key velocity to the mean flow velocity $v_{mean} = Q/(B \cdot y_u)$. For $P_{sed}/P = 0$ (Figure 8 a-c), the analysis reveals that $v/v_{mean} < 1$ in the lower half of the labyrinth weir, with a triangular region near the bottom where $v/v_{mean} < 0.5$. Towards the top, v/v_{mean} increase, indicating acceleration in both vertical and horizontal directions toward the weir crest. For $P_{sed}/P = 0.48$ (Figure 8 d-f) the velocity field remains similar to that observed at $P_{sed}/P = 0$, suggesting that sedimentation does not significantly influence flow dynamics up to this height. In both cases, flow detachment occurs at the front area of the crest, indicating that the boundary streamline adheres to the vertical wall of the labyrinth weir. For $P_{sed}/P = 0.93$ (Figure 8 g-h), a distinct modification of the velocity field is evident due to sedimentation. Acceleration is observed along the sediment geometry, with the highest velocities occurring closer to the water surface compared to the previous sedimentation levels. Additionally, the detachment point shifts slightly downstream.

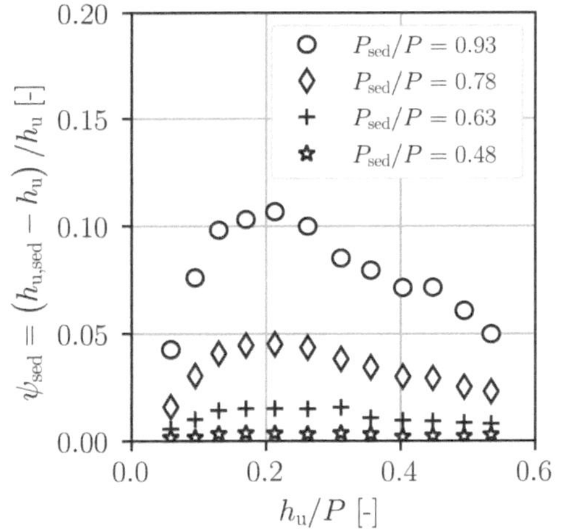

Figure 9. Increase of overflow height depending on relative sediment height.

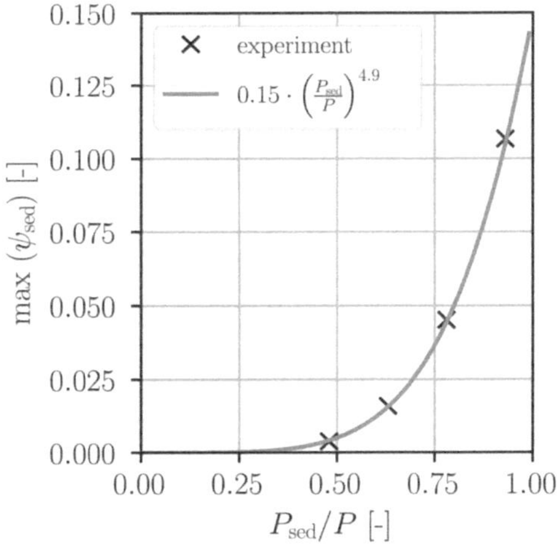

Figure 10. Maximum increase of overflow height depending on the relative height of sediment.

For the hydraulic design at inland waterways the upstream water depth is usually the dominating parameter. The overflow height with sedimentation $h_{u,sed}$ was compared to the overflow height without sedimentation h_u, expressed as a relative increase in upstream water level $\psi_{sed} = (h_{u,sed} - h_u)/h_u$.

Figure 9 presents the results, indicating that as sedimentation height P_{sed}/P increases, the upstream water level also rises. The relative increase ψ_{sed} reaches a maximum at approximately $h_u/P = 0.2$ and subsequently decreases for higher overflow heights. This maximum is more pronounced for greater sediment heights, while lower sediment heights yield a plateau in the range of $h_u/P \approx 0.15 - 0.25$ for $P_{sed}/P = 0.78$ and $h_u/P \approx 0.10 - 0.30$ for $P_{sed}/P = 0.63$.

Figure 10 illustrates the maximum increase of overflow height $\max(\psi_{sed})$ as a function of relative sediment height. A power function was fitted to the data.

$$\max(\psi_{sed}) = 0.15 \cdot \left(\frac{P_{sed}}{P}\right)^{4.9} \tag{2}$$

The fitting was performed using the Trust Region Reflective algorithm (Branch *et al.*, 1999), yielding a Pearson correlation coefficient of r = 0.99. Evaluation of this equation indicates that for $P_{sed}/P = 0.5$, the upstream water level increases by less than 1%, suggesting minimal impact. The highest sedimentation level modeled was $P_{sed}/P = 0.93$. Extrapolating to 100 % sedimentation ($P_{sed}/P = 1$) predicts a maximum increase in overflow height of approximately 15 %.

4 DISCUSSION

For validation purposes, the results for free flow without sedimentation were compared with laboratory experiments, revealing a strong agreement. Current numerical flow solvers typically yield accurate results; however, discrepancies often arise from insufficient discretization or from the inaccurate modeling of geometry, boundary conditions, or because the parameters used for evaluation were not measured at the same location or were derived indirectly. The numerical model developed in this study was designed to accurately represent the upstream water level. Consequently, a precise modeling of the flow structures in the vicinity of the labyrinth weir geometry, particularly at the crest, was essential. However, this study does not provide predictions regarding the necessary grid resolution for modeling other effects around a labyrinth weir, such as energy dissipation or nappe behavior.

The aeration of the nappe does not appear over all the investigated discharges. It is possible that this process has a coincidental component. The aeration condition could not be determined either by known literature or by comparison with the laboratory model, so that simulations with and without aeration had to be carried out first. In order to guarantee certain hydraulic conditions, the authors therefore recommend always providing artificial aeration.

The investigation of sedimentation effects indicates that as long as the sediment ramp is within a region where $v/v_{mean} < 1$, sedimentation does not significantly influence the hydraulic capacity of the weir. The analysis of the velocity field reveals a triangular region with $v/v_{mean} < 1$ with a height of about half of the height of the labyrinth. As sedimentation levels increase further, the sediment begins to direct the flow, leading to a vertical alignment of the streamlines and an increase in hydraulic loss. The presence of the nearly triangular region with $v/v_{mean} < 1$ supports the assumption regarding the shape of sedimentation in this study and in Gebhardt *et al.* (2017). Nevertheless, the sediment shape used in this study may not fully represent reality, so the applicability of the results may be limited. Further investigations into sedimentation processes in the inlet keys of labyrinth weirs are needed to provide insights into a realistic shape of the sediment bed. In a preliminary step, the angle and length of the diagonal sediment bed should be varied.

In contrast to the present study, Wilmore (2004) simulated a fully sedimented upstream region by reducing the weir height P, corresponding to a horizontal sedimentation profile. This approach led to an increase in upstream water levels by more than 25 % for a sediment height of 50 % of the weir height. In the present study, a triangular sediment wedge was assumed, representing a more realistic and localized form of sediment deposition. For the same relative sediment height (P_{sed}/P=50 %), the resulting increase in upstream head was only about 1 %. This difference highlights the significant influence of the assumed sediment geometry. Moreover, full horizontal sedimentation seems unlikely in natural river systems, which further limits the applicability of Wilmore's results in the context of inland waterways.

5 CONCLUSION

This study provides valuable insights into the hydraulic performance of trapezoidal labyrinth weirs under different sedimentation conditions. The strong agreement between numerical simulations and experimental data confirms the reliability of the applied modeling approach. The results indicate that moderate sedimentation has little impact on hydraulic capacity, whereas full sedimentation can significantly increase upstream water levels.

The findings emphasize the necessity of integrating sedimentation effects into the design and maintenance strategies of labyrinth weirs, particularly in non-core waterways where maintenance efforts are reduced. Further investigations should focus on realistic sediment deposition patterns, long-term sediment transport processes, and potential mitigation measures to ensure the efficient operation of fixed weirs in inland waterways. In addition, a deeper understanding of the mechanisms driving the physical aeration conditions in this application should be developed and further correlated with CFD outcomes.

ACKNOWLEDGEMENTS

The authors thank Udo Pfrommer and Timo Krebs for their great support in designing and constructing the physical model, and Rick Crocoll for the structured execution of the physical experiments and for producing high quality data.

REFERENCES

Belzner, F., Merkel, J., Gebhardt, M. & Thorenz, C. (2017) Piano Key and Labyrinth Weirs at German waterways: Recent and future research of the BAW. In: Erpicum, S., Laugier, F., Khanh, M.H.T. & Pfister, M. (eds.) *Labyrinth and Piano Key Weirs III: PKW 2017*. Boca Raton, CRC Press. pp. 167–174.

Branch, M.A., Coleman, T.F. & Li, Y. (1999) A Subspace, Interior, and Conjugate Gradient Method for Large-Scale Bound-Constrained Minimization Problems. *SIAM Journal on Scientific Computing*. [Online] 21 (1), 1–23. Available from doi:10.1137/S1064827595289108.

Crookston, B.M. (2010) *Labyrinth Weirs*.

Crookston, B.M. & Tullis, B.P. (2013) Hydraulic Design and Analysis of Labyrinth Weirs. I: Discharge Relationships. *Journal of Irrigation and Drainage Engineering*, 139 (5), 363–370.

Gebhardt, M. & Belzner, F. (2025) Physical model investigations on nappe instabilities at labyrinth weirs: unpublished. In: United States Society on Dams (ed.) *International Workshop on Labyrinth and Piano Key Weirs 2025: unpublished.*

Gebhardt, M., Gerstner, N. & Thorenz, C. (2013) Der Einfluss Über- und unterströmter WehrverschlÜsse auf den Sedimenttransport durch Stauhaltungen. *Wasserwirtschaft*. [Online]. 103 (2013), 16. Available from: https://hdl.handle.net/20.500.11970/100679.

Gebhardt, M., Herbst, J., Merkel, J. & Belzner, F. (2018) Sedimentation at labyrinth weirs – an experimental study of the self-cleaning process. *Journal of Hydraulic Research*. [Online] 90, 1–12. Available from doi:10.1080/00221686.2018.1494053.

Gebhardt, M., Merkel, J., Belzner, F. & Thorenz, C. (2017) A comparison of side weirs and labyrinth weirs at Ilmenau river. In: Erpicum, S., Laugier, F., Khanh, M.H.T. & Pfister, M. (eds.) *Labyrinth and Piano Key Weirs III: PKW 2017*. Boca Raton, CRC Press. pp. 221–228.

Houston, K. (1983) *Hydraulic Model Study of Hyrum Dam Auxiliary Labyrinth Spillway*. [Online]. Available from: https://digitalcommons.usu.edu/govdocs/159 Accessed [7th March 2025].

Noseda, M., Stojnic, I., Pfister, M. & Schleiss, A.J. (2019) Upstream Erosion and Sediment Passage at Piano Key Weirs. *Journal of Hydraulic Engineering*. [Online] 145 (8), 4019029. Available from doi:10.1061/(ASCE)HY.1943-7900.0001616.

Rusche, H. (2002) Computational Fluid Dynamics of Dispersed Two-Phase Flows at High Phase Fractions. [Online]. London. Available from: http://powerlab.fsb.hr/ped/kturbo/OpenFOAM/docs/HenrikRuschePhD2002.pdf Accessed [2nd July 2019].

Thorenz, C. (2024) *Boundary Conditions for Hydraulic Structures Modelling with OpenFOAM*.

Wilmore, C.M. (2004) *Hydraulic Characteristics of Labyrinth Weirs*. MS Thesis Utah State University, Logan, Utah.

I.4 Numerical modeling

Labyrinth and Piano Key Weirs IV – Erpicum, Crookston & Paxson (Eds)
© 2026 The Author(s), ISBN 978-1-041-10527-5

Sensitivity analysis of piano-key weirs: Optimizing discharge performance through CFD modeling and empirical insights

F. Oukaili and Y. Bercovitz
Laboratoire National d'Hydraulique et Environnement (LNHE), EDF R&D, Chatou, France

T. Fonty
Département Mécanique des Fluides, Energie et Environnement (MFEE), EDF R&D, Chatou, France

B. Blancher
Centre d'Ingénierie Hydraulique (CIH), Electricité De France, France

ABSTRACT: Piano-key weirs (PKWs) are widely recognized for their efficiency in enhancing spillway discharge capacity while maintaining a compact footprint. This study focuses on a parametric and sensitivity analysis of PKW discharge performance, exploring the influence of key design parameters on hydraulic behavior. By integrating experimental data with advanced computational fluid dynamics (CFD) modeling using open-source software code_saturne, to aim to refine predictive capabilities and establish guidelines for effective PKW design. The analysis emphasizes understanding the interplay between PKW geometry and flow characteristics. The analysis considers a range of geometrical configurations, including variations in key width ratios, and inlet-to-outlet dimensions, to quantify their impact on discharge efficiency. The results of the parametric study offer insights into optimal design strategies for maximizing performance under various hydraulic conditions. The outcomes of the study aim to support the development of design guidelines that improve both functionality and sustainability in spillway applications.

1 INTRODUCTION

Numerous studies have aimed to understand the underlying physical phenomena of the flow and quantify the impact of each parameter on discharge capacity to provide design recommendations for engineers. Ouamane and Lempérière (2006) were the first to investigate the influence of key parameters on weir efficiency, identifying the ratio (L/W) as the most influential parameter. This finding was later confirmed by Leite Ribero et al. (2012). An optimal ratio of 5 was suggested by Ouamane and Lempérière (2006). Laugier et al. (2005) and Hien et al. (2006) proposed an optimal ratio between 5 and 7, depending on the head discharge. The weir height (P) greatly influences the weir discharge capacity. Machiels et al. (2014) demonstrated that increasing the inlet key height (P) enhances weir capacity, with an optimal hydraulic ratio of P/W=1.3, though a ratio of 0.5 was recommended due to construction constraints in rehabilitation projects.

The inlet-to-outlet key width ratio (W_i/W_o) and overhang portions (B_i/B_o) were identified as critical parameters for a given (L/W) and (P/W_u). Extensive experimental and numerical studies have suggested optimal values for (W_i/W_o) ranging from 1.25 to 1.6, with a preference for larger inlet keys. Ouamane and Lempérière ($B_i/B_o > 1$) has higher efficiency. Cicero et al. (2016) concluded that Type A PKWs are more efficient than Type C but less efficient than Type B. Machiels et al. (2014) showed that the optimal value depends on the (P/W_u) ratio, with high PKWs height (P/W_u = 1.3) benefiting from increased upstream overhang portions, whereas for low PKWs height (P/W_u= 0.5), symmetric overhangs have optimal performances. The numerical investigation by Pralong et al. (2011b) suggested a ratio (B_i/B_o = 3), and Athmani (2019) demonstrated that the optimal value depends on head discharge and the (Wi/Wo) ratio.

DOI: 10.1201/9781003655541-12

In nearly all studies focusing on the (W_i/W_o) ratio, they assumed symmetric overhangs. Athmani (2019) addressed this limitation by simultaneously varying both (W_i/W_o) and (B_i/B_o) ratios, though his study considered only three values for the ratio. Additionally, the ratio (B_b), representing the weir base portion, was fixed at 1/3 in all reviewed studies. The current study aims to fill this gap by allowing the (B_b/B) ratio to vary, providing a comprehensive assessment of its influence on discharge capacity.

Thus, the research focuses on numerically investigating the impact of these parameters by varying them simultaneously. The analysis is completed with experimental results obtained at ULiège, EPFL, and EDF over the years.

2 METHODS AND TOOLS

2.1 *Fundamental PKW design parameters*

The standardized notations introduced by Pralong et al. (2011a) are recalled in Figure 1 which presents the key design parameters of a PKW, in cross-section on the left and plane view on the right.

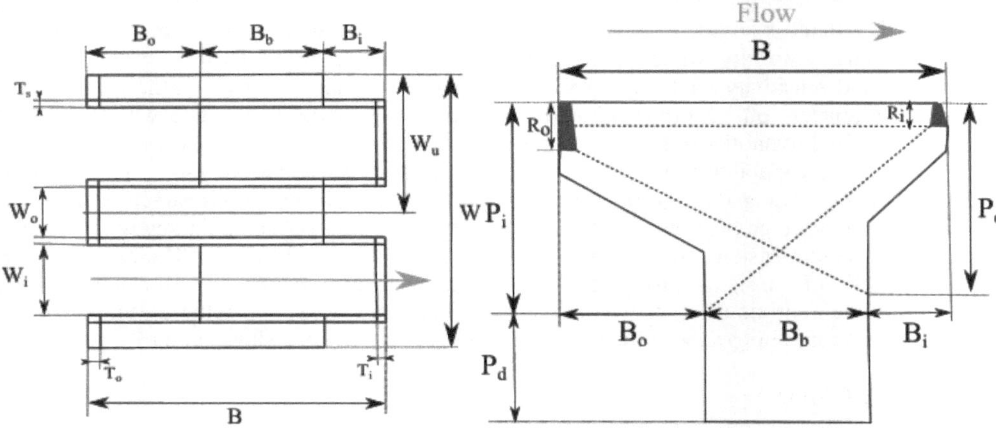

Figure 1. Fundamental PKW design parameters: cross-section (left), plane view (right), adapted from Pralong et al. (2011).

Dimensionless parameters allow a comparison independent of the actual size. In the literature, ratio key dimensionless parameters have been examined for each (H/P), namely the magnification ratio (L/W), PKW height to unit width ratio (P/W_u), inlet-to-outlet key width ratio W_i/W_o, overhang portions B_o/B_i. Other parameters, such as crest thickness (T_s, T_i, T_o) and its shape, were also investigated, along with the parapet walls (R_i, R_o). Additionally, the dam height (P_d) was also studied.

Previous studies have identified the ratio between the developed crest length (L) and the weir width (W) as the primary factor influencing the hydraulic efficiency of PKWs. This parameter also plays a crucial role in the complexity and cost of construction, as it is related to the side crest length (B) and the overhang dimensions. Lempérière et al. (2011) suggested that an (L/W) ratio of 5 represents an optimal trade-off between hydraulic performance and economic feasibility. Consequently, this ratio will not be investigated in this study which results in a fixed value of B and W (and indirectly W_u). Additionally, the inlet height-to-unit width ratio P_i/W_u has been identified as the second most influential parameter and was investigated experimentally by Machiels et al. (2014). It will also be kept constant throughout the present study to alleviate the analysis.

Furthermore, previous research has shown that reducing the wall thickness (T_s) increases the discharge coefficient. An experimental study conducted at LNHE in 2015 demonstrated that the shape of the crest walls has a negligible effect on discharge capacity (Cicero et al. (2016)). Numerical simulations by Pralong et al. (2011b) also highlighted the influence of dam height (P_d) and parapet walls (R_i, R_o) on discharge. Given that these parameters are not the primary focus of the present study, they will not be considered, though they may be explored in future work. Therefore, all numerical tests consider a flat-topped crest with no parapet walls.

Consequently, the influence of the four geometric parameters W_i, W_o, B_i, B_o and B_b is investigated for given values of W_u and B. To simplify the analysis, the following dimensionless ratios are redefined: W_i/W_u, W_o/W_u, B_o/B, B_i/B, and B_b/B. Considering the geometrical constraints, three independent parameters remain: W_o/W_u, B_o/B, and B_b/B to describe the variations in geometry while preserving the fixed values of (W_u) and (B).

The discharge capacity of the PKW is estimated by the discharge coefficient, computed from the discharge per PKW-unit width flow rate (q). The discharge coefficient is given by Equation (1), with g being the acceleration due to gravity, taken as 9.81 m/s^2.

$$C_d = \frac{q}{\sqrt{2gH^3}} \tag{1}$$

2.2 *Experimental database*

The experimental dataset used in this study comes from ULiège (Machiels et al. 2014), EPFL (Le Doucen et al. 2009), and EDF (Cicero et al. 2016). It includes over 9,000 tests conducted in a controlled flume environment, covering more than 40 different PKW configurations. The dataset contains key geometric parameters like crest dimensions, inlet and outlet widths, structural ratios, and flow characteristics such as discharge rates. The data was standardized to ensure consistency and relevant dimensionless ratios—like relative widths and crest efficiencies—were computed. This dataset is used to validate the numerical model and support the analysis to better understand how geometric parameters influence hydraulic performance.

2.3 *Numerical model description*

Numerical simulations were carried out using code_saturne v8.2, an open-source CFD solver developed by EDF R&D. The Volume of Fluid (VOF) method was employed to simulate free-surface flows, incorporating a compressive interface-capturing scheme based on Deshpande et al. (2012) and Harris et al. (2024), implemented via the β-limiter. The solver resolves the Navier–Stokes equations for a fictitious fluid, with properties defined through a mixture law and an additional transport equation for the void fraction. All numerical parameters were kept at their default settings.

The computational domain was based on the geometry of a reduced-scale physical model tested at LNHE. Using a scaled model allowed for lower wall distances at the weir ($35 < y^+ < 85$), which would require significantly finer meshes in a full-scale setup. To further reduce computational cost, only half of a PKW unit was modeled, applying symmetry boundary conditions along the lateral planes. The fixed geometric parameters are summarized in Table 1.

Table 1. Fixed geometric parameters of the PKW.

Parameters	B	W_u	P_i	P_o	T_s	T_o	T_i
Value (m)	0.5732	0.306	0.222	0.222	0.020	0.0206	0.0206

Fluid domain dimensions and boundary conditions are illustrated in Figure 2. An imposed hydrostatic pressure condition was applied at the outlet to stabilize the simulation. An incoming flow rate was applied at the inlet boundary condition, presented in blue in Figure 2, with a uniform velocity profile. The SMESH module in Salome (https://www.salome-platform.org) was used to mesh the fluid domain. An adapted quadrangle mapping mesh was applied. Figure 3 shows the generated mesh with a zoomed-in view around the PKW geometry. The mesh has a discretization size of 5 mm for a simulated head discharge H = [0.05 m; 0.15 m] and contains 1 million cells.

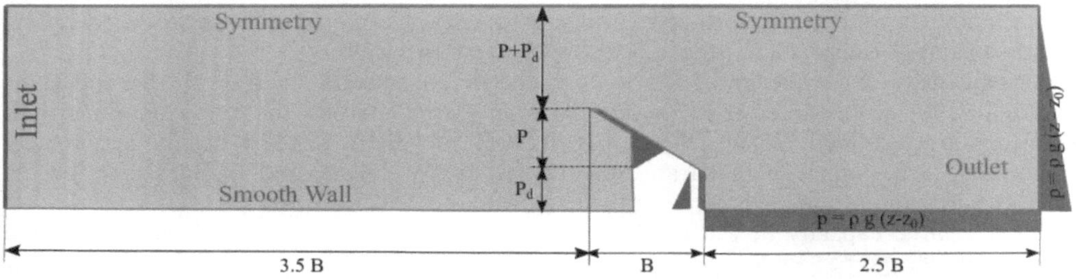

Figure 2. Fluid domain dimensions as a function of the PKW parameters and boundary conditions in code_saturne.

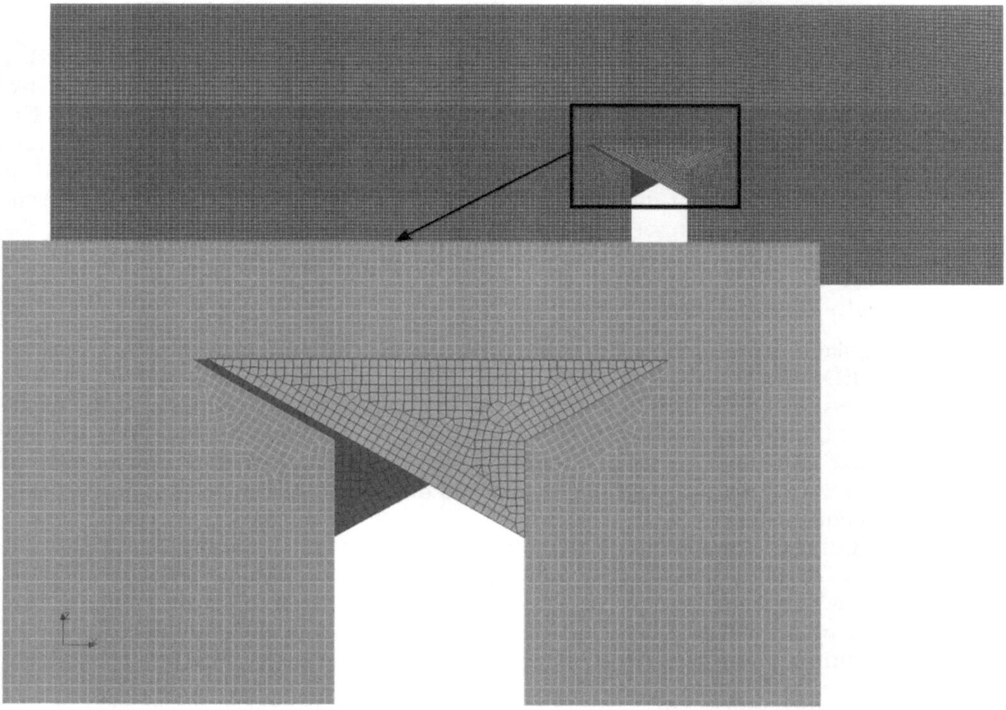

Figure 3. Domain mesh and zoom in around the PKW geometry.

2.4 *Numerical model validation*

A mesh convergence analysis was conducted on the Lempérière and Ouamane (2003) weir geometry to ensure the reliability of the numerical simulations. As illustrated in Figure 4, the relative error decreases consistently with mesh refinement, confirming convergence. The error stabilizes below 5% for finer meshes, indicating that further refinement yields diminishing returns in accuracy. Based on this trend, the fourth mesh level was selected as an optimal balance between computational cost and solution accuracy.

Subsequently, as shown in Figure 6, simulations were performed at multiple head levels using the selected mesh. The simulation results closely follow the experimental trend, with relative absolute errors remaining below 5% for all tested conditions, thus validating the model's predictive capability across a range of flow conditions while maintaining computational efficiency.

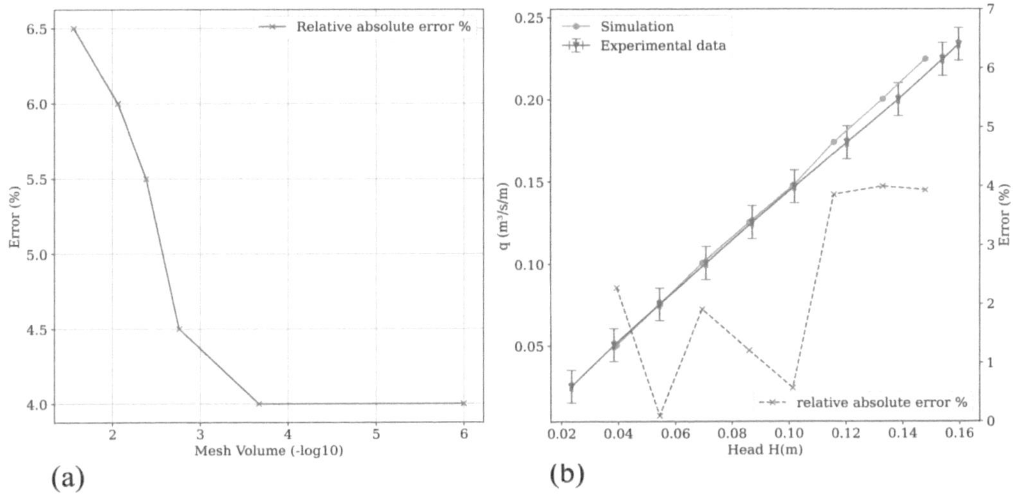

Figure 4. Comparison between numerical results obtained with code_saturne and experimental data on the reduced model tested in 2015 at EDF LNHE of PKW Type A.

3 RESULTS AND DISCUSSION

A series of simulations were conducted using eight Cartesian grids to investigate the combined effect of the three key geometric ratios (W_o/W_u, B_o/B, and B_b/B). A discrete set of B_b/B values ranging from 0.1 to 1.0 (step size 0.1) was first defined. For each value $(B_b/B)_k$, a Cartesian grid was constructed with nine equidistant values of (W_o/W_u) within [0.1, 0.8]. The corresponding range of B_o/B values was defined as $[0, 1-(B_b/B)_k]$, discretized into $N_k=10(1- (B_b/B)_k)+1$ points. Each grid configuration was evaluated using code_saturne simulations at three flow rates: q = {0.05,0.1,0.15} m3/s/m, corresponding to head-to-weir height ratios H/P ≈ {0.22,0.45,0.67} for a Type A PKW. These scenarios represent low, moderate, and high discharge conditions.

3.1 *The combined influence of key widths and overhang portions*

To better understand the influence of outlet geometry on PKW efficiency, contour plots were generated showing the discharge coefficient (Cd) as a function of (Wo/Wu) and (Bo/B), for a fixed base ratio Bb/B=0.3. This value was chosen to match previous studies, where the base typically represents one-third of the total weir length.

At low discharge (q=0.05 m³/s/m, Figure 5a), the results show a clear optimal zone around W_o/W_u=0.3–0.4 and B_o/B=0.6–0.7, where (C_d) reaches approximately 1.73. This suggests that a moderate outlet width, combined with a longer upstream overhang, enhances flow concentration and nappe development. Outside this range, discharge efficiency drops, particularly when W_o/W_u>0.5 or B_o/B<0.2, likely due to nappe interference or unstable flow patterns. For moderate discharge (q=0.10 m³/s/m, Figure 5b), the optimal configuration shifts slightly toward wider outlet keys, with peak C_d≈1.56 observed around W_o/W_u=0.35–0.4 and B_o/B=0.6. This indicates that the effect of the outlet key width becomes more significant as flow increases. At higher discharge (q=0.15 m³/s/m, Figure 5c), similar trends are observed, though the maximum Cd drops slightly to ≈ 1.37. The best-performing region remains around Wo/Wu=0.35 and Bo/B=0.6, confirming the robustness of this geometry across a range of flow conditions.

The results confirm that asymmetric geometries with long upstream overhangs and moderate outlet widths offer the best hydraulic efficiency. This aligns with previous experimental and numerical findings from Cicero et al. (2016) and Machiels et al. (2012). Furthermore, the asymmetry of the contour lines, especially near the lower and upper boundaries, further emphasizes the nonlinear interaction between the geometric ratios.

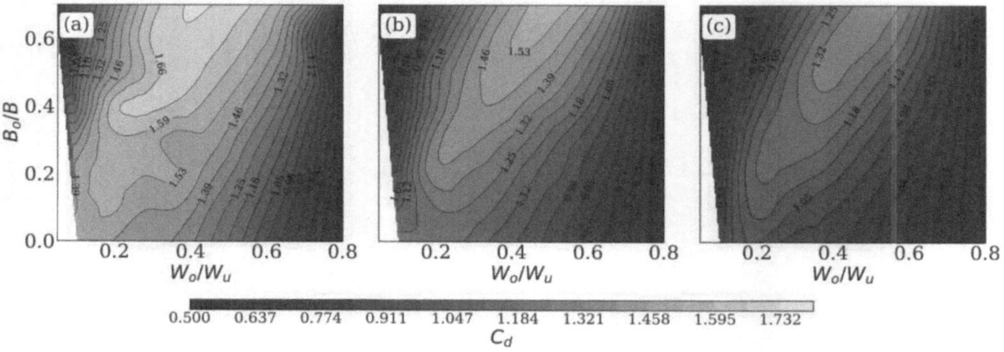

Figure 5. Effect of key widths and overhang portions on the discharge coefficient for q=0.05 m³/s/m (a), q=0.10 m³/s/m (b), and q=0.15 m³/s/m (c).

3.2 Influence of weir base length portion B_b/B

Additional simulations were conducted to evaluate the effect of the weir base length portion on PKW performance. The cases for $B_b/B=0.1$ and $B_b/B=0.8$ are presented here, complementing the baseline case of $B_b/B=0.3$ (Figure 6b). While some configurations may not be practical, they help reveal performance trends. Only results for moderate discharge (q=0.1 m3/s/m) are presented in Figure 6 to ensure clarity.

For $B_b/B=0.1$ (Figure 6a), the discharge coefficient (C_d) reaches a peak of approximately 1.7, significantly higher than in other cases. The optimal region is broader and occurs at higher B_o/B values, indicating that shorter base lengths promote more effective flow concentration and nappe formation across a broader range of geometries. In contrast, for $B_b/B=0.8$ (Figure 6c), the maximum (C_d) drops to around 1.12, with efficient performance confined to a narrow band at low (B_o /B) and moderate (W_o/W_u). This suggests that longer base lengths limit flow acceleration and reduce overall efficiency. These results highlight the strong influence of base length on PKW hydraulics, with shorter bases offering superior performance under the tested conditions.

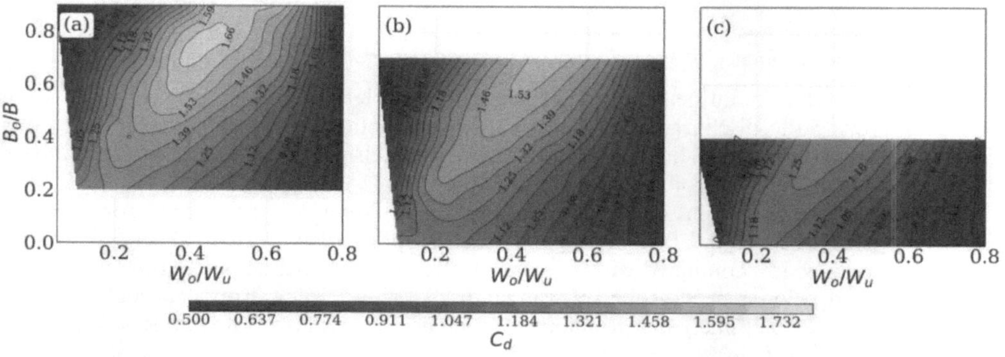

Figure 6. Effect of key widths and overhang portions on the discharge coefficient for moderate discharge head (q=0.1 m3/s/m) for weir base length portion Bb/B=0.1 (a), Bb/B=0.3 (b), and Bb/B=0.8 (c).

3.3 *Comparison with experimental results*

Figure 7 compares numerical and experimental discharge coefficients C_d as a function of the outlet width ratio W_o/W_u, for a fixed head-to-weir height ratio $H/P \approx 0.24$. The experimental dataset, filtered to match the numerical conditions (fixed L/W, P/P_d, and $P/W_u=0.5$), was sourced from Machiels et al. (2014). The results show good agreement across the tested range, with both datasets capturing similar trends and peak values. This consistency supports the validity of the numerical model under low-head flow conditions.

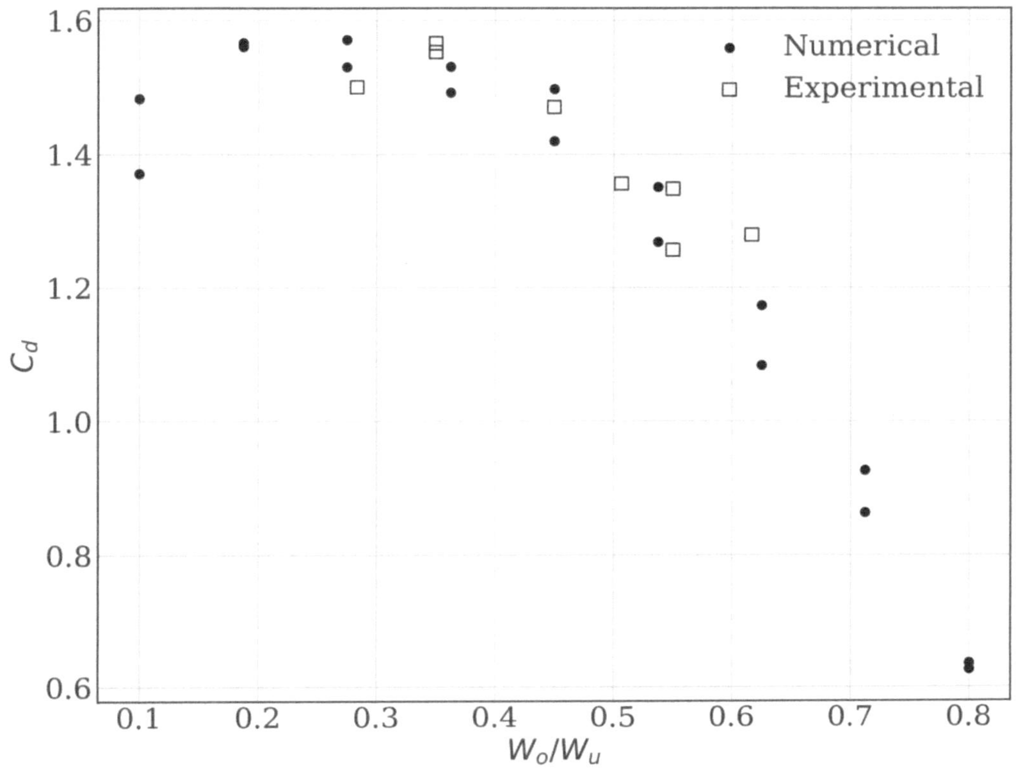

Figure 7. Comparison between experimental and numerical data.

4 CONCLUSIONS AND PERSPECTIVES

This study investigated the combined influence of key widths and overhang portions on the discharge efficiency of Piano Key Weirs (PKWs). The findings indicate that, for low-head discharge, the maximum discharge coefficient is achieved with symmetric inlet/outlet keys. However, as the hydraulic head increases, the optimal ratio shifts towards larger outlet keys. Longer upstream overhangs maximize discharge efficiency. The optimal overhang ratio depends on the head discharge and decreases slightly with increasing head. The numerical results align well with the experimental data. Allowing the base width ratio to vary in the study provided additional insights into its impact on discharge performance, extending beyond traditional assumptions in the literature. In future work, it could be interesting to investigate how these findings apply to different heights, widths, and lengths.

ACKNOWLEDGMENT

The writers would like to thank professor S. Erpicum at uliège for sharing the collected experimental database.

REFERENCES

Athmani, B. 2019. Optimisation de la forme du déversoir en touches de piano (PK-Weir) par voies numérique et expérimentale. (Doctoral dissertation, Université Mohamed Khider-Biskra).

Cicero, G.M., Vermeulen, J. & Laugier, F. 2016. Influence of some geometrical parameters on Piano Key Weir discharge efficiency. In: 6th International Symposium on Hydraulic Structures and Water System Management, Portland, Oregon, USA, 27–30 June 2016. ISBN 978-1-884575-75-4.

Deshpande, S.S., Anumolu, L. & Trujillo, M.F. 2012. Evaluating the performance of the two-phase flow solver interFoam. Computational Science & Discovery 5(1): 014016.

Fonty, T., Ferrand, F., Leroy, A., Joly, A. & Violeau, D. 2018. Mixture model for two-phase flows with high density ratios: A conservative realizable SPH formulation. International Journal of Multiphase Flow 111: 158–174.

Harris, J.C., Ferrand, M., Le Coupanec, E., Fonty, T. & Battle Martin, M. 2024. Validation of a volume of fluid approach for modeling water waves. *Submitted to International Journal for Numerical Methods in Fluids*.

Hien, T.C., Son, H.T. & Khanh, M.H.T. 2006. Results of some Piano Key Weir hydraulic model tests in Vietnam. In: Proceedings of the 22nd Congress of ICOLD, Barcelona, Spain.

Le Doucen, O., Leite Ribeiro, M., Boillat, J.L., Schleiss, A. & Laugier, F. 2009. Etude paramétrique de la capacité des PK-Weirs. Modèles Physiques Hydrauliques: Outils indispensables du XXIe siècle: 155–160.

Lempérière, F. & Ouamane, A. 2003. The Piano Key Weir: a new cost-effective solution for spillways. Hydropower Dams 10(5): 144–149.

Lempérière, F., Vigny, J.P. & Ouamane, A. 2011. General comments on Labyrinth and Piano Key Weirs: The past and present. In: Proceedings of the International Conference on Labyrinth and Piano Key Weirs (PKW 2011): 17–24. CRC Press, Taylor & Francis Group.

Leite Ribeiro, M., Pfister, M., Schleiss, A.J. & Boillat, J.L. 2012. Hydraulic design of A-type Piano Key Weirs. J. Hydraul. Res. 50(4): 400–408.

Machiels, O., Pirotton, M., Archambeau, P., Dewals, B. & Erpicum, S. 2014. Experimental parametric study and design of Piano Key Weirs. J. Hydraul. Res. 52(3): 326–335.

Ouamane, A. & Lempérière, F. 2006. Design of a new economic shape of weir. In: Proceedings of the International Symposium on Dams in the Societies of the 21st Century: 463–470. Barcelona, Spain.

Pralong, J., Vermeulen, J., Blancher, B., Laugier, F., Erpicum, S., Machiels, O., Pirotton, M., Boillat, J.L., Leite Ribeiro, M. & Schleiss, A.J. 2011a. A naming convention for the Piano Key Weirs geometrical parameters. In: Proceedings of the International Conference on Labyrinth and Piano Key Weirs (PKW 2011): 271–278. CRC Press, Taylor & Francis Group.

Pralong, J., Montarros, F., Blancher, B. & Laugier, F. 2011b. A sensitivity analysis of Piano Key Weirs geometrical parameters based on 3D numerical modeling. In: Labyrinth and Piano Key Weirs – Proceedings of the International Conference on Labyrinth and Piano Key Weirs (PKW 2011): 133–139.

Labyrinth and Piano Key Weirs IV – Erpicum, Crookston & Paxson (Eds)

Experimental and numerical investigation of A-type rectangular side Piano Key Weir

S. Hussain and Z. Ahmad
IIT Roorkee, Roorkee, Uttarakhand, India

ABSTRACT: A side weir is a structure installed along the side wall of a channel, serving multiple purposes, such as raising water levels for irrigation and drainage and diverting excess flow from rivers or channels during floods. Most studies have focused so far on linear rectangular and labyrinth side weirs, with relatively limited research conducted on side Piano key weirs. In this paper, an experimental and numerical simulation was performed on an A-type rectangular side piano key weir and a rectangular sharp-crested side weir. The diversion discharge of Piano key side weirs was within the range of ± 10% with numerical results. The diverted discharge over the rectangular side piano key weir was 1.5 to 6.7 times greater than the linear rectangular side weir. Water surface profile and velocity variations along the rectangular side Piano key weirs were also investigated experimentally and numerically.

1 INTRODUCTION

Side weirs are hydraulic structures frequently used in water management and engineering to divert surplus flow from a primary channel into a lateral or drainage system. They are commonly installed at diversion structures or intake points to regulate water surface elevation and control discharge direction. Side weirs have been extensively investigated due to their wide range of hydraulic and environmental engineering applications. Most of the existing research has concentrated on linear rectangular and labyrinth side weirs, while studies on side piano key weirs remain relatively limited. Mehri et al. (2018) conducted an experimental investigation on a C-type rectangular side piano key weir (RSPKW), comparing the coefficient of discharge for channel curvature angles of 30° and 120°. Karimi et al. (2018) simultaneously conducted an experimental study on a C-type rectangular side piano key weir (RSPKW) within a straight channel. Saghari et al. (2019) conducted an experimental study to evaluate the coefficient of discharge for an A-type trapezoidal side piano key weir (TSPKW) installed in a curved channel. In a subsequent study, Mehri et al. (2020) later identified P/h□ as the most significant parameter influencing the coefficient of discharge for A, B, C, and D-type side piano key weirs within a 120° curved channel. Zeyneb Kilic and Emin Emiroglu (2022) performed an experimental investigation on trapezoidal side piano key weirs (TSPKW) in a straight channel. Hussain et al. (2024) did a numerical simulation on an A-type rectangular side piano key weir. However, a review of the existing literature indicates that no researchers have yet derived a coefficient of discharge equation specifically for an A-type rectangular side piano key weir (SPKW) in a straight channel.

1.1 *Theoretical background*

The discharge per unit length of a side weir can be calculated using the formula (Subramanya, 2009)

DOI: 10.1201/9781003655541-13

$$\frac{dQ}{dx} = \frac{2}{3} C_d \sqrt{2g} h^{3/2} \tag{1}$$

C_d, g, and h represent De Marchi's discharge coefficient, acceleration due to gravity, and piezometric head ($h = y - P$) over the weir. Here, y and P represent the depth of flow and weir height, respectively. Along the side weir, De Marchi first presented the equation for the closed-form solution of the water surface differential, assuming constant energy in the main channel.

$$x = \frac{3}{2} \frac{B}{C_d} (\phi_2 - \phi_1) \tag{2}$$

where x is the distance from the beginning of the side weir, B is the channel width, and ϕ_1 and ϕ_2 are parameter functions at the upstream and downstream, defined as

$$\phi_1 = \frac{2E_1 - 3P}{E_1 - P} \sqrt{\frac{E_1 - y_1}{y_1 - P}} - 3\sin^{-1} \sqrt{\frac{E_1 - y_1}{E_1 - P}} \tag{3}$$

$$\phi_2 = \frac{2E_2 - 3P}{E_2 - P} \sqrt{\frac{E_2 - y_2}{y_2 - P}} - 3\sin^{-1} \sqrt{\frac{E_2 - y_2}{E_2 - P}} \tag{4}$$

Here, E_1 and E_2 represent specific energy at the upstream and downstream ends of the side weir, and y_1 and y_2 represent flow depth at the upstream and downstream. The relation between the weir opening (L) and other parameters of flow is

$$L = \frac{3}{2} \frac{B}{C_d} (\phi_2 - \phi_1) \tag{5}$$

L is calculated as the difference between positionsx_2 and x_1 (x_1 and x_2 represent the upstream and downstream ends of the side weir, respectively).

The objective of this study was to investigate water surface profile and velocity variations along the rectangular side Piano key weirs experimentally and numerically, and the diversion efficiency comparison of rectangular side Piano key weirs (SPKW) with linear rectangular side weirs (LSW).

2 EXPERIMENTAL SETUP & PROCEDURE

The experiment was performed in a hydraulic laboratory of IIT Roorkee, India. The setup consists of a 20 m long channel with a height of 0.5 m and a width of 0.67 m, as shown in Figure1. Lateral flow was introduced through a 0.67 m opening in the side wall of the main channel, starting 6.2m from the inlet end. The diversion channel's length, height, and width were 4 m, 0.8 m, and 0.59 m, respectively. The linear side weir's height, length, and thickness were 0.2 m, 0.67 m, and 0.004 m, respectively. The inlet and outlet key widths of the rectangular side Piano key weir were 0.16 m, and the thickness of the key was 0.006 m. The length of a key for an A-type piano key weir was 0.36 m (0.18 m base length and 0.18 m overhang length). The bed level of the diversion channel was 0.3 m below the bed of the main channel. To minimize flow turbulence, a honeycomb wall (grid) was placed at the upstream end of the main channel, and a tailgate was used to regulate the water level in the main channel. A rectangular, sharp-crested weir was provided at the end of the diversion channel to measure diverted discharge. Table 1 shows the range of experimental data. The experimental setup used in this study was comparatively larger than those reported in previous investigations. As a result, scale effects associated with viscous forces and surface tension are minimized due to higher Reynolds numbers. The larger scale also facilitates improved replication of free-surface behavior and air-water interactions under Froude similarity. Hence, the observed flow features, such as separation zones and reverse flow, were considered representative of prototype conditions, with negligible distortion due to scale limitations.

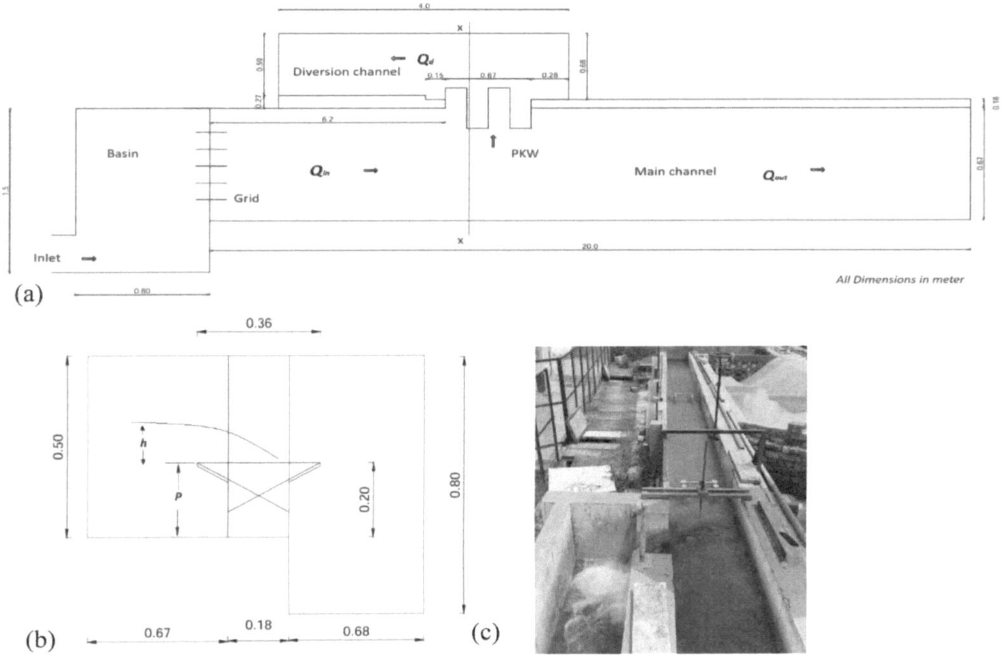

Figure 1. (a) Plan view, (b) x-x sectional view, and (c) photographic view of the experiment setup.

Table1. A range of experimental data was collected.

Types of weir	Q_{in} (m^3/s)	Q_d (m^3/s)	Froude (F_1) (-)	Height of weir (P) (m)	Effective length of weir (L_e) (m)	h_1/P -	h_1/L_e -
SPKW	0.1-0.15	0.003-0.118	0.25-0.64	0.2	2.08	0.114-0.836	0.0109-0.0803

3 NUMERICAL METHODOLOGY

Numerical simulations were conducted using ANSYS Fluent to investigate the hydraulic behavior of flow over a side Piano Key weir (PKW) in an open channel. The simulations were based on the Reynolds-Averaged Navier–Stokes (RANS) equations for incompressible flow. The Volume of Fluid (VOF) method was employed to model the multiphase interface between water and air, and turbulence was represented using the standard k–ε model. The PKW geometry was developed in ICEM-CFD based on experimental dimensions. A structured mesh consisting of approximately 3.96 million elements was generated, ensuring sufficient resolution near the weir crest and channel boundaries to capture key flow features while maintaining computational efficiency. Boundary conditions were specified as follows: the inlet was defined as a mass flow inlet, while both the main channel and side channel outlets were set as pressure outlets. The channel bed was treated as the reference elevation. A symmetry boundary condition was applied at the top surface to simulate the atmospheric interface, and no-slip conditions were applied to all solid walls, including the side walls and weir surface. The model was validated against experimental data for water surface profiles, flow depth, and velocity distribution. Numerical results showed good agreement with the experimental observations, with acceptable deviations, thereby confirming the accuracy and reliability of the simulation approach.

4 RESULTS AND DISCUSSION

4.1 *Empirical equation for C_d*

Most of the studies used the De Marchi approach to compute the coefficient of discharge (Ahmad et al., 2023). A similar approach was used to find observed C_d, and using the experimental parameters in Table 1, the Generalized Reduced Gradient (GRG) nonlinear Excel Solver method was used to compute C_d. As a result, the equation (6) for C_d was derived for a rectangular side Piano key weir.

$$C_d = \left[0.1034 - 0.599(F_1)^{1.5427} - 5.463\left(\frac{h_1}{l_e}\right)^{-0.0173} + 6.692\left(\frac{h_1}{p}\right)^{-0.0279}\right]^{3.094} \tag{6}$$

Figure 2(a) shows the comparison between the observed coefficient of discharge and the computed coefficient of discharge, with a good line of agreement with ±10% error. Due to the limitations in experimentation, some points were outside the range.

4.2 *Diversion flow efficiency*

It was observed that the diverted discharge (Q_d) obtained both experimentally and numerically showed a good level of agreement for the Piano Key side weir, with deviations found to be within ±10% error, as shown in Figure 2(b).

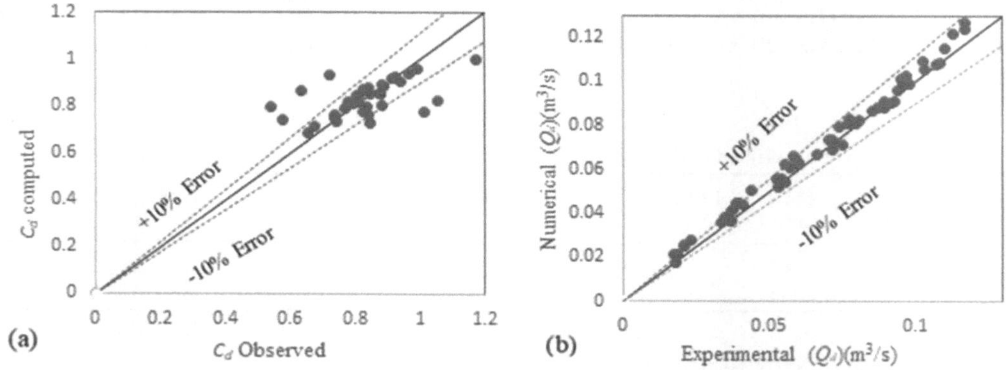

Figure 2. (a) Comparison of C_d observed and computed, and (b) Comparison of Q_d experimentally and numerically.

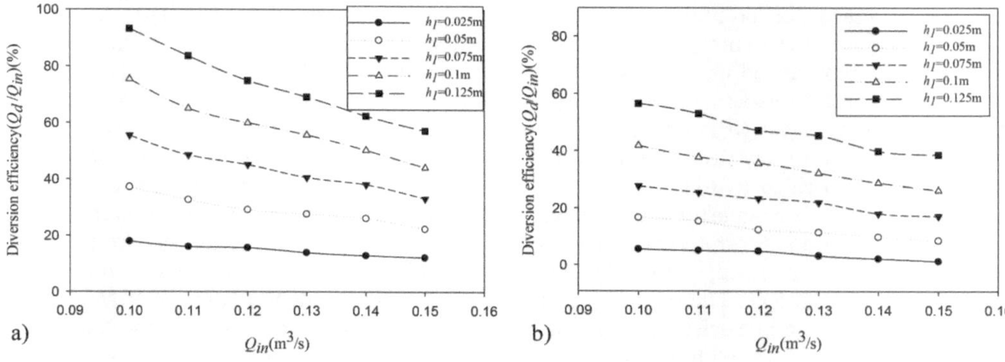

Figure 3. Variation of inlet discharge with diversion efficiency of different flow depths of (a)SPKW and (b) LSW.

Diversion efficiency ($\eta = Q_d/Q_{in}$) decreases with an increase in inlet discharge at a constant flow depth for both linear side weirs and side Piano key weirs. The observations indicate that the rate of decline was steeper for the rectangular side piano key weir, as compared to the linear side weir shown in Figure 3. When comparing the side piano key weir with the linear side weir under the same inlet head and discharge conditions, the side piano key weir achieves higher diversion rates ranging from 1.5 to 6.7 times greater than the linear rectangular side weir, as shown in Figure 4.

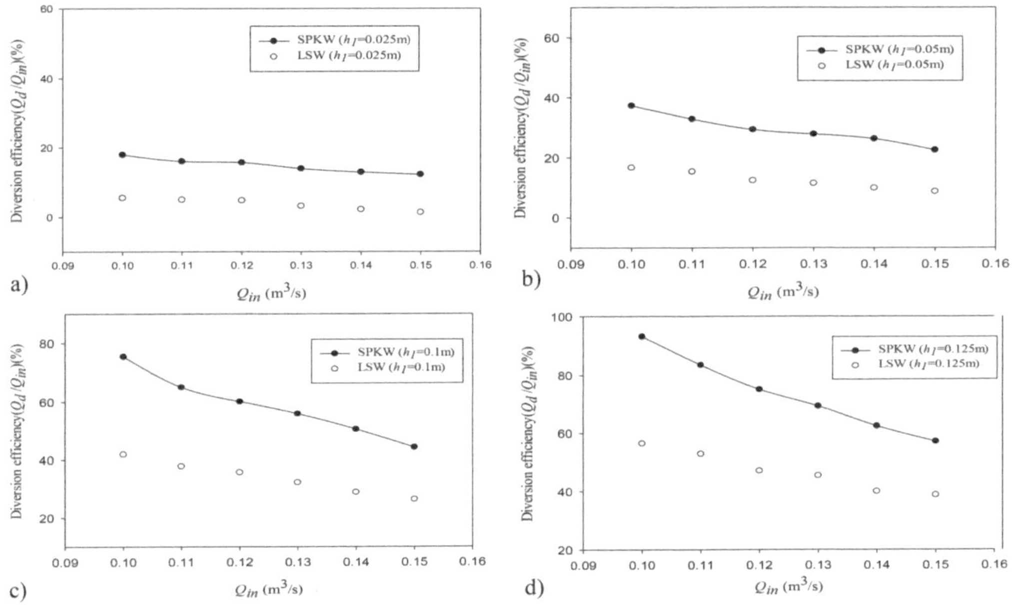

Figure 4. Comparison of SPKW and LSW diversion efficiency with inlet discharge at (a) h_l=0.025 m, (b) h_l=0.05 m, (c) h_l=0.1 m, and (d) h_l=0.125 m.

4.3 *Water surface profile*

In this study, a two-cycle piano key side weir was utilized to observe water surface profiles using a pointer gauge. Flow depths were recorded at intervals of 0.045 m from the starting point of the side piano key weir and 0.09 m beyond the key for each cycle, as well as the inlet and outlet keys. Figure 5 illustrates the variations in water surface profiles for both inlet and outlet keys under different discharge conditions while maintaining a constant inlet head (h_l=0.025 m); the results demonstrate a similar trend in water surface profiles for both the first and second inlet keys across all discharge conditions. However, the outlet keys exhibit rapid changes as discharge increases. Notably, the second outlet key delivers reduced water, decreasing the profile.

Figure 6 illustrates the changes in water surface profiles at both the inlet and outlet keys under varying inlet head conditions, with a constant discharge (Q_{in}=0.12 m^3/s). It demonstrates that the water surface profile exhibited reduced non-linearity as the inlet head increased.

Numerical simulations were conducted to analyze the behavior of the water surface profile under varying inlet head conditions while maintaining a constant discharge. Mass flow inlet boundary conditions were applied at the inlet, while pressure outlet conditions were imposed at both the main channel and the diversion channel outlets. Figure 7 shows the numerically simulated water surface profile, demonstrating a trend consistent with the experimental observations.

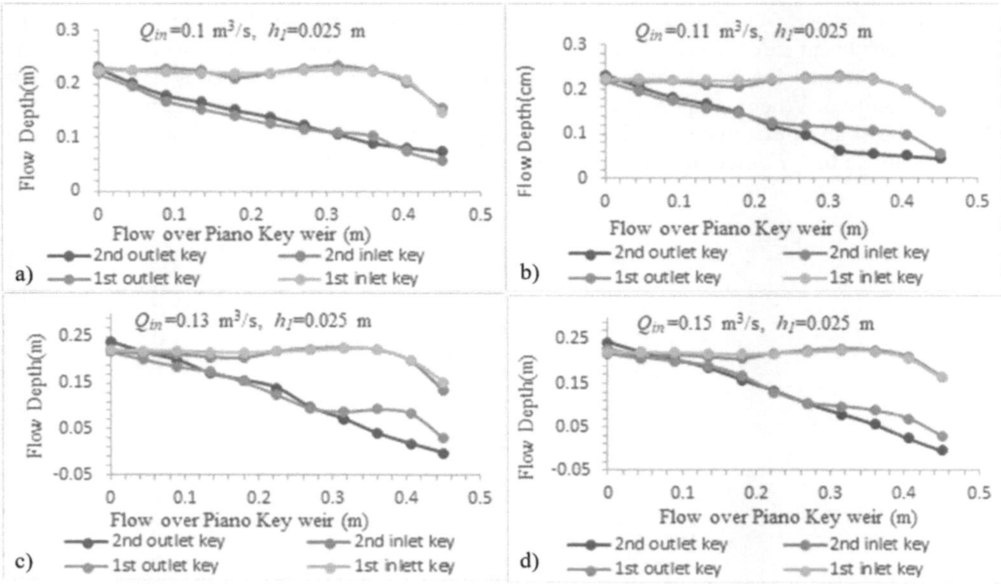

Figure 5. Water surface profile variations for inlet and outlet keys under (a) Q_{in} =0.1 m^3/s, (b) Q_{in} =0.11 m^3/s, (c) Q_{in} =0.13 m^3/s, and (d) Q_{in} =0.15 m^3/s at a constant inlet head (h_1=0.025 m).

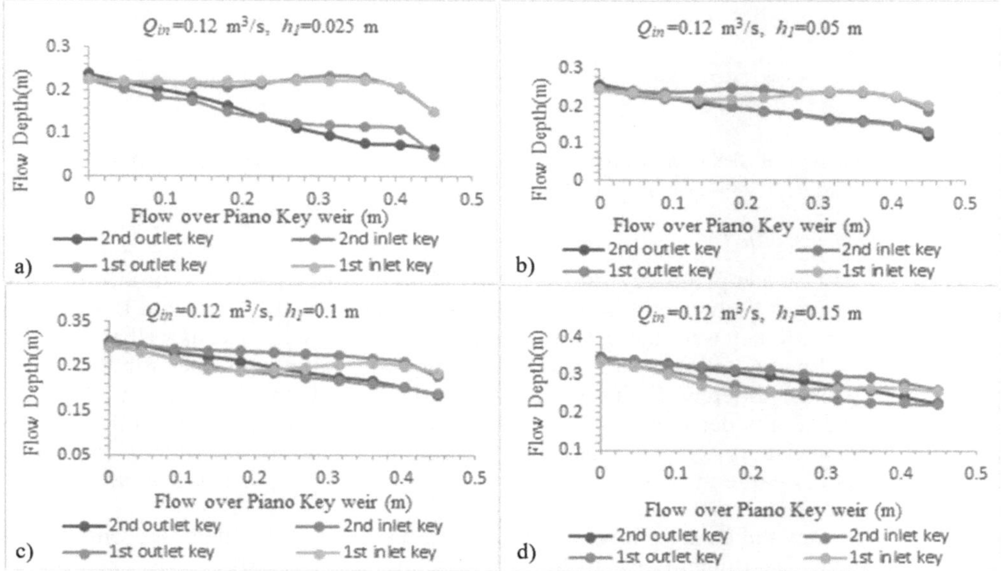

Figure 6. Water surface profile variations for inlet and outlet keys under (a) h_1=0.025m,(b) h_1=0.05m, (c) h_1=0.1m, and (d) h_1=0.15 mat aconstant discharge (Q_{in} =0.12 m^3/s).

4.4 *Velocity variation along the side Piano key weir at varying depths*

To measure the three-dimensional velocity in a horizontal plane passing through the Piano key side weirs using a flowtracker2 acoustic Doppler velocity meter (ADV), the total grid coverage extended 0.167 m along the flow direction and 0.67 m in the lateral direction. Measurements were taken at each grid point, resulting in 153 points across the plane generating velocity contours

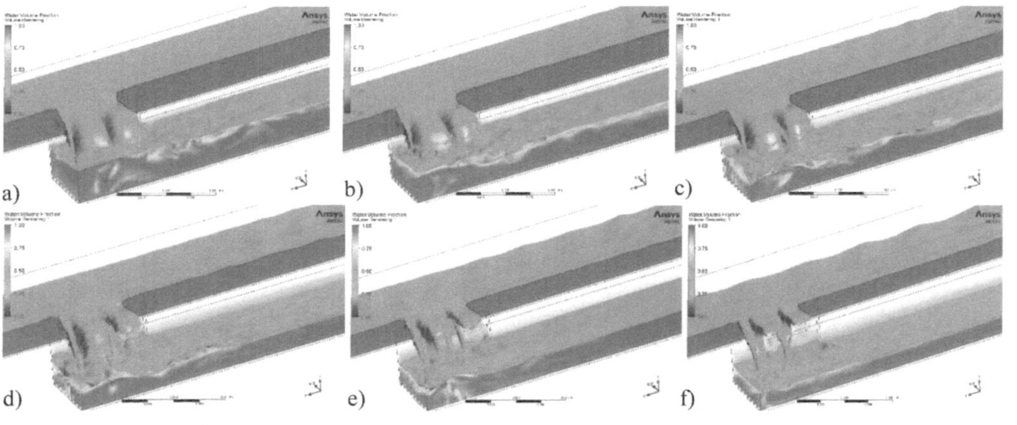

Figure 7. Water surface profile over a SPKW for 0.12 m^3/s discharge and flow depth (a) 0.35 m, (b) 0.325 m, (c) 0.30 m, (d) 0.275 m, (e) 0.25 m, and (f) 0.225 m.

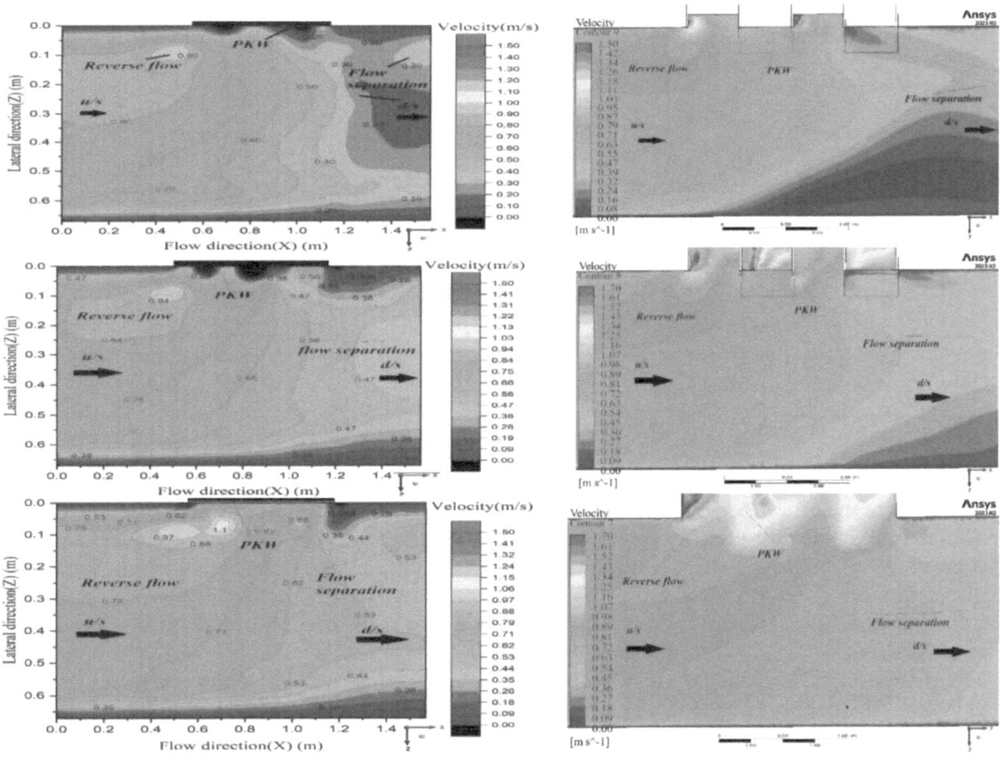

Figure 8. Velocity contour of SPKW for (a) y^*=0.25, (b) y^*=0.50, (c) y^*=0.75 obtained experimentally, and (d) y^*=0.25, (e) y^*=0.50, (f) y^*=0.75 obtained numerically at Q=0.15 m^3/s and y=0.3m.

experimentally. Three depths at plane y^* (measured depth/total depth) =0.25, 0.50, and 0.75 were selected to analyze the velocity variation at a discharge (Q_{in}) of 0.15 m^3/s, with a flow depth (y) maintained at 0.3 m. The results revealed flow separation at the downstream end of the side weir, with the effect being more pronounced at y^*=0.25. Furthermore, the reverse flow was observed near the upstream end of the side weir, as shown in Figure 8 (a), (b), and (c) experimentally and Figure8 (d), (e), and (f) numerically. Numerical simulations demonstrated similar flow behavior patterns, validated by the experimental findings.

4.5 *Model Validation*

The numerical results obtained from the CFD simulations were validated against experimental velocity measurements at varying flow depths for an inlet discharge of 0.12 m^3/s at the upstream end of the SPKW. As shown in Figure 9, the CFD model accurately replicates the experimental trend, particularly at lower flow depths (up to approximately 0.05 m), where the agreement was strong, with an average error of 7%. At higher flow depths, the CFD predictions slightly underestimate the velocity; however, the overall flow behavior remains consistent with experimental observations. These minor deviations can be attributed to idealized assumptions in the numerical model, such as turbulence closure and boundary condition specifications. Despite these limitations, the close correspondence between the datasets confirms the reliability of the CFD approach in modeling the hydraulic characteristics of the flow under the given conditions.

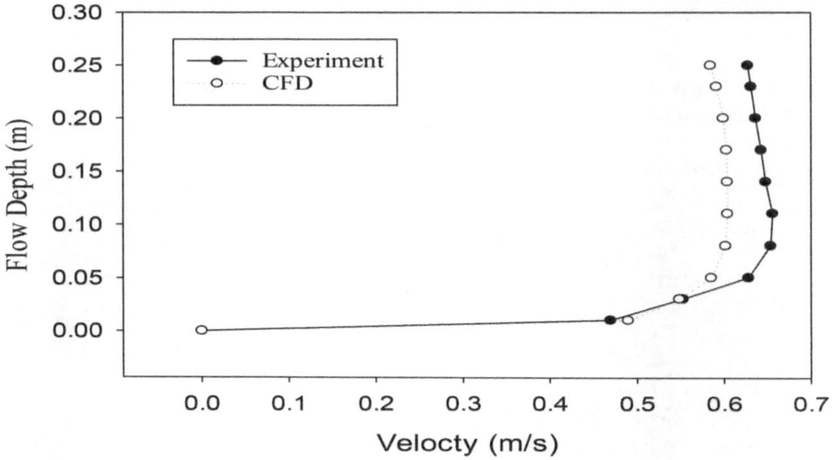

Figure 9. Comparison of velocity profiles obtained experimentally and CFD simulation at varying flow depths.

5 CONCLUSIONS

This study presents a comprehensive hydraulic investigation of a Side Piano Key Weir (SPKW), integrating experimental observations with numerical simulations to analyze flow behavior under varying conditions. An empirical equation for the discharge coefficient (C_d) was developed using experimental data, capturing the influence of key hydraulic and geometrical parameters. The proposed equation demonstrated strong predictive capability, with high correlation and low error metrics, making it a practical tool for preliminary design and analysis of SPKWs.

The diversion flow efficiency was evaluated across a range of discharges, showing that the SPKW effectively enhances flow diversion compared to LSW due to its increased crest length and unique geometry. Water surface profiles were accurately captured along the weir channel, indicating a gradual drawdown over the weir length and validating the free-surface behavior predicted by the numerical model.

Three-dimensional velocity measurements were conducted experimentally at different relative depths (y^* = 0.25, 0.50, and 0.75) using an Acoustic Doppler Velocimeter (ADV). The velocity distribution along the SPKW revealed key flow features, including separation zones at the downstream end and reverse flow near the upstream inlet. These experimental results were in good agreement with the numerical simulations carried out using ANSYS Fluent, thereby validating the accuracy of the computational model.

Overall, the study confirms that the SPKW offers improved hydraulic performance and flow control capabilities. The combined use of experimental and numerical approaches strengthens the reliability of the findings and provides a validated framework for further optimization and design of side piano key weir systems under varying flow conditions.

REFERENCES

Ranga Raju, K. G., Gupta, S. K., & Prasad, B. (1979). Side weir in rectangular channel. *Journal of the Hydraulics Division, 105(5)*, 547-554.

Emiroglu, M. E., & Kaya, N. (2011). Discharge coefficient for trapezoidal labyrinth side weir in subcritical flow. *Water Resources Management*, *25*(3), 1037–1058.

M.Emin Emiroglu,M. Cihan Aydin & Nihat Kaya(2014). Discharge Characteristics of a Trapezoidal Labyrinth Side Weir with One and Two Cycles in Subcritical Flow. *J. Irrig. Drain Eng., 2014, 140(5)*: 04014007.

Shariq, A., Hussain, A., & Ansari, M. A. (2018). Lateral flow through the sharp crested side rectangular weirs in open channels. *Flow Measurement and Instrumentation*, *59*, 8-17.

Mahmoud Karimi, Jalal Attari, Mojtaba Saneie, & Mohammad Reza Jalili Ghazizadeh(2018), Side Weir Flow Characteristics: Comparison of Piano Key, Labyrinth, and Linear Types *J. Hydraul. Eng., 2018, 144(12)*: 04018075.

Abolfazl Saghari, Khosrow Hosseini & Mojtaba Saneie (2019). Experimental study of trapezoidal piano key side weirs in a curved channel. *Flow Measurement and Instrumentation 70 (2019)* 101640.

Mehri, Y., Esmaeili, S., & Soltani, J. (2020). Experimental study and performance comparison on various types of rectangular piano key side weirs at a 120 section of a 180 curved channel. *Applied Water Science*, *10*, 1-13.

Kilic, Z., & Emin Emiroglu, M. (2022). Study of hydraulic characteristics of trapezoidal piano key side weir using different approaches. Water Supply, 22(8), 6672–6691.

Ahmad, F., Hussain, A., & Ansari, M. A. (2023). Development of ANN model for the prediction of discharge coefficient of an arced labyrinth side weir. *Modeling Earth Systems and Environment*, *9*(2), 1835–1842.

Hussain, S. (2024). Discharge Characteristics of Piano Key Side Weirs. In *Proceedings of the 10th International Symposium on Hydraulic Structures (ISHS) 2024*. ETH Zurich.

Labyrinth and Piano Key Weirs IV – Erpicum, Crookston & Paxson (Eds)

Key width ratio influence on discharge coefficients of trapezoidal Piano Key Weirs via OpenFOAM simulations

L. Lüddecke and M. Oertel
Helmut-Schmidt-University, Hydraulic Engineering Section, Hamburg, Germany

ABSTRACT: Piano key weirs are characterized by complex three-dimensional flow structures with multiple parameters describing their geometry. The variation of these parameters can affect several flow characteristics such as efficiency, energy dissipation, and air entrainment. Identifying the optimal geometry for enhanced hydraulic performance presents a significant challenge. Physical or numerical modeling of multiple geometric variants is necessary to identify the relevant optimal parameters. The key width ratio (between an inlet and an outlet key) is one of the main parameters that influences the hydraulic behavior of the piano key weir. In particular, it affects the discharge coefficient, which in turn affects the overall efficiency of the weir. Moreover, the impact of this parameter varies depending on the specific discharge. Accordingly, a comprehensive study has been conducted on a high-performance computer using the CFD software OpenFOAM to find an optimum range of inlet/outlet key width ratios for a trapezoidal Piano key weir under various discharges. In total, 45 different geometries of Piano key weirs (a weir height of 0.5 m and a width of 1.5 m) with varying key width ratios have been simulated over a range of 60 discharges resulting in 2,700 numerical model runs. The results were evaluated, and a best-fitting function is introduced that describes the efficiency of the weir as a function of key width ratio and discharge.

1 INTRODUCTION

1.1 *Background*

The Piano Key Weir (PKW) is a non-linear hydraulic structure specifically designed to achieve enhanced discharge efficiency compared to conventional linear weirs. This improved performance is primarily attributed to its folded geometry, which increases the effective overfall length. Due to its compact footprint, the PKW can be installed atop dams, offering a distinct advantage over its predecessor, the Labyrinth Weir.

The geometric configuration of the PKW is defined by multiple design parameters (Pralong, 2011), which can be adjusted in various ways and combinations, significantly influencing its flow discharge efficiency. Prior research has examined the influence of these parameters not only on discharge efficiency but also on structural stability and energy dissipation (Leite Ribeiro, 2009; Shen, 2023). Additionally, these parameters have been classified into primary and secondary groups based on their relative impact on discharge capacity. The most influential parameter has been identified as the magnification ratio (L/W), followed by the weir height (P) (Cicero, 2016; Erpicum, 2014). Leite Ribeiro (2012) classified the key width ratio of the inlet and outlet key (W_i/W_o), the relative overhang lengths (B_i/B_o), and the relative height of the parapet walls as secondary parameters with minor but not negligible effect on the discharge efficiency. The complexity of optimizing PKW design arises not only from the interplay of multiple geometric configurations but also from the influence of varying discharge conditions.

DOI: 10.1201/9781003655541-14

The discharge capacity (Q) of a Piano Key Weir is calculated a standard form of the weir equation by Du Buat:

$$Q = \frac{2}{3} c_D L \sqrt{2g} H_t^{1.5} \tag{1}$$

where c_D is the discharge coefficient, g is the acceleration due to gravity and H_t is the total upstream head ($H_t = v^2/2g + h_t$). The key width ratio (W_i/W_o) is a parameter of the Piano Key Weir that has been subject to many studies to find an optimum value in combination with the discharge. In existing studies, the recommended value for the key width ratio is between 1 and 1.67. Most researchers suggest that an optimal key width ratio ranging between 1.25 and 1.5 (Anderson & Tullis, 2013; Machiels et al., 2012; Erpicum et al., 2013). Empirical findings generally indicate that Piano Key Weirs (PKWs) with a key width ratio (W_i/W_o) exceeding unity ($W_i/W_o > 1$) exhibit enhanced discharge capacity relative to symmetric configurations ($W_i/W_o = 1$). Conversely, configurations characterized by a key width ratio below unity ($W_i/W_o < 1$) demonstrate diminished discharge efficiency (Li et al., 2019; Shen & Oertel, 2021). However, it has also been noted that a single optimal value does not exist (Machiels, 2014), and accurately characterizing the relationship between PKW geometry, discharge, and the discharge coefficient remains highly complex and has yet to be fully established.

1.2 *Objectives*

Although extensive research and data collection have been conducted to evaluate the influence of the key width ratio on the efficiency of Piano Key Weirs, the available data remains insufficient to establish a definitive relationship between this ratio, the discharge and discharge coefficients. This study seeks to build upon previous investigations by generating a comprehensive dataset of upstream flow depths and discharge coefficients for a range of PKW geometries with varying key width ratios and discharge conditions (see Figure 1). From this dataset, a predictive equation will be derived to describe the efficiency of the weir as a function of the key width ratio and discharge. This study functions as a preliminary investigation into the possibility of identifying a theoretical relationship between the specified parameters.

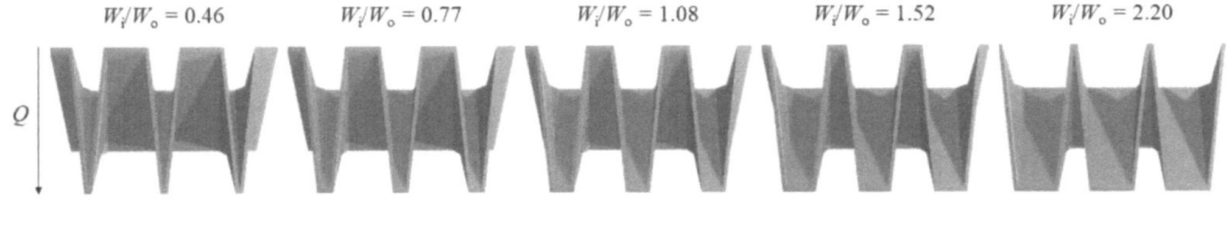

Figure 1. Examples of PKW geometries with different key width ratios.

2 NUMERICAL MODEL

2.1 *Model design*

The numerical model was developed using the open-source computational fluid dynamics (CFD) software OpenFOAM. The simulation setup was based on established boundary conditions for hydraulic structure modeling in OpenFOAM, as outlined in the study by Thorenz (2024). The model represents a scaled physical system, featuring a channel width of 1.5 m and a height of 1.0 m. A weir with a height of P = 0.5 m was installed within the channel. The computational domain extended 20 × P upstream and 10 × P downstream of the weir's upstream crest, resulting in a total domain length of 15 meters. The chosen base cell size of 0.1 m represented a balance between computational efficiency and data accuracy, given the large number of simulations conducted. To ensure an accurate geometric representation of the weir, double mesh refinements were applied locally around the weir structure. These refinements extended upstream to a distance of 3 × P, corresponding to the region where flow depth data were extracted. To ensure smooth transitions and maintain numerical stability, a series of single-level refinements were applied between the base mesh and the double-refined zones. The computational mesh was generated using blockMesh and snappyHexMesh, consisting of approximately 560,000 cells, varying depending on the weir geometry. By only refining the critical regions of interest and reducing the cell density in less significant areas, computational efficiency is significantly improved. This optimization reduces the computing time per run, enabling a larger number of simulations and facilitating the generation of a more extensive dataset. Boundary conditions were set according to Thorenz (2024) with inflow rate at the upstream boundary (x_{min}) and downstream water level at the outflow boundary (x_{max}) serving as key parameters. Flow rates were assigned according to the values listed in Table 1. The downstream water depth was maintained at a constant level that did not influence the overflow conditions at the weir, as downstream flow dynamics were beyond the scope of this investigation. A roughness height of K_s = 0.001 m was specified for the flume bottom (z_{min}) to represent a hydraulically smooth boundary.

The interFOAM solver, which employs the Volume of Fluid (VOF) method, was used to simulate free-surface flows. Turbulence was modeled using the Reynolds-Averaged Navier-Stokes (RANS) equations in combination with the k-ω turbulence model. The simulation time was set to 100 seconds to simulate a near-steady state. A pressure probe was positioned at a distance of 3 × P from the weir crest. Pressure data were recorded at 0.5-second intervals. Flow depths were derived from the recorded pressure values, taking into account the specific location of the probe. To mitigate the influence of surface fluctuations and improve the reliability of the results, the mean flow depth was calculated over the final five seconds of the simulation. All simulations were executed in parallel computing mode on a high-performance computing (HPC) cluster at Helmut Schmidt University, Hamburg, Germany. The cluster comprises 571 compute nodes, each equipped with 256 GB of RAM and two Intel Ice Lake sockets. Each socket features an Intel® Xeon® Platinum 8360Y processor with 36 cores, yielding a total of 72 cores per node. The simulation time for each run was set to at least 50 seconds and was increased for very small and very large discharges to ensure steady-state conditions.

2.2 *Investigation program*

The dataset consists of 45 trapezoidal Piano Key Weir geometries, each varying in its key width ratio (W_i/W_o), and 60 different discharge values. For each geometry, all 60 discharges were simulated, resulting in a total of 2,700 numerical model runs. The weir models have a uniform centerline crest length L of 6.77 m, a width W of 1.5 m, a wall thickness T_s of 2.5 cm and an angle $\alpha = 6°$. B_b is set to 1.00 m while B_i and B_o are 0.25 m. The key width ratio ranged from 0.46 to 2.20. Detailed geometry specifications are provided in Table 2 and Figure 2. Computation times varied between 4 and 24 hours. Resulting file sizes are approximately 75 MB. Total simulation file sizes including mesh data are considerably larger.

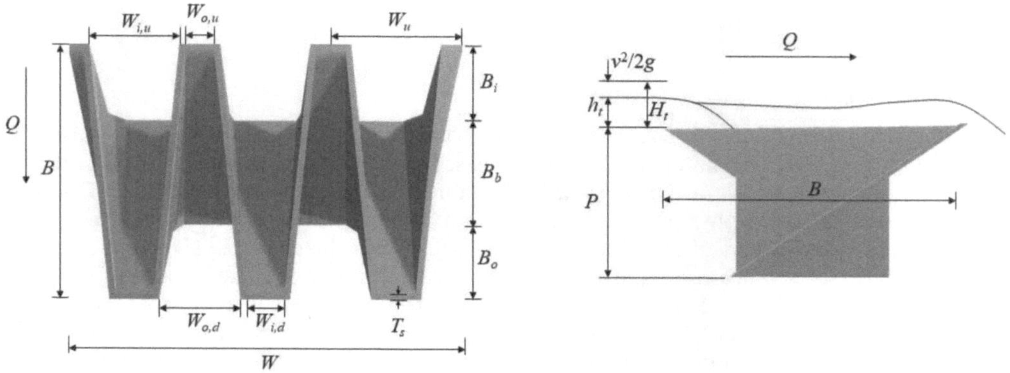

Figure 2. Geometric parameters of the trapezoidal Piano Key Weir.

Table 1. Simulated discharges.

Q [l/s]											
55	80	105	130	155	180	220	300	425	550	750	1200
60	85	110	135	160	185	230	325	450	575	800	1300
65	90	115	140	165	190	240	350	475	600	900	1400
70	95	120	145	170	200	250	375	500	650	1000	1500
75	100	125	150	175	210	275	400	525	700	1100	1600

Table 2. Geometry specifications of Piano Key Weir models.

Model	W_i/W_o [-]	$W_{i,u}$ [mm]	$W_{i,d}$ [mm]	$W_{o,u}$ [mm]	$W_{o,d}$ [mm]	Model	W_i/W_o [-]	$W_{i,u}$ [mm]	$W_{i,d}$ [mm]	$W_{o,u}$ [mm]	$W_{o,d}$ [mm]
R046	0.46	205.0	5.0	250.0	450.0	R096	0.96	320.0	120.0	135.0	335.0
R047	0.47	210.0	10.0	245.0	445.0	R098	0.98	325.0	125.0	130.0	330.0
R049	0.49	215.0	15.0	240.0	440.0	R102	1.02	330.0	130.0	125.0	325.0
R051	0.51	220.0	20.0	235.0	435.0	R105	1.05	335.0	135.0	120.0	320.0
R052	0.52	225.0	25.0	230.0	430.0	R108	1.08	340.0	140.0	115.0	315.0
R054	0.54	230.0	30.0	225.0	425.0	R111	1.11	345.0	145.0	110.0	310.0
R056	0.56	235.0	35.0	220.0	420.0	R115	1.15	350.0	150.0	105.0	305.0
R058	0.58	240.0	40.0	215.0	415.0	R118	1.18	355.0	155.0	100.0	300.0
R060	0.60	245.0	45.0	210.0	410.0	R122	1.22	360.0	160.0	95.0	295.0
R062	0.62	250.0	50.0	205.0	405.0	R126	1.26	365.0	165.0	90.0	290.0
R064	0.64	255.0	55.0	200.0	400.0	R130	1.30	370.0	170.0	85.0	285.0
R066	0.66	260.0	60.0	195.0	395.0	R134	1.34	375.0	175.0	80.0	280.0
R068	0.68	265.0	65.0	190.0	390.0	R138	1.38	380.0	180.0	75.0	275.0
R070	0.70	270.0	70.0	185.0	385.0	R143	1.43	385.0	185.0	70.0	270.0
R072	0.72	275.0	75.0	180.0	380.0	R147	1.47	390.0	190.0	65.0	265.0
R075	0.75	280.0	80.0	175.0	375.0	R152	1.52	395.0	195.0	60.0	260.0
R077	0.77	285.0	85.0	170.0	370.0	R157	1.57	400.0	200.0	55.0	255.0
R079	0.79	290.0	90.0	165.0	365.0	R162	1.62	405.0	205.0	50.0	250.0
R082	0.82	295.0	95.0	160.0	360.0	R173	1.73	415.0	215.0	40.0	240.0
R085	0.85	300.0	100.0	155.0	355.0	R191	1.91	430.0	230.0	25.0	225.0
R087	0.87	305.0	105.0	150.0	350.0	R212	2.12	445.0	245.0	10.0	210.0
R090	0.90	310.0	110.0	145.0	345.0	R220	2.20	450.0	250.0	5.0	205.0
R093	0.93	315.0	115.0	140.0	340.0						

3 RESULTS

3.1 *Discharge coefficients*

The relation between the efficiency of the different weir geometries and the discharge Q was investigated by calculating discharge coefficients c_D using Eq. 1. The results are illustrated in Figure 3 as dimensionless discharge coefficients as a function of dimensionless upstream head H_t/P and dimensionless key width ratio W_i/W_o. The data shows non-linear behavior with two main trends. A single fitting function could not capture both regions with equal accuracy, necessitating the implementation of two distinct fittings to achieve the desired level of precision. The data was fitted using polynomial functions of degree 53 (see Eq. 2) for $H_t/P \leq 0.27$ (see parameters in Table 3) and for $H_t/P > 0.27$ (see parameters in Table 4) with $x = H_t/P$ and $y = W_i/W_o$. The complex polynomial fit demonstrates a high degree of accuracy, with an R^2 value of 0.988 and 0.999 respectively. The discharge coefficient curves are observed to align with established trends (Anderson & Tullis 2012, Oertel & Tullis 2014), exhibiting maximum efficiency within a range of approximately $0.07 < H_t/P < 0.14$. Beyond this range, the curves demonstrate a consistent decline in discharge coefficients.

$$C_d(x,y) = a + b\cdot x + c\cdot y + d\cdot x^2 + e\cdot x\cdot y + f\cdot y^2 + g\cdot x^3 + h\cdot x^2\cdot y + i\cdot x\cdot y^2 + j\cdot y^3 + k\cdot x^4 + l\cdot x^3\cdot y + m\cdot x^2\cdot y^2 + n\cdot x\cdot y^3 + o\cdot x^5 + p\cdot x^4\cdot y + q\cdot x^3\cdot y^2 + r\cdot x^2\cdot y^3 \tag{2}$$

Table 3. Parameters for Eq. 2 ($H_t/P < 0.27$).

a	=	-2.183	*d*	=	-961.1	*g*	=	5644	*j*	=	0.050	*m*	=	-5.485	*p*	=	-198.7
b	=	78.77	*e*	=	2.960	*h*	=	-24.65	*k*	=	-16190	*n*	=	-0.390	*q*	=	0.1939
c	=	0.465	*f*	=	-0.282	*i*	=	-1.582	*l*	=	107	*o*	=	18200	*r*	=	-1.352

Table 4. Parameters for Eq. 2 ($H_t/P > 0.27$).

a	=	0.517	*d*	=	3.759	*g*	=	-2.151	*j*	=	0.031	*m*	=	1.024	*p*	=	-1.038
b	=	-2.01	*e*	=	1.212	*h*	=	-3.756	*k*	=	-0.828	*n*	=	0.105	*q*	=	-0.331
c	=	0.368	*f*	=	-0.203	*i*	=	-0.539	*l*	=	3.277	*o*	=	0.947	*r*	=	-0.126

Figure 4 illustrates the percentage deviation between the calculated discharge coefficients and those predicted by Eq. 2. The results indicate a strong agreement within the range of $2.2 > W_i/W_o > 0.52$ and $0.99 > H_t/P > 0.08$, where the deviation remains below 5%. Within these boundaries, the prediction of discharge coefficients as a function of discharge and key width ratio is highly accurate. The discharge capacity for small heads but also for flood events can be estimated using this approach. However, data points outside these boundaries exhibit deviations of up to 11%, suggesting that predictions of discharge efficiency beyond these ranges may be less reliable when using the given equation. Additionally, it should be noted that Eq. 2 applies exclusively to the specific Piano Key Weir geometry considered in this study, where wall thickness, weir height, base and overhang lengths, and wall angle remain constant, with only the key width ratio varying.

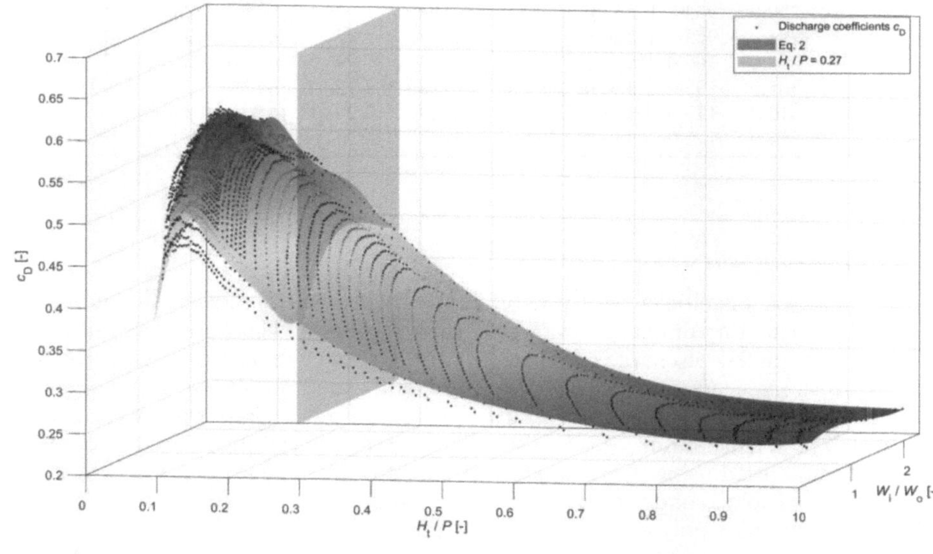

Figure 3. Discharge coefficients as a function of key width ratio and discharge.

Moreover, the optimal combination of geometry and discharge can be determined by identifying the maximum value of the function. The peak indicates that a Piano Key weir with a key width ratio of $W_i/W_o = 1.34$ in combination with a discharge of $H_t/P = 0.091$ achieves the highest

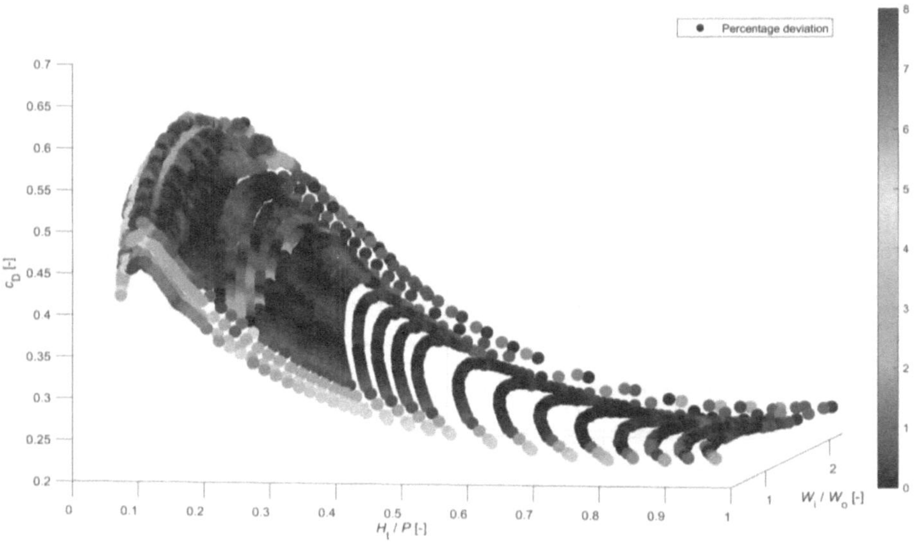

Figure 4. Percentage deviation between calculated discharge coefficients and Eq. 2.

efficiency, with a discharge coefficient c_D of 0.589. When calculating the maximum discharge coefficient for each geometry, the highest possible efficiency for every geometry variation can be described using a multivariate polynomial regression, resulting in Eq. 3. This function describes the maximum c_D values with determination coefficients $R_x^2 = 0.57$, $R_y^2 = 0.99$ and $R_z^2 = 0.99$.

$$\begin{aligned} x(t) &= 0.015 \cdot t^2 - 0.007 \cdot t + 0.092 \\ y(t) &= -2.366 \cdot t^3 + 4.920 \cdot t^2 - 4.222 \cdot t + 2.080 \\ z(t) &= -0.852 \cdot t^4 + 1.701 \cdot t^3 - 1.332 \cdot t^2 + 0.395 \cdot t + 0.548 \end{aligned} \tag{3}$$

With x describing the dimensionless discharge H_t/P, y represents the key width ratio W_i/W_o and z describes the discharge coefficient c_D. Previous studies have identified optimal W_i/W_o ratios ranging from 1.0 to 1.67. As shown in Figure 5, which represents the fitting of the peak discharge coefficients, the highest discharge coefficient values in this study also fall within this range. The top 25% of discharge coefficients are highlighted, which corresponds to range of $0.98 < W_i/W_o < 1.52$ where discharges range between $H_t/P = 0.089$ and $H_t/P = 0.095$.

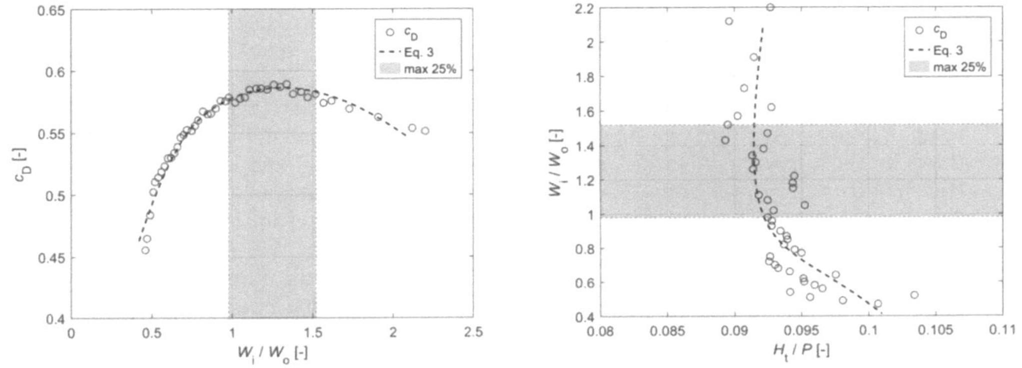

Figure 5. Peak discharge coefficients for varying key width ratios.

3.2 *Data validation*

To validate the dataset and the equation presented, experimental data from a similar study was used. Shen & Oertel (2021) tested the influence of key width ratio of Piano Key Weirs with the same geometry for a smaller scale. The experimental data was collected in a 30 cm wide flume with a weir height of $P = 0.1$ m. The weir geometries tested in their study differ from those in the present study only in scale, with a scaling factor of $\lambda = 5$. Figure 6 presents the percentage deviation of the experimental data from Shen & Oertel (2021) and Eq. 2. Within the parameter ranges where the equation demonstrates the highest accuracy ($2.2 > W_i/W_o > 0.52$ and $0.99 > H_t/P > 0.08$), the experimental data exhibits only minor deviations, remaining below 10%. These deviations may be attributed to scale effects in the small weir model or to inaccuracies related to the numerical model and the selected mesh size.

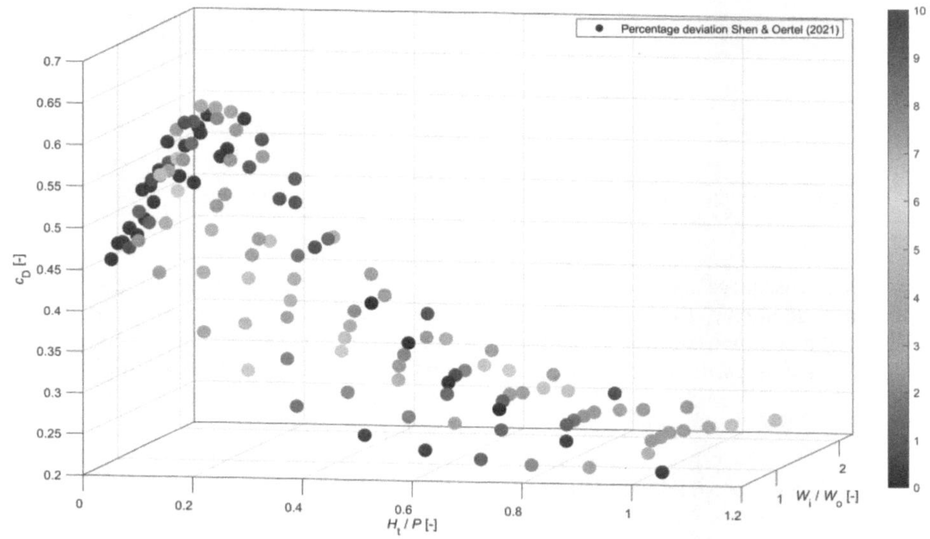

Figure 6. Percentage deviation between Eq. 2 and experimental data from Shen & Oertel (2021).

4 CONCLUSIONS AND OUTLOOK

In this study, a comprehensive dataset of upstream flow depth of a Piano Key Weir with varying key width ratios was generated consisting of 2,700 numerical simulations. From this dataset, discharge coefficients were calculated and displayed as a function of discharge and key width ratio. Based on this relationship, a best-fitting function was identified, thereby enabling the determination of the discharge coefficient for any flow rates and key width ratios within the boundaries of $2.2 > W_i/W_o > 0.52$ and $0.99 > H_t/P > 0.08$. However, it should be noted that the application of the function presented is limited to PKW that share the same geometric parameters as those presented in this study, with the exception of the key width ratio. The best efficiency (top 25%) was found for key width ratios in the range of 0.98 to 1.52, which corresponds to the average results from former studies. The corresponding discharge varies between $H_t/P = 0.089$ and $H_t/P = 0.095$.

Future research should consider finer cell sizes in order to further improve the accuracy of the function. Validation with experimental data of the same scale could also optimize the numerical model and the resulting predictions. As a preliminary study, this investigation could identify a theoretical relationship between key width ratio, discharge and efficiency of a Piano Key Weir. To develop an approach for a practical application, further investigations are recommended. A comprehensive investigation into the relationship between all geometry parameters, discharge, and efficiency of piano key weirs requires the analysis of extensive datasets that cover a wide range of parameter variations. This approach, when combined with the application of artificial intelligence, could enable deeper understanding of the relationships between the parameters. A diverse range of geometries can be generated through parametric design, a topic currently under investigation at Helmut-Schmidt University.

ACKNOWLEDGENTS

Computational resources (HPC cluster HSUper) have been provided by the project hpc.bw, funded by dtec.bw – Digitalization and Technology Research Center of the Bundeswehr. dtec. bw is funded by the European Union – NextGenerationEU.

REFERENCES

Anderson, R. M. & Tullis, B. P. (2012): Comparison of Piano Key and Rectangular Labyrinth Weir Hydraulics. In: *Journal of Hydraulic Engineering.*

Anderson, R. M. & Tullis, B. P. (2013): Piano Key Weir Hydraulics and Labyrinth Weir Comparison. In: *Journal of Irrigation and Drainage Engineering.*

Cicero, G. M.; Vermeulen, J. & Laugier, F. (2016): Influence of Some Geometrical Parameters on Piano Key Weir Discharge Efficiency. In: *International Symposium on Hydraulic Structures.*

Erpicum, S.; Archambeau, P.; Pirotton, M. & Dewals, B. (2014): Geometric parameters influence on Piano Key Weir hydraulic performances.

Erpicum, S.; Dewals, B.; Archambeau, P.; Pirotton, M. & Machiels, O. (2013): Considerations about the optimum design of Piano Key Weirs. In: *Erpicum, Laugier et al. (Eds.) – Labyrinth and Piano Key Weirs.*

Leite Ribeiro, M.; Bieri, M.; Boillat, J. L.; Schleiss, A. J.; Delorme, F. & Laugier, F. (2009): Hydraulic Capacity Improvement of Existing Spillways - Design of Piano Key Weirs.

Leite Ribeiro, M.; Pfister, M.; Schleiss, A. J. & Boillat, J. L. (2012): Hydraulic design of A-type Piano Key Weirs. In: *Journal of Hydraulic Research.*

Li, G.; Li, S. & Hu, Y. (2019): The effect of the inlet/outlet width ratio on the discharge of piano key weirs. In: *Journal of Hydraulic Research.*

Machiels, O.; Pirotton, M.; Archambeau, P.; Dewals, B. & Erpicum, S. (2014): Experimental parametric study and design of Piano Key Weirs. In: *Journal of Hydraulic Research.*

Machiels, O.; Erpicum, S.; Dewals, B.; Archambeau, P. & Pirotton, M. (2012): Influence of the Relative Alveoli Widths on Piano Key Weirs Efficiency for Varied Weir Heights. In: *Proceedings of the International Symposium on Dams for a Changing World.*

Oertel, M. & Tullis, B. P. (2014): Comparison of Piano Key Weir Discharge Coefficients from Experimental and Numerical Models. In: *Proceedings of the 3rd IAHR Europe Congress.*

Pralong, J.; Vermeulen, J.; Blancher, B.; Laugier, F.; Erpicum, S.; Machiels, O. et al. (2011): A Naming Convention for the Piano Key Weirs Geometrical Parameters. In: *S. Erpicum, F. Laugier, J. L. Boillat, M. Pirotton, B. Reverchon, A. J. Schleiss (Eds.) – Labyrinth and Piano Key Weirs – PKW. 2011, pp. 293–300.*

Shen, X. & Oertel, M. (2021): Comparative Study of Nonsymmetrical Trapezoidal and Rectangular Piano Key Weirs with Varying Key Width Ratios. In: *Journal of Hydraulic Engineering.*

Shen, X. & Oertel, M. (2023): Influence of Piano Key Weir Crest Shapes on Flow Characteristics, Scale Effects, and Energy Dissipation for In-Channel Application. In: *Journal of Hydraulic Engineering.*

Thorenz, C. (2024): Boundary Conditions for Hydraulic Structures Modelling with OpenFOAM. In: *Proceedings of the 10th International Symposium on Hydraulic Structures.*

Labyrinth and Piano Key Weirs IV – Erpicum, Crookston & Paxson (Eds)
© 2026 The Author(s), ISBN 978-1-041-10527-5

Investigation of flow field near a type-A Piano Key Weir

M.S. Quamar and Z. Ahmad
Indian Institute of Technology Roorkee, India

ABSTRACT: Piano Key Weirs (PKWs) enhance spillway capacity and induce complex flow patterns around their inlet and outlet keys. This study investigates the flow near a Type-A PKW using Computational Fluid Dynamics (CFD). A three-dimensional model was developed in Ansys Fluent to simulate flow characteristics for various discharges. The velocity components—streamwise, vertical, and lateral—were analyzed in detail. Results show that as flow approaches the inlet key, all velocity components increase, with the highest velocity observed near the inlet. In contrast, the streamwise velocity decreases near the outlet key while vertical and lateral velocities increase. Turbulent kinetic energy (TKE) was also found to be the highest downstream of the inlet key. The numerical model was validated using experimental velocity measurements, demonstrating good agreement. This confirms the accuracy of the CFD simulations and provides valuable insights into the hydrodynamics of PKWs, supporting improved design and erosion mitigation.

1 INTRODUCTION

Efficient water management is a critical challenge in modern hydrology, and innovative spillway designs are needed to maximize flow capacity in constrained spaces. Piano Key Weirs (PKWs) are a type of non-linear weir that can significantly increase spillway discharge compared to traditional linear weirs. The distinctive piano-key shape of PKWs allows for an elongated crest length within a given channel width, yielding discharge capacities more than four times that of conventional weirs (Ouamane and Lempérière 2006).

PKWs are generally classified into four geometrical types—A, B, C, and D—based on their shape and configuration (Ribeiro et al. 2012). Among these, Type-A PKWs are the most used due to their structural stability. The Type-A design is characterized by alternating inlet and outlet keys, which induce a complex three-dimensional flow field around the weir. Several studies have previously examined the hydrodynamics of flow around PKWs. For example, Denys and Basson (2018) investigated the transient flow behaviour of PKWs and observed that vortex shedding occurs at the inlet key due to pressure fluctuations, while negative pressure develops at the outlet key due to flow interactions. Kumar and Ahmad (2022) studied the three-dimensional flow field around a Type-A PKW (with a nose on the upstream side of the outlet keys) and reported that as flow approaches the inlet key, velocities in all three directions increase. They also noted that the vertical velocity component is higher in front of the outlet key than in front of the inlet key. Tiwari and Sharma (2015) focused on the flow near an inlet key and found only minor variations in the mean streamwise velocity, but significant changes in the lateral velocity component. Several other researchers have contributed to characterizing PKW flow behaviour (e.g., Machiels 2012; Erpicum et al. 2013; Denys et al. 2017; Kumar et al. 2024).

Understanding the turbulence characteristics of the PKW flow is crucial for predicting sediment transport and scour around these weirs. Turbulent flow structures near the keys can influence how sediment is mobilized and where erosion occurs. Therefore, a detailed investigation of the flow

DOI: 10.1201/9781003655541-15

behaviour near PKWs is essential for optimizing their hydraulic performance and assessing their impact on sediment dynamics. The aim of this study is to investigate the three-dimensional velocity field (streamwise, lateral, and vertical components) along the inlet and outlet keys of a Type-A PKW. This analysis provides better insight into the flow mechanics relevant to sediment transport over PKWs and informs more effective design and operation of such weir.

2 METHODOLOGY

2.1 *Numerical simulation*

The complex flow in the vicinity of the PKW was simulated using a CFD model in ANSYS Fluent version 2023 R2. The model solves the Reynolds-Averaged Navier–Stokes (RANS) equations for incompressible flow, which consist of the continuity and momentum equations. The governing equations for the three-dimensional simulation are given as follows:

Continuity Equation:

$$\frac{\partial \bar{u}_l}{\partial x_i} = 0 \tag{1}$$

Momentum Equation:

$$\frac{\partial \bar{u}_i}{\partial t} + \frac{u \partial u}{\partial x} + \bar{u}_j \frac{\partial \bar{u}_i}{\partial x_j} = -\frac{1}{\rho}\frac{\partial \bar{p}}{\partial x_i} + \nu \frac{\partial^2 \bar{u}_i}{\partial x_j x_j} + \frac{\overline{\partial u'_i u'_j}}{\partial x_j} + f_i \tag{2}$$

where, $\overline{u_i}$ mean velocity components in x, y and z directions; $\bar{p}$ mean pressure; ν = kinematic viscosity; ρ = density of fluid; f_i Body force per unit volume (e.g., gravity); $\overline{\partial u'_i u'_j}$ = Reynolds stress tensor, representing turbulent fluctuation.

To predict the highly turbulent flow over the PKW (especially near the inlet key and the walls), the k–ω Shear Stress Transport (SST) turbulence model was adopted. This two-equation turbulence model is well-suited for capturing flow separation and adverse pressure gradients in complex flows (Lynch and Smith 2011). It introduces additional transport equations for Turbulent Kinetic Energy (TKE) and the specific dissipation rate (ω), which are solved alongside the RANS equations. For completeness, the transport equations of the SST model are:

$$\frac{\partial k}{at} + \overline{u}_j \frac{\partial \overline{u_i}}{\partial x_j} = P_k - \beta' k\omega + \frac{\partial}{\partial x_j}\left[(\nu + \sigma_k \nu_t)\frac{\partial \omega}{\partial x_j}\right] \tag{3}$$

$$\frac{\partial \omega}{at} + \overline{u}_j \frac{\partial \omega}{\partial x_j} = \alpha \frac{\omega}{k} P_k - \beta \omega^2 + \frac{\partial}{\partial x_j}\left[(\nu + \sigma_\omega \nu_t)\frac{\partial \omega}{\partial x_j}\right] \tag{4}$$

where P_k is the turbulence production term and ν_t is the eddy viscosity, which is computed as

$$\nu_t = \frac{k}{\omega} \tag{5}$$

where, k is the turbulent kinetic energy (m^2/s^2), ω is the specific dissipation rate (1/s), and α is a model coefficient related to turbulent diffusivity. β is the coefficient governing the dissipation of specific dissipation rate (ω), and β'is the coefficient governing the dissipation of TKE.

Figure 1 shows the computational domain and the PKW geometry used in the CFD model. The simulation was performed for two flow rates: 12.0×10^{-3} m^3/s and 16.80×10^{-3} m^3/s. The rectangular computational domain was 2.0 m long with a cross-section of 0.39 m (width) by

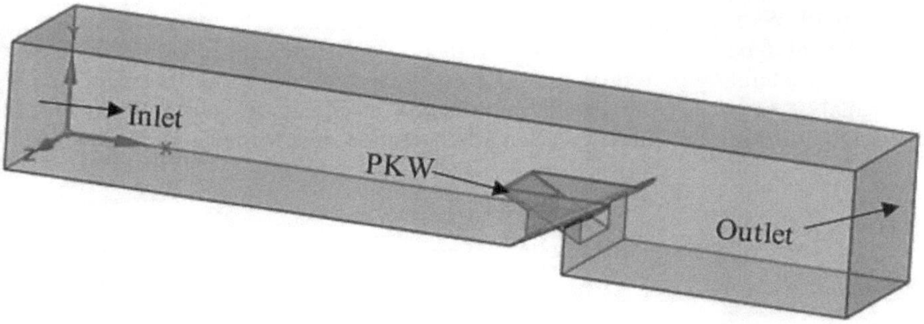

Figure 1. Fluent computational domain with the Type-A PKW model.

0.30 m (height), matching the dimensions of the experimental flume. A grid independence study was carried out by generating three meshes of varying densities using Fluent Meshing. To assess mesh sensitivity, the streamwise velocity component (u) was extracted along the vertical direction (depth of flow) at a representative point located 0.128 m upstream of the inlet key, measured along the channel centerline. Figure 2 shows the comparison of velocity profiles for the coarse, medium, and fine meshes, along with experimental measurements. The fine mesh provides the best agreement with the experimental data throughout the flow depth, especially near the surface where gradients are higher. Based on this analysis, the fine mesh was adopted for all subsequent simulations. The total cell count for the selected mesh was approximately 1.5 million, and the results were confirmed to be grid independent.

A transient solver was employed, simulating 20 s of real flow time with a time step of 0.01 s. An open-channel multiphase flow approach was used to capture the water–air interface dynamics, using the Volume of Fluid (VOF) method (Hirt and Nichols 1981). Boundary conditions were defined as a constant mass-flow inlet at the upstream end, a pressure outlet at the downstream end, and a symmetry condition at the top free surface. The governing equations were discretized using the finite volume method.

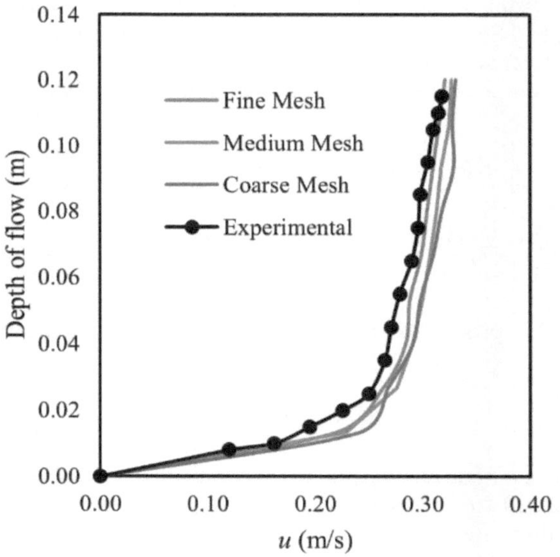

Figure 2. Streamwise velocity profiles (u) along the depth of flow at a point located 0.128 m upstream of the inlet key, comparing three mesh densities with experimental data.

3 VALIDATION OF THE NUMERICAL MODEL

To validate the numerical model, an experimental study was conducted in the hydraulic laboratory at the Indian Institute of Technology Roorkee. The experiments were carried out in a 15.0 m long, 0.39 m wide, and 0.50 m deep rectangular flume. A physical model of a Type-A PKW (one-cycle configuration) made of 6 mm thick acrylic sheet was installed in the flume, replicating the geometry used in the simulation. The plan and longitudinal view of the model are shown in Figure 3, where W_i and W_o are the width of inlet and outlet keys respectively; B_i and B_o are the inlet and outlet overhang lengths respectively; B_b is the base length; B is the sidewall overflowing length; P is the height of weir; W is the channel width and T is the thickness of side wall. The dimensions of the physical model are given in Table 1.

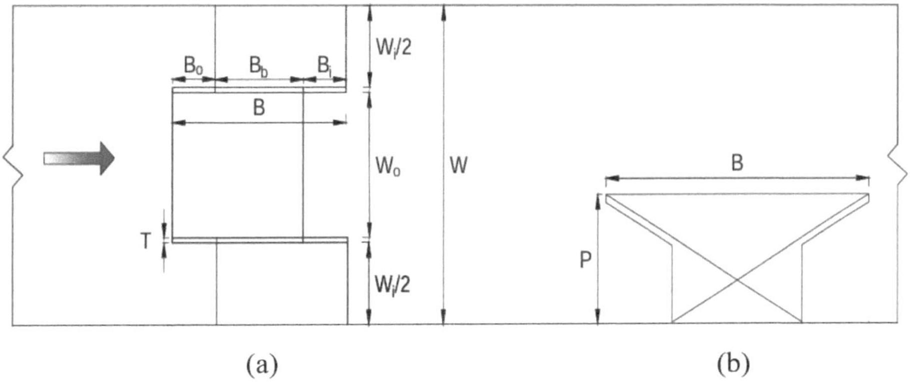

Figure 3. (a) Plan view and (b) longitudinal view of PKW model.

A continuous water supply from a recirculating tank provided a steady flow through the flume. The discharge was set to $12.0 \times 10^{-3} m^3/s$ and measured with an ultrasonic flowmeter (accuracy ±1%). The flow rate was regulated using a valve in the supply line. The water depth in the flume was monitored with a point gauge having a least count of 0.1 mm. Average velocity (U) was measured at 0.21 m of the upstream side of PKW which was 0.252 m/s. An Acoustic Doppler Velocimeter (ADV, 50 MHz, Nortek) was used to measure instantaneous velocity components at specific locations in the flow. Each point was sampled at 50 Hz for 60 seconds, resulting in 3000 samples per point. The time-averaged velocity was then computed for each location. In the numerical simulations, velocity data were extracted from selected points corresponding to the experimental locations. The time-averaged velocity was obtained from steady-state simulations using the k–ω SST turbulence model, and spatial averaging was applied where appropriate

Table 1. Dimensions of PKW model.

Model	No. of cycle	L(m)	W(m)	P(m)	W_i(m)	W_o(m)	B_b(m)	B_i=B_o(m)	S_i=S_o
PKW	One	0.898	0.39	0.105	0.2	0.178	0.125	0.064	5(V):9(H)

Figure 4. shows the measurement locations (grid points) where velocities were recorded. Five points (labeled A, B, C, D, E) were located along the centerline upstream of the half inlet key, and four points (F, G, H, I) were positioned upstream of the outlet key. At each of these points, the ADV was used to collect vertical profiles of the streamwise velocity component for comparison with the CFD results. The numerical model was validated by comparing the simulated and measured streamwise velocity (u) profiles at these locations (Figure 5). The velocity

profiles from the simulation closely match the experimental data across the flow depth, showing strong agreement between the numerical and experimental results. This agreement confirms that the CFD model can reliably reproduce the flow field around the PKW keys.

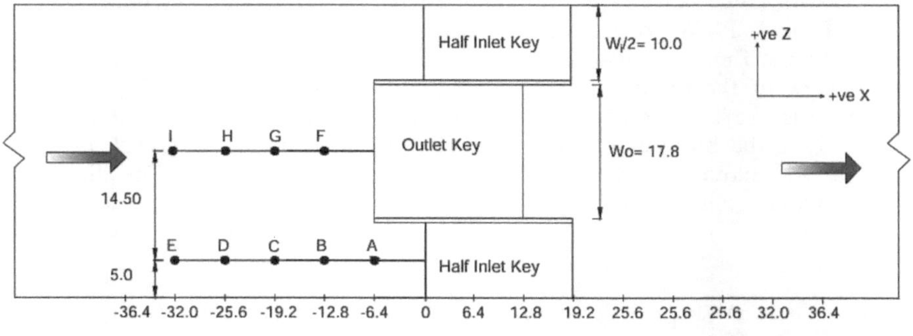

Figure 4. Schematic of measurement points (A–E near the inlet key, and F–I near the outlet key) for velocity validation.

4 RESULTS AND DISCUSSION

4.1 *Flow field in (X-Y) plane of inlet key*

The flow field in a vertical longitudinal plane (*x–y* plane) along the centerline of the half inlet key was obtained from the CFD simulation (Figure 6a). As the discharge increases, the nappe (free-surface jet) on the downstream side of the inlet key becomes more pronounced. Additionally, a vortex forms along the downstream vertical wall of the inlet key. Figures 6b–6d present contours of the mean velocity components (streamwise u, vertical v, and lateral w, respectively) in the *x-y* plane. As the flow approaches the inlet key, the streamwise velocity (u) remains nearly uniform upstream of the key. Upon entering the inlet key, streamwise velocity increases sharply due to flow contraction. The flow accelerates over the inlet key crest, reaching its highest speed near the downstream edge of the crest where flow separation occurs. In the region immediately downstream of the inlet key, a low-velocity recirculation zone is observed near the rear wall of the key. In this zone, streamwise velocity becomes negative, indicating reverse flow (backflow) associated with the vortex. Overall, the resultant velocity magnitude in the inlet key plane (Figure 6e) reflects these patterns: the velocity increases toward the inlet key and peaks at the downstream outlet of the inlet key, while a marked decrease appears in the recirculation region below the overhang part of the inlet key.

The lateral velocity component (w) is nearly zero in the approach flow upstream of the inlet key, indicating no significant cross-stream motion before the flow reaches the structure. Within the inlet key, however, regions of both positive and negative lateral velocity develop. This pattern suggests localized transverse flow, where some water moves sideways from the inlet key toward the adjacent outlet key region. Near the channel bed, turbulence-induced fluctuations lead to small variations in the lateral velocity. The vertical velocity component (v) also remains approximately zero in the upstream flow and throughout much of the inlet key, confirming that the incoming flow is primarily horizontal. A slight downward vertical velocity is observed just upstream of the inlet crest (as flow clings to and drops into the inlet), and a small upward vertical velocity occurs just past the crest. These vertical velocity deviations are indicative of flow separation and an adverse pressure gradient in the crest region of the inlet key.

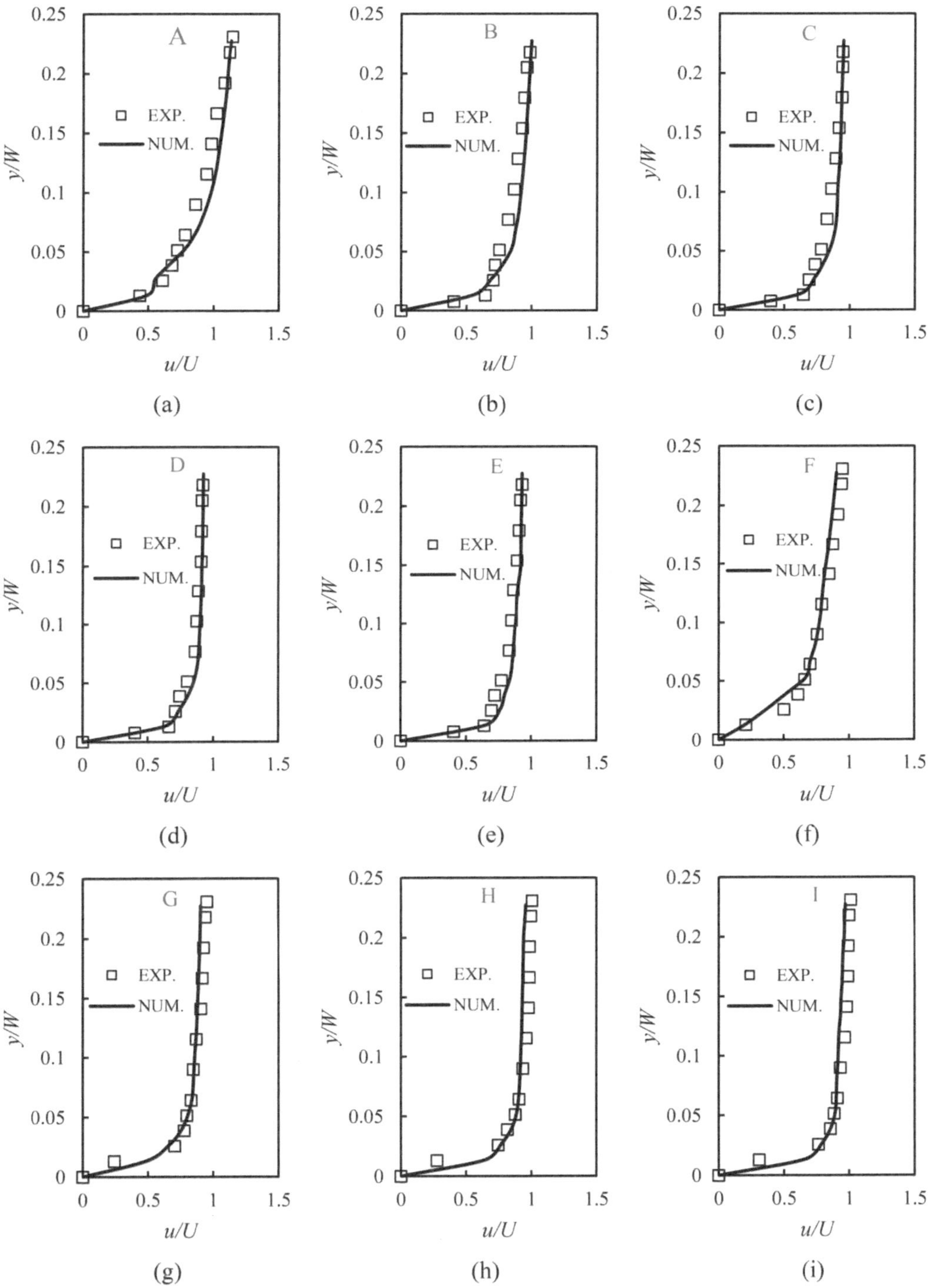

Figure 5. Measured vs. simulated streamwise velocity profiles at various depths for $Q = 12.0 \times 10^{-3}$ m³/s. Profiles (a–e) correspond to points A–E in front of the inlet key, and (f–i) correspond to points F–I in front of the outlet key.

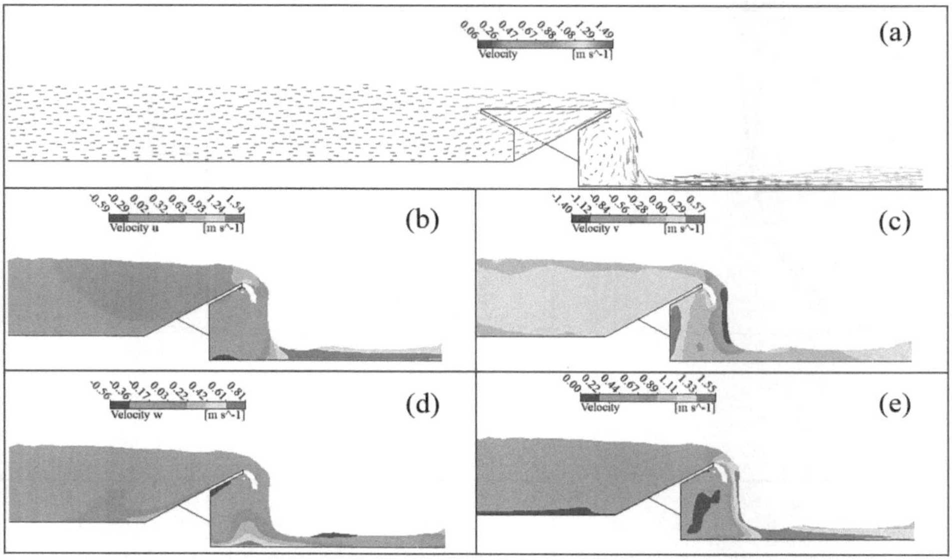

Figure 6. Velocity distribution in x-y plane along the centerline of half inlet key: (a) velocity vectors; (b) streamwise velocity (u); (c) vertical velocity (v); (d) lateral velocity (w); and (e) resultant velocity contours for Q=16.80 x10^{-3} m^3/s.

4.2 *Flow field in (X-Y) plan at outlet key*

Figure 7a shows the flow field (velocity vectors) in the vertical mid-plane through the outlet key. The simulation reveals a weak vortex forming beneath the upstream overhang of the outlet key and a circulatory flow region just downstream, immediately below the outlet key. Figure 7b presents the contour of streamwise velocity in the same plane (along the outlet key centerline). Upstream of the outlet key, the streamwise velocity is uniform. As the flow approaches the outlet key, streamwise velocity decreases in front of the key (i.e. in the region just upstream of the outlet entrance, up to the height of the PKW crest). Upon entering the outlet key, the flow contracts and streamwise velocity increases again. The streamwise velocity continues to accelerate through the outlet passage, reaching the maximum immediately downstream of the outlet key.

A low-velocity region is observed below the outlet key, attributable to the recirculating reverse flow in that area. Figure 7c shows the vertical velocity of the outlet key plane. Upstream of the outlet key, indicating minimal vertical motion in the approach flow. As the water drops into the outlet bay, vertical velocity becomes negative (downward), a result of the adverse pressure gradient pulling flow down into the outlet key. The lateral velocity (Figure 7d) is essentially zero before the flow reaches the outlet key, but lateral velocity begins to vary near and past the outlet key due to turbulence and the spreading of flow laterally as it exits the constricted outlet. Figure 7e depicts the resultant velocity magnitude in the outlet key plane. The overall flow deceleration in front of the outlet is evident, with the lowest speeds occurring up to about the crest height (approx. 0.105 m, which is the height of the PKW).

Beyond this point, inside the outlet and as the flow goes over the downstream slope of the key, the velocity increases again due to flow contraction and acceleration down the slope. The highest velocity occurs just downstream of the outlet key, where the flow separates from the structure. A pronounced low-velocity zone persists immediately below the outlet key, in the wake region behind the lower corner of the outlet key. The results obtained for both discharge rates showed similar trends.

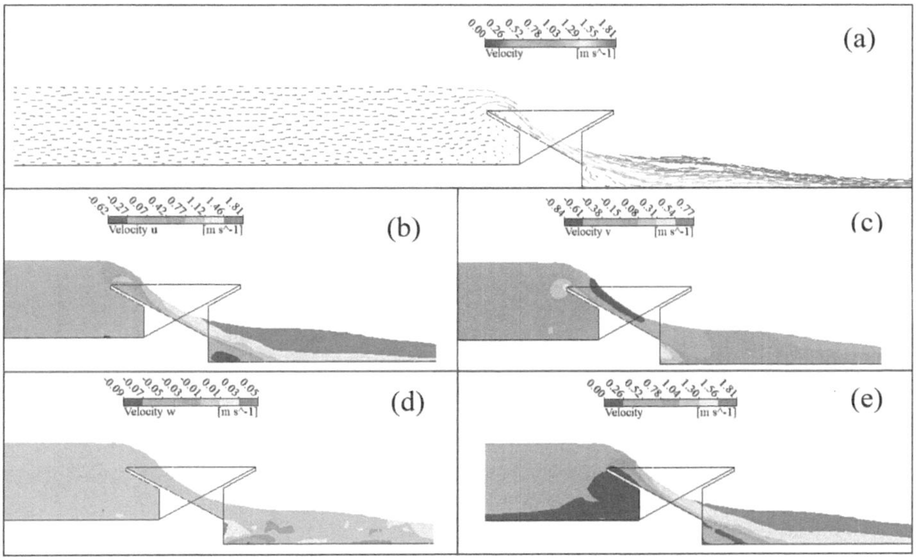

Figure 7. Velocity distribution in x-y plane along the centerline of outlet key: (a) velocity vectors; (b) streamwise velocity (u); (c) vertical velocity (v); (d) lateral velocity (w); and (e) resultant velocity contours for Q=16.80 x10^{-3} m^3/s.

4.3 *Turbulent kinetic energy (TKE)*

The turbulent kinetic energy field was also examined in the vertical centerline planes of a representative inlet key and outlet key (Figure 8). In general, the upstream sides of both the inlet and outlet keys exhibit very low TKE, since the approach flow is relatively uniform and undisturbed. On the downstream side of the inlet key, however, the TKE increases sharply due to the intense turbulence generated by flow separation and the recirculation zone in that area. In contrast, the TKE on the downstream side of the outlet key remains comparatively low. There is a slight increase in TKE near the channel bed just downstream of the outlet key (relative to the TKE near the outlet key free-surface region), likely because the recirculating flow at the bottom produces additional turbulence.

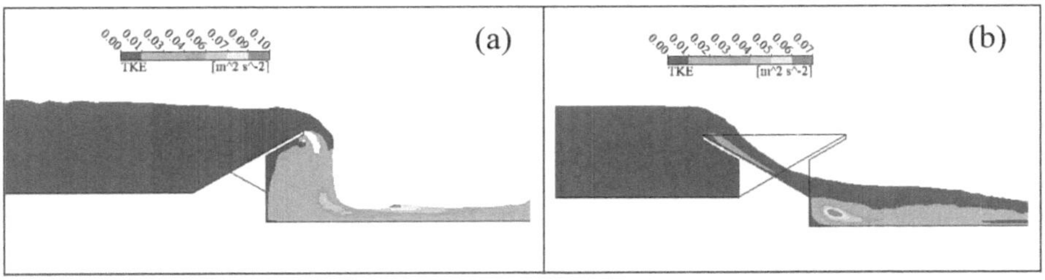

Figure 8. Contours of turbulent kinetic energy in x-y plane along the centerline of; (a) half inlet key; and (b) outlet key for Q=16.80 x10^{-3} m^3/s.

The highest TKE is found in the wake of the inlet key, where it is roughly 50% greater than the maximum TKE observed in the outlet key wake. These results indicate that the inlet key generates stronger turbulent fluctuations than the outlet key under the same flow conditions.

5 CONCLUSION

This study provides a detailed investigation of the flow field near the inlet and outlet keys of a Type-A Piano Key Weir. The inlet key region was found to experience markedly higher velocities and turbulence compared to the outlet key region. As flow approaches the inlet key, it accelerates and produces a strong recirculation zone on the downstream side of the inlet key. In contrast, flow deceleration occurs in front of the outlet key, and although the flow re-accelerates within and beyond the outlet key, the recirculation zone downstream of the outlet key is weaker than that of the inlet key. The turbulent kinetic energy is significantly higher in the downstream region of the inlet key, on the order of 50% higher than in the outlet key's wake, highlighting more vigorous turbulent mixing near the inlet key. The numerical simulation results showed good agreement with experimental velocity measurements in the upstream region, providing confidence in the CFD model's ability to represent the near-inlet flow behaviour of the PKW under the tested conditions. The insights gained from this work advance the understanding of the complex flow dynamics around PKWs. These findings can inform the hydraulic design and optimization of PKWs, enhancing sediment transport capacity and minimizing potential scour by better managing high-velocity and high-turbulence regions around the weir.

REFERENCES

Denys, F. J. M., Basson, G.R., & Strasheim, J. 2017. Fluid structure interaction of Piano Key Weirs. *In Proc., 3rd Int. Workshop Labyrinth and Piano Key Weirs (PKW 2017), 119–126.* Leiden, Netherlands: CRC Press.

Denys, F.J.M., & Basson, G.R. 2018. Transient Hydrodynamics of Piano Key Weirs. *In Proceedings of the 7th International Symposium on Hydraulic Structures*, Aachen, Germany, 15–18 May 2018.

Erpicum, S., Laugier, F., Pfister, M., Pirotton, M., Cicéro, G.M., & Schleiss, A. (Eds.). 2013. *Labyrinth and piano key weirs II-PKW 2013*, CRC Press, London, pp. 93–100.

Hirt, C.W., and Nichols, B.D. 1981. "Volume of fluid (VOF) method for the dynamics of free boundaries." *J. Comput. Phys., 39(1), 201–225.*

Kumar, B., Pandey, M., & Ahmad, Z. 2024. Flow field and sediment passing capacity of type-a piano key weirs. *International Journal of Sediment Research, 39*(4), 540–551.

Kumar, B., & Ahmad, Z. 2022. Experimental Study on Three-Dimensional Velocity Near Piano Key Weirs. *Proceedings, 9th IAHR international symposium on hydraulic structures (9th ISHS)*, October 24–27, 2022, Roorkee, India (2022).

Lynch, C.E., and Smith, M.J. 2011. "Extension and exploration of a hybrid turbulence Model on unstructured grids." *Aiaa J., 49(11), 2585–2591.*

Machiels, O. 2012. "Experimental study of the hydraulic behaviour of Piano Key Weirs." *Ph.D. thesis, Dept. of Urban and Environmental Engineering*, Univ. of Liège.

Ouamane, A., and Lempérière, F. 2006. "Design of a new economic shape of weir." *International Symposium on Dams in the Societies of the 21st Century; Dams and Reservoirs, Societies and Environment in the 21st Century ICOLD-SPANCOLD; 2006;* Barcelona, Spain, Taylor & Francis Group, Spain, 2006: pp. 463–470.

Ribeiro, M.L., Pfister, M., Schleiss, A.J. & Boillat, J.L. 2012. Hydraulic Design of A-Type Piano Key Weirs. *Journal of Hydraulic Research, 50, 400–408.*

Tiwari, H., & Sharma, N. 2015. Flow hydrodynamics near inlet key of Piano Key Weir (PKW). *Sadhana 40, 2253–2261 (2015).*

II Projects

II.1 Lessons learned and specific applications

Labyrinth and Piano Key Weirs IV – Erpicum, Crookston & Paxson (Eds)

Lessons learned from the hydraulic design of the Prado Dam labyrinth spillway

Y. Diaz, M.A. Phillips and J.D. Allen
US Army Corps of Engineers Risk Management Center, Lakewood, Colorado, USA

B.M. Crookston
Utah State University Water Research Laboratory, Logan Utah, USA

ABSTRACT: The U.S. Army Corps of Engineers has completed the design of a new labyrinth weir spillway at Prado Dam, in southern California. The proposed labyrinth weir would consist of an arced configuration, 7-cycles, with varying weir wall height from 28 ft to 41 ft (8.5 to 12.5 m) with upstream head over the weir cycles of up to 30 ft (9 m). This paper highlights the key lessons learned from the hydraulic design of the new labyrinth weir, including significant fluctuating negative pressures at the weir wall resulting in hydrodynamic loads that should be considered during structural design and submergence of the weir at high heads.

1 INTRODUCTION

1.1 *Background*

Prado Dam is located on the Santa Ana River, approximately 30 miles (50 km) upstream of the Pacific Ocean. The dam is in Riverside County, California, approximately 2 miles (3.2 km) west of the City of Corona. The U.S. Army Corps of Engineers (USACE) completed the existing Prado Dam in April 1941. The existing spillway consists of a 1,000 ft (305 m) wide, 13-feet (4 m) high concrete ogee crest, a converging concrete spillway channel, and a flip bucket.

1.2 *Proposed modifications*

Spillway modifications are necessary to increase the Prado Dam reservoir storage capacity and to pass outflow discharges as estimated from a recent Probable Maximum Flood (PMF) study. The new design discharge for the labyrinth spillway is estimated at approximately 500,000 cfs (14,000 m^3/s) and an upper estimate of 613,000 cfs (17,800 m^3/s). This is considerably higher than the original 1941 design discharge of 178,000 cfs (5,000 m^3/s).

The proposed spillway modifications, as shown in Figure 1, consists of:

- Replacing the existing 1000-ft (305 m) wide ogee crest with a new labyrinth weir structure that would project into the reservoir and be generally oriented with the flow direction from the reservoir.
- Converging the existing spillway chute that extends from the labyrinth control structure at 640 ft to 480 ft (195 m to 146 m) wide.
- Determining the need to upgrade the existing flip bucket and whether additional scour protection is required downstream of the flip bucket.

This paper focuses on the control structure modifications.

DOI: 10.1201/9781003655541-16

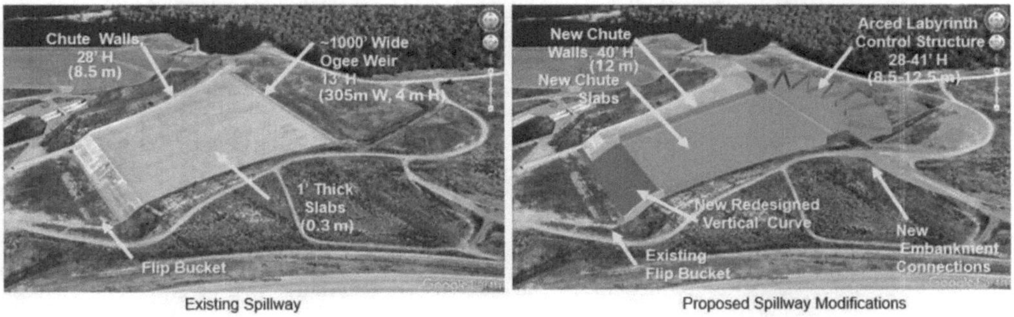

Figure 1. Existing spillway at Prado Dam (left) and proposed spillway features to be hydraulically evaluated and designed (right).

2 NEW LABYRINTH CONTROL STRUCTURE

2.1 *Design*

A new labyrinth weir was initially hydraulically designed using standard methodologies from Falvey (2003) and Tullis, et al (2013). The discharge rating curve was verified using both computational fluid dynamics (CFD) modeling and physical modeling. The new labyrinth weir control structure, as shown in Figure 2, will consist of:

- Replacing the existing 1000-ft (305 m) wide ogee crest with a new labyrinth weir structure 640-ft (195 m) wide arced, 7-cycle, multi-level labyrinth weir.
- Labyrinth weir walls 28 - 41 ft (8.5 - 12.5 m) high passing up to 30 ft (9 m) of head.
- Raising the spillway crest by 20 ft (6 m) for additional flood storage attenuation.

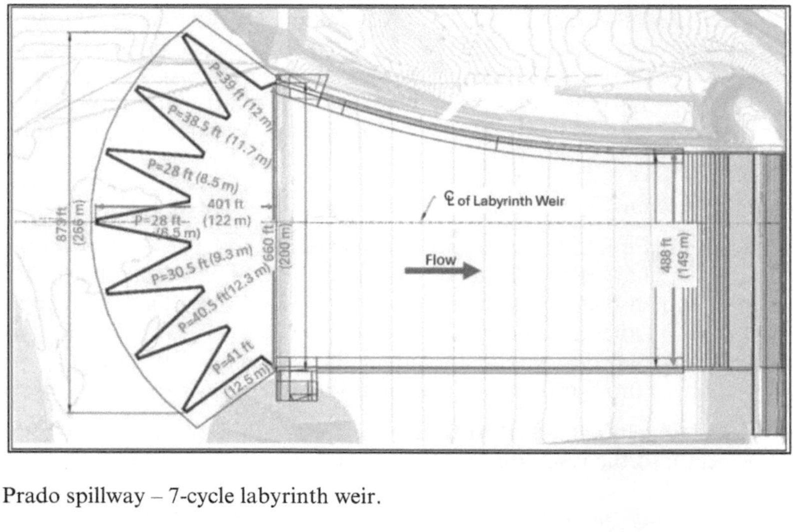

Figure 2. Prado spillway – 7-cycle labyrinth weir.

3 PHYSICAL MODELLING INVESTIGATIONS

Two hydraulic outcomes were modeled extensively in two scale physical hydraulic models to refine the hydraulic and structural design of the weir:

- Submergence of the labyrinth weir at high heads
- Hydrodynamic pressure fluctuations on the labyrinth weir crest at varying heads that affect the structural design of the weir walls and base.

The physical modeling efforts for the Prado Dam Spillway were conducted at Utah State University (USU) at the Utah Water Research Laboratory (UWRL) located in Logan, Utah.

3.1 *Submergence of the labyrinth weir*

Submergence of the lower labyrinth weir cycles was expected due to the flat weir apron, convergence of the arc and the mild chute slope. CFD model results showed that the submergence effects were more pronounced at the higher. This was investigated further using a physical hydraulic model for the spillway upgrade.

The General Model (see Figure 3) included a portion of the reservoir, the proposed labyrinth crest, chute, flip bucket, and a large downstream area to consider the downstream flow field, scour potential, waves, tailwater, etc. The purpose of this model was to confirm hydraulic performance and test design options, and provide detailed hydraulic information about the interaction between water, structure, and terrain.

A range of flow rates from were tested (75,000 - 613,000 cfs or 2,100 - 17,400 m^3/s) with reservoir elevation, downstream velocity and water surface elevation as measurements for submergence. A comparison of the measured reservoir elevation – discharge curve from the physical model was compared to the results from the CFD modeling. The results showed very similar results for the full range of flow rates, including at submerged conditions for the higher discharges, as shown in Table 1 and Figure 4.

Figure 3. 1:38 Scale physical hydraulic model at UWRL.

Table 1. Summary reservoir elevation – discharge for CFD model and physical model.

CFD Model Results		1:38 Scale Physical Model Results	
Elevation, ft (m)	Discharge, cfs (m^3/s)	Elevation, ft (m)	Discharge, cfs (m^3/s)
566.60 (172.5)	30,000 (850)	566.60 (172.5)	30,000 (850)
569.88 (173.7)	75,000 (2,100)	569.88 (173.7)	75,000 (2,100)
576.12 (175.6)	175,000 (5,000)	576.12 (175.6)	175,000 (5,000)
587.27 (179.0)	500,000 (14,200)	580.38 (176.9)	500,000 (14,200)
590.88 (180.1)	575,000 (16,300)	583.99 (178.0)	575,000 (16,300)
592.85(180.7)	630,000 (17,800)	586.94 (178.9)	630,000 (17,800)
595.47 (181.50)	680,000 (19,255)	591.86 (180.40)	680,000 (19,255)

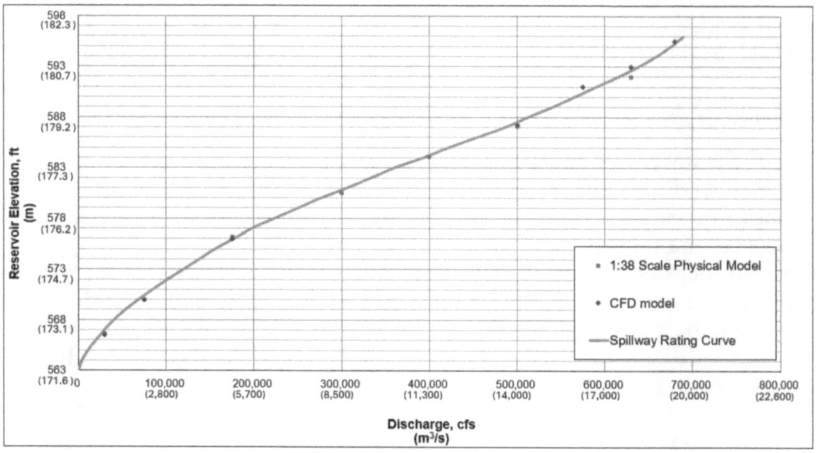

Figure 4. Reservoir elevation – discharge rating curve (CFD and physical model).

Physical modeling showed that the cycles begin to submerge at different discharges, as shown in Figure 5. Cycle 3, 4 and 5 submerge at approx. 300,000 cfs (8,500 m^3/s); cycle 6 and 7 at approx. 500,000 cfs (14,200 m^3/s); and cycles 1 and 2 did not submerge for the flows tested.

Figure 5. Prado Dam physical model photos at select discharges.

Hydraulic data from the physical and CFD models at discharges of 500,000 cfs (14,200 m^3/s) and 613,000 cfs (17,400 m^3/s) were utilized to develop H_d/H_o and H^*/H_o ratios for each cycle as shown in Figure 6 and Table 2. The unsubmerged total upstream head, H_o, was calculated using the discharge at each cycle from the CFD modeling, then using the labyrinth weir design procedure from Crookston and Tullis (2013), the height of the weir cycle was increased until a hypothetical H_o/P ratio of around 0.4 was achieved while at the same time iterating on the reservoir level until the submerged discharge was achieved.

The calculated ratios of H_d/H_o and H^*/H_o are summarized in Table 3 and plotted on the data from Tullis et al (2007) as shown in Figure 7. The results show that the Prado spillway data generally follows the research data from Tullis et al (2007), except for cycle 6 and cycle 7 at a discharge of 613,000 cfs (17,400 m^3/s), where their position is much higher than the research data. The difference in parameters is most likely a result of the variability in the upstream topography as opposed to an ideal condition in a laboratory flume model.

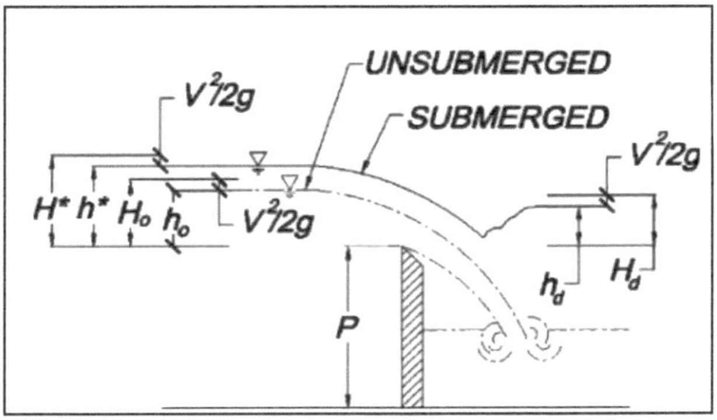

Figure 6. Submerged and free-flow weir parameters (Tullis et al 2007).

Table 2. Submerged and unsubmerged labyrinth weir calculations.

									Labyrinth design spreadsheet (US Customary Units)				
Cycle #	Cycle Crest El. (ft)	P (ft)	H^* (ft)	Q (cfs)	Tailwater Depth @ 3+35 (ft)	h_d (ft)	Average downstream velocity @ 11+00 (ft/s)	$V_d^2/2g$ (ft)	H_d (ft)	H_d/H_o	Reservoir El., Unsubmerged (ft)	H_o, Unsubmerged (ft)	H^*/H_o
Reservoir El. 587.3 ft; 500,000 cfs													
3	563.5	28.5	23.8	85,000	36.0	7.5	25	9.70	17.20	0.94	581.9	18.4	1.29
4	563.0	28.0	24.3	87,000	36.0	8.0	25	9.70	17.70	0.94	581.9	18.9	1.29
5	563.0	28.0	24.3	87,000	37.0	9.0	25	9.70	18.70	0.94	581.9	18.9	1.29
Reservoir El. 593.0 ft; 630,000 cfs													
3	563.5	28.5	29.5	110,000	40.0	11.5	28	12.17	23.67	0.99	587.3	23.8	1.24
4	563.0	28.0	30.0	118,000	40.0	12.0	28	12.17	24.17	0.99	587.3	24.3	1.23
5	563.0	28.0	30.0	118,000	40.0	12.0	28	12.17	24.17	0.99	587.3	24.3	1.23
6	573.5	38.5	19.5	52,000	40.0	1.5	28	12.17	13.67	1.13	585.6	12.1	1.61
7	574.0	39.0	19.0	51,000	40.0	1.0	28	12.17	13.17	1.12	585.8	11.8	1.61

									Labyrinth design spreadsheet (SI Units)				
Cycle #	Cycle Crest EL. (m)	P (m)	H^* (m)	Q (m_3/s)	Tailwater Depth @Sta 3+35 (m)	h_d (m)	Average downstream velocity @ 11+00 (m/s)	$V_d^2/2g$ (m)	H_d (m)	H_d/H_o	Reservoir EL., Unsubmerged (m)	H_o, Unsubmerged (m)	H^*/H_o
Reservoir EL 179.0 m; 14,160 m^3/s													
3	171.8	8.7	7.3	2407	11.0	2.286	7.62	2.96	5.24	0.94	177.36	5.61	1.29
4	171.6	8.5	7.4	2464	11.0	2.4384	7.62	2.96	5.40	0.94	177.36	5.76	1.29
5	171.6	8.5	7.4	2464	11.3	2.7432	7.62	2.96	5.70	0.99	177.36	5.76	1.29
Reservoir EL 180.7 m; 17,840 m^3/s													
3	171.8	8.7	9.0	3115	12.2	3.51	8.53	3.71	7.22	0.99	179.0	7.25	0.38
4	171.6	8.5	9.1	3341	12.2	3.66	8.53	3.71	7.37	0.99	179.0	7.41	0.38
5	171.6	8.5	9.1	3341	12.2	3.66	8.53	3.71	7.37	0.99	179.0	7.41	0.38
6	174.8	11.7	5.9	1472	12.2	0.46	8.53	3.71	4.17	1.13	178.5	3.69	0.49
7	175.0	11.9	5.8	1444	12.2	0.30	8.53	3.71	4.02	1.12	178.6	3.60	0.49

Table 3. Summary of calculated prado labyrinth H_d/H_o and H^*/H_o ratios.

	Q = 500,000 cfs (14,200 m^3/s)		Q = 630,000 cfs (17,800 m^3/s)	
Cycle #	H^*/H_o	H_d/H_o	H^*/H_o	H_d/H_o
Cycle 3	1.29	0.94	1.24	0.99
Cycle 4	1.29	0.94	1.23	0.99
Cycle 5	1.29	0.99	1.23	0.99
Cycle 6	-	-	1.61	1.13
Cycle 7	-	-	1.61	1.12

3.2 *Hydrodynamic pressure fluctuations on the labyrinth weir*

Limited available research shows that high head labyrinth weirs can experience significant fluctuating negative pressures at the weir crest that increase the loading on weir walls,

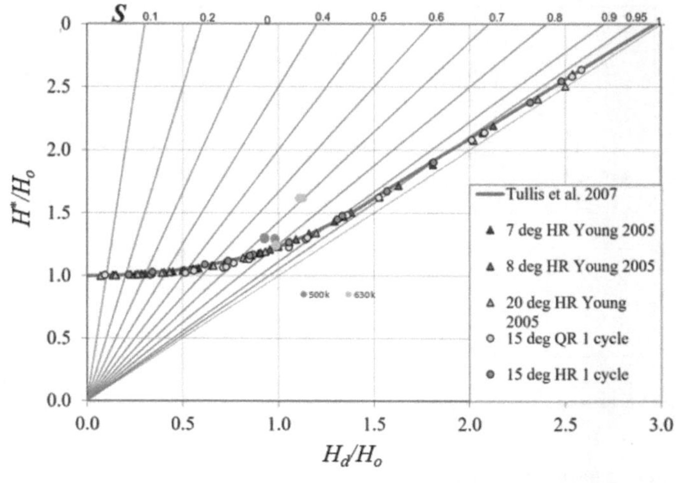

Figure 7. Calculated H_d/H_o and H^*/H_o ratios compared to Tullis et al (2007) research for a submerged, unvented, labyrinth weir.

resulting in high hydrodynamic loads that should be considered during structural design. Therefore, a single cycle physical model was constructed to investigate this phenomenon, as shown in Figure 8. The physical modelling efforts for the Prado Dam Spillway were conducted at UWRL in parallel with the general model.

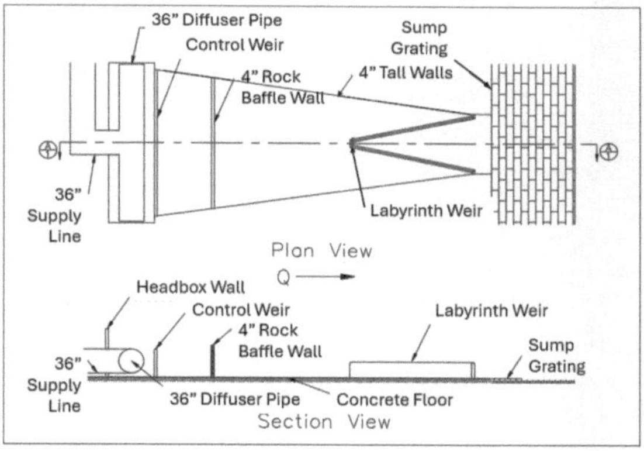

Figure 8. Single-cycle labyrinth weir model setup.

Piezometric and dynamic pressure readings were taken at 5 locations along the weir cycle, with 4 pressure taps on the downstream face at each location, for a total of 20 pressure taps, as shown in Figure 9a. Dynamic pressures were recorded for a minimum of 300 seconds (model scale) at 100 kHz to estimate the min, max, and average pressure. Observations of the nappe and flow in the downstream section of the labyrinth cycle were noted, including aeration of the nappe, flow shifting from sided, and surging, as seen in Figure 9b.

Key observations from the measured pressures are summarized below:

- Nappe aeration and stability observations were:
  - Clinging flow generally observed at 300,000 cfs (8,500 m^3/s) ($H_o/P = 0.11$)
- Partially aerated nappe at 400,000 cfs (11,300 m^3/s) ($H_o/P = 0.20$)
  - Unstable partially aerated nappe and surging at 500,000 cfs (14,200 m^3/s) ($H_o/P = 0.27$)

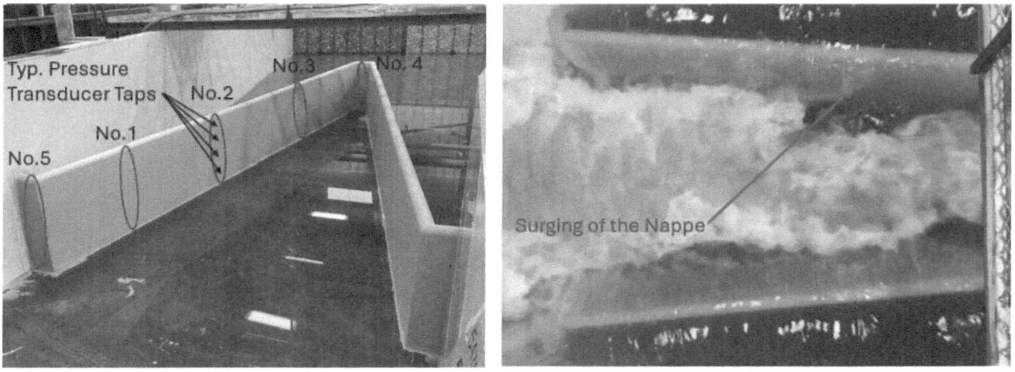

(a) (b)

Figure 9. (Left) Single cycle model and pressure tap locations. (Right) Example of surging of the nappe at the downstream end of the cycle.

  - Submerged nappe was observed at 613,000 cfs (17,400 m^3/s) ($H_o/P = 0.39$)

- High reservoir head and low pressures at the nappe occurs at (refer to Figure 12):
  - Location 4 at 300,000 cfs (8,500 m^3/s) of total spillway flow
  - Location 1, 2, and 5 at 400,000 cfs (11,300 m^3/s) of total spillway flow
  - Location 3 at 500,000 cfs (14,200 m^3/s) of total spillway flow

- The ratio of the minimum pressure at the crest, p_{min}, and the total head over the weir crest, H_o, showed that at the flows of 400,000 cfs (11,300 m^3/s) and 500,000 cfs (14,200 m^3/s), the resulting negative pressure at the crest were 2 to 3 times higher in magnitude.
- Dynamic pressure averages closely matched the measured piezometric pressures in the model (not shown in graph) and closely matched the CFD model results.
- The minimum dynamic pressure near the crest of the weir wall were 5 ft (1.5m) to 18 ft (5.5 m) lower than the measured average pressures.

Table 4. Minimum pressures at weir crest.

Pressure Tapping Location	Discharge w/ minimum pressure at crest, cfs (m)	Total head, H_o,ft (m)	Minimum pressures at crest, p_{min} (ft)	p_{min}/H_o
1	400,000 (11,300)	7.94 (2.42)	-13.78 (4.20)	-1.74
2	400,000 (11,300)	7.94 (2.42)	-14.76 (4.50)	-1.86
3	500,000 (14,200)	11.09 (3.38)	-15.09 (4.60)	-1.33
4	300,000 (8,500)	4.27 (1.61)	-11.48 (3.50)	-2.69
5	400,000 (11,300)	7.94 (2.42)	-18.04 (5.50)	-2.27

Nappe aeration and stability observations included:

- Clinging flow at the weir walls was generally observed $H_o/P = 0.11$
- A partially aerated nappe was observed at $H_o/P = 0.20$
- An unstable partially aerated nappe and surging was observed at $H_o/P = 0.27$
- A submerged nappe was observed at $H_o/P = 0.39$
- Surging of the nappe and in the downstream areas resulted in larger negative pressures due to the oscillating tailwater depths.

Pressure profiles for the labyrinth weir apexes and weir walls were developed for design with -15 ft (-4.6 m) and -18 ft (-5.5 m) pressure at the crest, respectively.

4 CONCLUSIONS

The hydraulic design of the proposed replacement of the Prado Dam Spillway includes the design of a new labyrinth weir control structure. The hydraulic design process, with the aid of CFD and two physical models resulted in the following observations and conclusions:

- The labyrinth weir generally follows standard design guidance for both free flowing and submerged conditions. Additional hydraulic analysis showed a non-symmetric approach flow, further reducing efficiency and exacerbating submergence. Practitioners should be careful when calculating the flow over a labyrinth weir if approach conditions are not ideal, as modelled in a laboratory.
- The physical modelling of the single cycle labyrinth weir showed that the measure dynamic pressure minimums on the downstream wall faces were lowest at H_0/P ratios of 0.20 to 0.30 and are generally 5 ft to 18 ft (1.5 m to 5.5 m) lower than hydrostatic pressure. Pressure profiles for the labyrinth weir apexes and weir walls were developed for design with -15 ft and -18 ft (-4.6 m to -5.5 m) pressure at the crest, respectively.

4.1 *Endnote*

- Of interest to designers of labyrinth weirs, for the Prado labyrinth weir, at the H_0/P ratios of 0.20 and 0.27, the magnitude of negative pressure under nappe at the crest was 2 to 3 times the incoming total head due to oscillating tailwater depths in the cycle areas.
- It is recommended that designers consider the hydrodynamic contributions to the loading diagram as this could lead to an unconservative structural design. Conversely, the assumption of -33.8 ft (-1 atm) at the top portion of the weir wall could lead to an overly conservative structural design. Further research should be conducted to assist hydraulic structure engineers in the development of pressure profiles on labyrinth weirs to ensure adequate structural design of the labyrinth weir features.

ACKNOWLEDGEMENTS

This work was conducted in collaboration with Utah State University (USU) at the Utah Water Research Laboratory (UWRL) located in Logan, Utah, USA. We appreciate their efforts in the physical modelling and extensive data collection for this important project.

REFERENCES

Crookston, B.M. and Tullis, B.P. (2013). *Hydraulic Design and Analysis of Labyrinth Weirs. I: Discharge Relationships.* J. Irrig. Drain. Eng., 139(5), 363–370.

Crookston, B.M., and Tullis, B.P. (2013). *Hydraulic design and analysis of labyrinth weirs. II: nappe aeration, instability, and vibration.* J. Irrig. Drain. Eng., 139(5), 371–377.

Crookston, B.M., and Tullis, B.P. (2012). *Arced labyrinth weirs.* J. Hydraul. Eng., 2012, 138 (6): 555–562.

Crookston, B.M., and Tullis, B.P. (2012). *Discharge efficiency of reservoir-application-specific labyrinth weirs.* J. Irrig. Drain Eng., 2012, 138(6): 564–568.

Crookston, B.M., et al (2012). *Hydraulic performance of labyrinth weirs for high headwater ratios.* Proc., 4th IAHR International Symposium on Hydraulic Structures, Porto, Portugal.

Falvey, Henry T., (2003). *Hydraulic Design of Labyrinth Weirs*, ASCE Press, Reston Virginia.

Tullis, B.P. et al. (2020). *Geometric effects on discharge relationships for labyrinth weirs.* J. Hydraul. Eng., 2020, 146(10): 04020066.

Tullis, B.P; Young, J.C. and Chandler, M.A. (2013). *Head-Discharge Relationships for Submerged Labyrinth Weirs.* J Hydraul. Eng., 133(3), 248–254.

U.S. Army Corps of Engineers. (2024). *DRAFT Prado Dam Spillway modifications – control structure hydrodynamic loading.* Memorandum for Record.

Labyrinth and Piano Key Weirs IV – Erpicum, Crookston & Paxson (Eds)
© 2026 The Author(s), ISBN 978-1-041-10527-5

Research and application of Piano Key Weirs in Vietnam (2025)

M. Ho Ta Khanh
VNCOLD - CFBR Independent Consultant, Vietnam - France

Dau Xuan Thuy
Power Engineering Consulting Joint Stock Company 2 (PECC2), Ho Chi Minh, Vietnam

ABSTRACT: Since the introduction of Piano Key Weirs (PKW) by F. Lempérière and A. Ouamane in 2003, many such structures have been built for irrigation, hydroelectricity and water regulation due to their outstanding efficiency. In Vietnam, PKWs have also attracted significant attention. There have been several laboratory hydraulic model studies carried out to test the discharge capacity, flow characteristics, and energy dissipation downstream of the spillway. Vietnam had also issued PKW design standard in 2018 to ease its application. This paper describes shortly 18 PKWs in operation in Vietnam among other existing structures. Due to lack of space, the corresponding drawings and details are not provided but can be gained from the authors.

1 RESEARCH AND DEVELOPMENT

In 2004-2006, the Ho Chi Minh City University of Technology (HCMUT) carried out physical model tests focusing on the discharge capacity of different types of PKW and a labyrinth weir. In 2007-2010, the Ministry of Agriculture and Rural Development (MARD) - Hydraulic laboratories and Southern Institute of Water Resources Research (SIWWR) carried out studies concerning the dissipation of flow energy at the spillway toe. In 2014, SIWRR carried out additional research related to the use of PKW for irrigation projects in Vietnam.

In 2018, the document titled "TCVN 12262:2018 – Hydraulic Structures – Spillway – Hydraulic Calculation of PKW" has been issued. It provides comprehensive guidelines for the hydraulic design and operation of PKWs. The guidelines include the following items: Scope of Application, Definitions and Classifications, Design Guidelines, Hydraulic Calculations, Steps for Designing a New PKW, Upgrading Existing Structures, Energy Dissipation and Downstream Protection, Management and Maintenance.

2 PKW UNDER OPERATION IN VIETNAM

Up to now, PKW has been used in many irrigation and hydropower projects, including new construction projects and reservoir rehabilitation. Most of the PKWs are implemented on small and medium size reservoirs. In Vietnam, there are many PKWs in operation and other projects are under design and construction. In the following, 18 of the projects that have PKW in operation in Vietnam are briefly described. Most of the PKWs are type A (U/S and D/S overhangs), type B (U/S overhang only) and less frequently type C (D/S overhang only).

2.1 *Dak Mi 4B*

Location: 15°27′32″N 107°54′44″E. The Dak Mi 4B hydroelectric plant, located in Quang Nam province, was built in 2013. The project components include the main dam, a PKW spillway, a water intake, a pressure tunnel, and a powerhouse.

DOI: 10.1201/9781003655541-17

This design is a very special combination of labyrinth weir (1st stage) and PKW (2nd stage) due to the required increase of the spillway capacity during the dam construction. The PKW width is 55.0 m and the total discharge capacity is 641 m^3/s. Figure 1 shows Dak Mi 4B PKW in operation.

Figure 1. Dak Mi 4B PKW in operation.

2.2 *Dak Mi 3*

Location: 15°19′09″N 107°49′32″E. The Dak Mi 3 is located in Quang Nam province. The spillway, designed as a PKW with an initial width of 144.0 m, is in operation since May 2017. From October 2020 to August 2021, the powerplant had to stop operation due to damage caused by an extreme flood during the October 2020 storm. A 28.0 m long section of the PKW was demolished and replaced by two orifice sluices. Thus, the PKW width has been reduced to 116 m (Figure 2). Maximum discharge capacity is 8,777 m^3/s, of which 5,354 m^3/s are released by the PKW.

Figure 2. Downstream view of Dak Mi 3 PKW after rehabilitation.

2.3 *Dak Rong 3*

Location: 16°33′38″N 106°57′13″E. Dak Rong 3 hydropower project was built in 2013. It is located in Quang Tri province. The Dak Rong 3 spillway is a type A PKW (Figure 3) with a total width across the river equal to 200.0 m. The maximum discharge of the check flood (Probability of occurrence of 0.2%) is 6,550 m^3/s and the design flood (Probability of occurrence of 1%) is 4,830 m^3/s.

2.4 *Da Dang 3*

Location: 11°40′03″N 108°14′00″E. Da Dang 3 hydropower project was built in 2016. The project is in Lam Dong, and has a capacity of 12 MW. The total volume of the reservoir is 7.6 hm^3. The main barrage is a PKW type B (Figure 4) with a width of 144 m and a maximum discharge capacity of 7,280 m^3/s.

Figure 3. Dak Rong 3 PKW during a small flood.

Figure 4. Downstream view of the Da Dang 3 PKW.

2.5 *Van Phong*

Location: 13°56′23"N 108°52′29"E. The Van Phong barrage (Figure 5) is part of a large irrigation project. The barrage was built from 2010 to 2015 and has a total width of 474 m across the river. It consists of two PKWs (181 m on the right bank and 121 m on the left bank) and a gated spillway in the center part of 172 m with 10 radial gates. The 1000-year flood is 15,350 m^3/s for a maximum discharge over the PKW of 8,700 m^3/s.

Figure 5. Van Phong PKW releasing a small discharge.

2.6 *Xuan Minh*

Location: 19°52′52"N 105°20′20"E. The Xuan Minh Hydropower project (Figure 6), with a capacity of 15MW, is located on the Chu River, Thuong Xuan district, Thanh Hoa province.

The spillway consists of two 10x10 m gates on the left bank of the river and a 150 m wide PKW on the right side of the gates.

Figure 6. Downstream view of Xuan Minh project just after construction.

2.7 *Dak Trua 1*

Location: 14°42′55"N 108°01′49"E. Dak Trua 1 HPP was built in 2020 in Kon Tum province. The spillway is a PKW with downstream steps for energy dissipation (Figure 7). The PKW consists of 5 units with a total width of 25 m. The maximum flood discharge is 317 m^3/s, with a head of 1.54 m.

Figure 7. Dak Trua 1 PKW.

2.8 *Dak Trua 2*

Location: 14°42′47"N 108°00′14"E. Dak Trua 2 HPP was built in 2021 in Kon Tum province. The PKW spillway (Figure 8) consists of 9 units with a total width of 45 m and a maximum discharge capacity of 543 m^3/s with a head of 1.47 m.

Figure 8. Upstream view of Dak Trua 2 PKW.

2.9 *Dak Nghe*

Location: 14°36′16″N 108°13′51″E. Built in 2023, the Dak Nghe Hydropower Project is located on the main stream of Dak Nghe River in Kon Tum province. A PKW (Figure 9) was selected to increase the discharge capacity of the spillway. The design flood (Probability of occurrence of 1.5%) is 2,410 m^3/s and the check flood (Probability of occurrence of 0.5%) is 3,360 m^3/s for a head of 3.44 m.

Figure 9. Dak Nghe PKW.

2.10 *Dak Sor 2*

Location: 12°26′57″N 107°43′27″E. The Dak Sor-2 hydropower project, built in 2021, is located in Dak Nong. The dam (Figure 10) consists of 10 PKW units with a total width of 60 m. The spillway is designed for a design flood (Probability of occurrence of 1.5%) of 716 m^3/s and a check flood (Probability of occurrence of 0.5%) of 913 m^3/s.

Figure 10. Dak Sor 2 reservoir and PKW.

2.11 *Buon Triet*

Location: 12°22′17″N 108°04′34″E. The Buon Triet PKW (Figure 11) was built in 2019 as part of a dam upgrading and repair project. It is located in Buon Triet commune, Lak district, Dak Lak province. The purpose of the project was to increase the capacity of the spillway. The old rectilinear spillway with a Creager weir was demolished and replaced by a PKW. The PKW with 15 units is located on the left bank of the earthfill dam.

2.12 *Dong Do 1*

Location: 21°08′12″N 106°34′18″E. Built in 2019, the Dong Do 1 spillway is part of the Vanh Mam reservoir located in Binh Khe commune, Dong Trieu town, Quang Ninh province. The project involved the repair and upgrading of the Dong Do 1 reservoir, including demolition of the old ogee crest spillway and its replacement with a 27 m wide PKW (Figure 12).

Figure 11. Buon Triet PKW.

Figure 12. Dong Do 1 PKW.

2.13 *Dam Ha Dong reservoir*

Location: 21°24′04”N 107°34′04”E. Dam Ha Dong reservoir was built in 2010 in Dam Ha district, Quang Ninh province. In 2018, the project was upgraded and repaired, including the addition of a PKW on the left side of the existing gated spillway (Figure 13). The PKW width is 54 m.

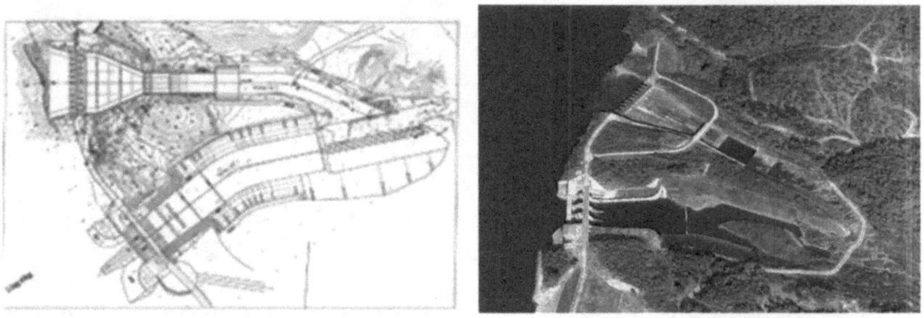

Figure 13. Dam Ha Dong PKW: plan view (left) and aerial view (right - Google Earth, © 2025 Airbus).

2.14 *Dong Nai 1*

Location: 11°39′49”N 108°12′23”E. The Project was built in 2022 in Lam Dong provice. It has a capacity of 15MW with a yearly output of 43.76 GWh. The spillway includes a type B PKW (Figure 14) with a total width of 108 m and a 10 m wide gate for sand flushing. The maximum flood discharge capacity of the spillway is 4,810 m^3/s with a head 4.77m.

Figure 14. Downstream view of Dong Nai 1 PKW just after construction.

2.15 *Canh Tang reservoir*

Location: 20°30′12"N 105°28′32"E. The Canh Tang project was built in 2022. It consists of a about 91 hm^3 reservoir for a large irrigation project in Hoa Binh province. The main dam is 53.23 m high and 850.5 m long; the auxiliary dam is 36.5 m high and 160.5 m long.

For safety reasons, PKWs were added on both side of the two radial gates (Figure 15) in order to mitigate the consequences of jamming of one of the gates.

Figure 15. Upstream view of the Canh Tang PKW just after construction.

2.16 *Thuong Song Ong*

Location: 11°46′18"N 108°47′56"E. The project was built in 2016 in Ninh Thuan province. The spillway is a long PKW type B 180 m wide (Figure 16). The maximum discharge capacity is 1,674 m^3/s with a head of 1.71m.

Figure 16. Thuong song ong PKW.

2.17 *Dak Rong 4*

Location: 16°35′46"N 106°52′55"E. The Dak Rong 4 hydropower plant is located in Quang Tri Province. The project with a 28 MW capacity was commissioned in 2019. The spillway is a 283 m wide type B PKW (Figure 17).

Figure 17. Downstream view of Dak Rong 4 PKW.

2.18 *My Son HPP*

Location: 11°43′40.″N 108°48′44″E. The My Son hydropower project was built in 2023. It is located in Ninh Thuan province. The spillway is a 350 m wide type A PKW (Figure 18). This may be the widest PKW constructed in Vietnam.

Figure 18. My Son PKW.

2.19 *Summary of main characteristics of the PKWs*

The Table 1 below summarizes the main characteristics of the 18 PKWs in operation in Vietnam, including PKW type and width, dam height, design head, discharge capacity, and the main reason for applying a PKW solution to the project.

Table 1. Summary of characteristics of the PKWs in Vietnam.

No.	Project	Type	Width (m)	Dam height (m)	Head (m)	Discharge of PKW (m^3/s)	Gated spillway or sluice	Reasons for applying a PKW solution
1	Dak Mi 4B	B	55	15	3.11	641		Alternative to ogee spillway to increase reservoir water level
2	Dak Mi 3	A	116	33	4.96	5,354	2 gates (9.5x12)	Alternative to ogee spillway to increase reservoir water level
3	Dak Rong 3	A	200	20	5.00	6,650		Alternative to gated spillway to reduce costs
4	Da Dang 3	B	144	23	5.70	7,280		Alternative to gated spillway to reduce costs
5	Van Phong	A	302	16	5.00	8,700	10 gates	Alternative to gated spillway to reduce costs
6	Xuan Minh	B	150	25	8.50	9,700	2 gates (10x10)	Alternative to gated spillway to reduce costs

(*Continued*)

Table 1. (*Continued*)

No.	Project	Type	Width (m)	Dam height (m)	Head (m)	Discharge of PKW (m^3/s)	Gated spillway or sluice	Reasons for applying a PKW solution
7	Dak Trua 1	A	25	11	1.54	317		Alternative to ogee spillway to increase reservoir water level
8	Dak Trua 2	A	45	19	1.47	543		Alternative to ogee spillway to increase reservoir water level
9	Dak Nghe	B	117	13	2.94	3,360		Alternative to ogee spillway to increase reservoir water level
10	Dak Sor 2	B	60	12	1.63	913		Alternative to ogee spillway to reduce the length of concrete dam
11	Buon Triet	A	50	9				Upgrade – Replacement of an ogee spillway to increase discharge capacity
12	Dong Do 1	A	27	10	1.17	97		Upgrade – Replacement of an ogee spillway to increase discharge capacity
13	Dam Ha Dong	A	54	8	2.98		3 gates	Upgrade – Additional structure to increase discharge capacity
14	Dong Nai 1	B	108	14	5.40		1 gate (10x12)	Alternative to gated spillway to reduce costs
15	Canh Tang	B	24	15			2 gates	For safety reasons in case of jamming of one of the radial gates
16	Thuong Song Ong	B	180	8	1.85	1,674	2 gates (2.5x5.4)	Alternative to ogee spillway to increase reservoir water level
17	Dak Rong 4	B	283	15				Alternative to gated spillway to reduce costs
18	My Son	A	350	14	2.27		2 gates (7x7)	Alternative to ogee spillway to increase reservoir water level

3 CONCLUSION

When a fixed crest spillway with high discharge capacity under relatively low heads is necessary, the labyrinth weirs are sometimes adopted. However, such structures often require large footprints, making it economically impractical to place them on the crest of high concrete dams or in rivers with thick alluvial bed.

The PKW solution, a kind of labyrinth with upstream and/or downstream overhangs and sloped keys, often addresses this issue by decreasing footprint requirements.

In Vietnam, PKWs have recently been very successfully implemented on dams in rivers, as demonstrated but the 18 examples presented in this report.

Vietnam has had early PKW design standards which have created favorable conditions for practical application, especially for small and medium-sized projects, because there is no need to carry out hydraulic model tests with cost and delay savings.

Up to now, there are nearly a few dozen PKWs operating at hydroelectric and irrigation reservoirs in Vietnam, including newly built and upgraded reservoirs. Some others are in the process of design and construction. During the operation of existing PKWs, some lessons have been drawn, namely the combination of PKW, gated spillway and sand discharge sluice to ensure an optimal operation, maintenance and repair of the future projects.

Most reasons for applying the PKW is to raise the normal water level of the reservoir while ensuring that the maximum flood level remains unchanged. This approach prevents an increase in the reservoir's flooded area, maintaining ecological balance and minimizing environmental impact. Additionally, some projects implement PKW to enhance flood discharge capacity, addressing the challenges of recent climate change and occurrence of unusually intense rainfall and flooding. This adaptation helps improve water management efficiency and strengthens infrastructure resilience against extreme weather events.

REFERENCES

Lempérière, F & Ouamane, A. 2003. "The Piano Keys Weir: a new cost-effective solution for spillway". The International Journal on Hydropower & Dams, Issue Four.

Laugier F. 2007. "Design and construction of the first Piano Key Weir (PKW) spillway at the Goulours dam", International Journal of Hydropower and Dams 14 (5), 94–101

Ho Ta Khanh. 2017. "History and Development of Piano Key Weirs in Vietnam (2004–2016)". Labyrinth and Piano Key Weirs III - PKW 2017, CRC Press/Balkema.

Ho Ta Khanh, M, T. Chi Hien & N. Thanh Hai. 2011. "Main results of the P.K weir model tests in Vietnam (2004 to 2010)". Labyrinth and Piano Key Weirs-PKW 2011. CRC Press/Balkema.

Ho Ta Khanh, M, Dinh Sy Quat & D. Xuan Thuy. 2011. "P.K weirs under design and construction in Vietnam (2010)". Labyrinth and Piano Key Weirs-PKW 2011. CRC Press/Balkema.

S. Erpicum. et al. 2017. "Hydraulics of Piano Key Weirs: A Review". Labyrinth and Piano Key Weirs III - PKW 2017, CRC Press/Balkema.

Reynolds, P.2011. "PK Weirs in Vietnam". International Waterpower & Dam Construction. 26 January 2011.

Labyrinth and Piano Key Weirs IV – Erpicum, Crookston & Paxson (Eds)

Honey, I shrunk the piano key weir: A case study of Burton Pond Dam

D. Bisio
Schnabel Engineering, Chadds Ford, Pennsylvania, USA

ABSTRACT: Between a low-head tidal environment, razor-thin hydraulic margins, and a long list of site restrictions, the Burton Pond Dam Rehabilitation project offered the chance to pursue a one-of-a-kind solution: a miniature stainless steel piano key weir as part of a new mixed active-passive spillway. This case study explores how established hydraulic designs were modified specifically for the site and verified with computational flow dynamics; how the innovative cantilevered geometry was analyzed structurally with finite element modeling; and how detailing, fabrication and constructability were considered throughout the design process. From conceptual-level design submissions to the final construction walkthrough, the whole project was made possible and successful by embracing Building Information Modeling software.

1 INTRODUCTION

1.1 *Site description*

Burton Pond Dam is located roughly 8 km southwest of Rehoboth Beach in Sussex County, Delaware, USA, along a tributary to Burton Prong, a brackish tidal inlet that flows into Rehoboth Bay. The earth embankment dam is 206 m in length but only 2.4 m in average height. Elevations within the project limits ranged from EL 0.0 m to 3.7 m. Despite its small stature, the upstream reservoir has a drainage area of 23.6 sq. km (Schnabel Engineering 2021). Additional hydrologic information related to the dam can be found in Table 1.

Table 1. Existing hydrologic conditions information for Burton Pond Dam.

	Elevation*	Storage Volume	Pool Area
Pool Level	m	m^3	km^2
Normal Pool	2.01	181,815	0.186
Top of Dam	2.94	471,807	0.481

* Elevations reference to NAVD88.

Delaware Route 24 (SR-24, John J. Williams Highway), a state highway and designated evacuation route, runs across the length of the dam and functions as its crest. Positioned near the northern abutment, a cast-in-place, two-cell (each cell being 3.7 m wide and 1.8 m high) concrete box culvert, designated as BR 3-709 by the Delaware Department of Transportation (DelDOT), serves as the dam's only spillway. Previously, 8 sets of 0.9-m wide wooden stoplog gates were mounted to the upstream side of the culvert and were used to control pool level and discharge through the structure. DelDOT is the owner and operator of the dam, spillway and SR-24 (Schnabel Engineering 2021).

DOI: 10.1201/9781003655541-18

1.2 *Project purpose and requirements*

Schnabel Engineering, LLC (Schnabel) was contracted by DelDOT to design rehabilitation measures for Burton Pond Dam. Under its previous configuration, Burton Pond Dam was unable to pass the spillway design flood (SDF), defined as the 100-year frequency storm, without active operation of the deteriorating stoplog gates. The dam was estimated to overtop from the 10-year frequency storm without gate operation (Schnabel Engineering 2015). The wooden stoplog boards leaked and the steel I-beams acting as guides were severely corroded (Pennoni Associates, Inc. 2018). Timber beams had been installed on the upstream face of each gate and each board had been fastened to the timber beam so that all of the boards moved as a single unit. The gates were operated from the road shoulder of the heavily trafficked SR-24, often in poor weather conditions. The previous spillway was unsafe, ineffective, and out of compliance with state regulations.

The proposed design needed to meet hydraulic requirements, as outlined below, from both DelDOT and the Delaware Department of Natural Resources and Environmental Control (DNREC), whose Dam Safety Program oversees, inspects, and regulates dams in the state (Schnabel Engineering 2021).

1. Pass the SDF (100-year frequency storm) with a maximum water surface elevation less than the Top of Dam elevation minus wave run-up height from the SDF (5103 Delaware Dam Safety Regulations, Code § 8.0).
2. Pass the 50-year frequency storm with at least 0.3 m of freeboard from the lowest edge of pavement.
3. Maintain existing Normal Pool water surface elevation (WSEL) in the upstream reservoir.
4. Increase passive spillway capacity relative to existing conditions.
5. Provide a low-level outlet (LLO) that will "allow draining of the reservoir in a reasonable amount of time", a rate that was determined to be the top 0.6 m of the reservoir (55% of Burton Pond's total storage) plus 70% of the maximum monthly baseflow (5103 Delaware Dam Safety Regulations, Code § 8.3.1).

Additionally, a new guardrail system and roadway rehabilitation measures were required for compliance purposes, along with armoring of the embankment slopes to improve erosion protection and slope stability.

1.3 *Site restrictions*

Potential designs to meet the hydraulic requirements were severely restricted by the low-head environment present at Burton Pond Dam. Normal Pool in the reservoir was to be maintained at EL 2.01, while the low chord of the culvert on the upstream face was at EL 2.16, leaving roughly 15 cm of reservoir attenuation before submergence of the structure. The lowest edge of pavement was surveyed at EL 2.64; wave run-up from the SDF was estimated to be 0.44 m. Thus, the 50-year frequency storm and SDF had to be passed with just 0.33 and 0.50 m of allowable head, respectively (Schnabel Engineering 2021). A proposed passive spillway would need to be hyper-efficient under low-head conditions, with a crest positioned as far upstream of the culvert as possible.

Utilities at the site also constrained design and construction means and methods. Subsurface exploration revealed a 138 kV transmission line 1.8 m upstream of the culvert face and approximately 3.0 m below the invert slab of the structure. Roughly 8.5 m upstream of the transmission line, a regional force main sewer line runs parallel to the dam through the reservoir at an unknown depth (Schnabel Engineering 2021). Neither line could be decommissioned or relocated, even temporarily. Thus, the ability to use piles or sheeting for a new spillway foundation, embankment protection, seepage control or control of water during construction was effectively eliminated. Overhead electric and communications lines directly above the spillway were able to be relocated roughly 4.0 m upstream of the culvert face but only 6.1 m of vertical clearance above the roadway could be guaranteed (Schnabel Engineering 2024). Any proposed structure or element that required picking, lifting, and placing would be limited in size.

SR-24 is a two-lane, state evacuation route and major artery near Delaware's southern beaches. During peak season (April 30 to October 1) both travel lanes are required to be open from 6 am until 9 pm; a single lane closure was permissible at night. During the off-season, a full road closure and detour were acceptable from 7 pm until 8 am. Additionally, no in-water work was permitted within the reservoir or downstream tidal channel from March 1 to July 31 to minimize impacts to endangered and spawning fish species (Schnabel Engineering 2021). A variance to allow in-water work until April 1 was eventually granted for the project. These temporal restrictions meant that the proposed spillway structure needed to be quick to install, cure, or set. Construction equipment would be limited to the width of a single lane, or else installation would need to take place at night during the colder months, or from a barge in the reservoir.

2 ANALYSIS AND DESIGN

2.1 *Proposed design*

It became clear early on in the alternatives analysis phase of the project that Burton Pond Dam required a unique solution to satisfy the presented requirements and constraints. Eventually, a mixed-use spillway design (see Figure 1) was proposed by Schnabel and approved by DelDOT.

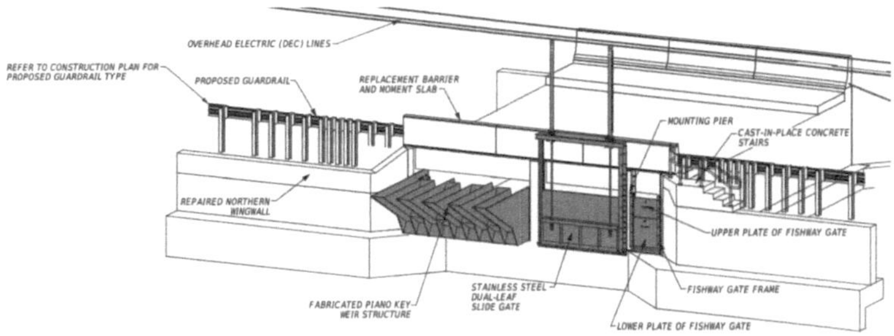

Figure 1. Proposed spillway design isometric (Schnabel Engineering 2024).

A "miniature" piano key weir (PKW) would extend over 2.7 m out into the reservoir from the upstream culvert face, providing passive hydraulic capacity. The cantilevered shape, a slightly modified version of a Type B PKW, offered several benefits: high-efficiency discharge under low-head conditions, sufficient nappe clearance and no need for modifications to the existing concrete culvert. Fabricated entirely from stainless steel plate ranging from 1.27 to 2.54 cm thick, the PKW structure would be attached to the interior of the existing left culvert bay with stainless steel epoxy anchors and grouted in place; no deep foundation elements were required (Schnabel Engineering 2021).

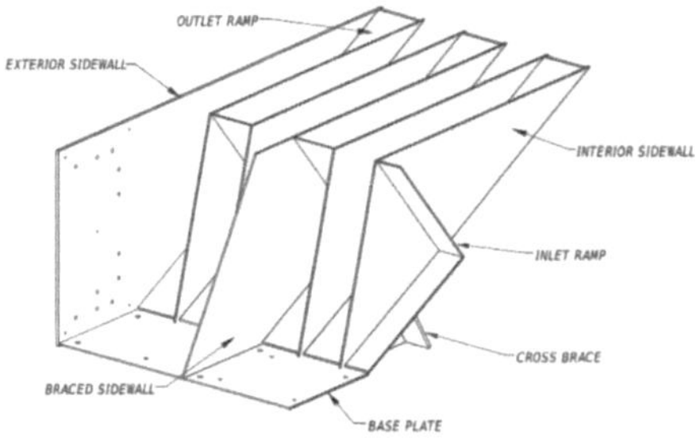

Figure 2. Section isometric view of proposed PKW (Schnabel Engineering 2024).

In the right culvert bay, a 2.4-m wide, stainless-steel split-leaf slide gate would be installed, providing active discharge capacity for larger storms and doubling as the reservoir's LLO. The gate would be more safely operated from a new concrete platform off of the road on the right up-stream wingwall. A set of removable plates on the right-most side of the culvert structure would allow for installation of a fishway or ladder in the future, and regulation of reservoir's normal pool level. The remainder of this case study focuses on the PKW.

2.2 *Hydraulic analysis*

The discharge rating curve of the cantilevered PKW was calculated using methodology presented by Khassaf et al (2016) due to the proposed geometry's resemblance to the Type B configuration. There was a concerted effort to choose geometric ratios that fell within the limits of the conducted empirical study, and that maximized discharge efficiency for the proposed design when feasible. Inlet (W_i) and outlet key (W_o) widths were set equal to each other at 0.38 m. The overhang of the outlet key (B_o) was set to 1.75 m which, along with a base width (B) of 0.76 m, produced B_o/B and B/P ratios that were found to maximize low-head discharge capacities (Khassaf et al. 2016). With 5 outlet and 4 inlet keys, the proposed PKW had a total weir crest length (L) of 28.58 m over a total transverse width (W) of 3.58 m (Schnabel Engineering 2021). Khassaf et al (2016) found that increasing the L/W ratio up to 8 increased hydraulic efficiency at low heads but had diminishing and even negative returns at higher H/P ratios. The design of the PKW emphasized discharge capacity under low-head conditions because it was acknowledged that the active portion of the spillway i.e. the dual-leaf slide gate, would need to be operated to pass the 50-yr frequency and SDF storms. There was not enough allowable head to pass larger inflows relying on the passive PKW capacity alone, given the other site constraints. Maximizing low-head efficiency meant less frequent operation of the gate would be needed, which reduces operational cost and safety risks for DelDOT personnel. A 10% reduction in discharge capacity was initially estimated due to slight variations between the proposed geometry and standard Type B weir configuration as well as impacts to flow when the adjacent gate was opened (Schnabel Engineering 2021).

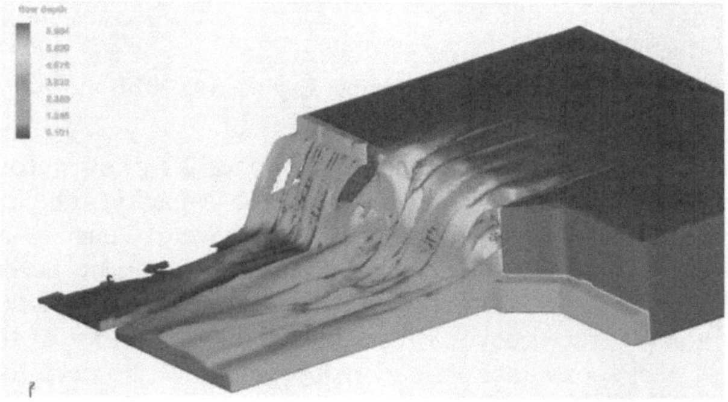

Figure 3. CFD model output – SDF with slide gate partially opened.

A computational fluid dynamics (CFD) model was developed in FLOW-3D v12.0 to validate the PKW's discharge rating curve. Geometry of the PKW, left culvert cell, half of the slide gate (in both opened and fully seated positions) and right culvert cell were imported. The initial water surface boundary was set at EL 2.51, the maximum allowable WSEL during the SDF. A convergence study was conducted with 3.04, 7.62, and 9.14 cm mesh sizing, which resulted in use of the 7.62 cm grid mesh (Schnabel Engineering 2021). The Renormalization Group (RNG) k-ε turbulence model was run for 30 seconds until steady-state conditions were achieved. Results from the CFD model matched the calculated discharge capacity within 1% when the slide gate was in the seated position. The model showed a 10.1% decrease in calculated discharge with the gate opened; thus the 10% reduction to the proposed PKW's discharge rating curve was applied only when the slide gate was operated (Schnabel Engineering 2021).

Two runs were conducted in HEC-HMS to determine peak reservoir elevations and discharge flows; a fully passive spillway system in which the slide gate remained closed, and a mixed active-passive spillway system in which the gate was assumed to be opened fully once the WSEL was 0.15 m above Normal Pool (EL 2.17). The HEC-HMS model results indicated that the proposed spillway was capable of passing the SDF plus the associated wave run-up with 4.6 cm of freeboard, and the 50-year frequency storm with an additional 11.3 cm of freeboard above the required 30.5 cm. Under fully passive conditions, the proposed PKW was able to pass the 25-year frequency storm while keeping the reservoir under the lowest edge of pavement, and the 50-year frequency storm without overtopping the dam (Schnabel Engineering 2021). Detailed results from the HEC-HMS model routing can be found in Table 2.

Table 2. HEC-HMS model results under proposed conditions.

		Gate Closed		Gate Opened (at WSEL 2.17)	
	Peak Inflow	Peak Outflow	Peak WSEL	Peak Outflow	Peak WSEL
Flood Event	m^3/s	m^3/s	m	m^3/s	m
2-year	2.0	1.9	2.09	1.9	2.09
5-year	4.2	4.1	2.18	4.3	2.17
10-year	6.8	6.3	2.30	6.8	2.17
25-year	11.3	9.8	2.56	11.3	2.19
50-year[1]	15.7	12.8	2.82	15.5	2.23
100-year[2]	21.1	17.8	3.07	19.9	2.46

1. Lowest edge of pavement (EL 2.64) – 0.30 m required freeboard = EL 2.34 (max allowable WSEL).
2. Top of dam (EL 2.94) – wave run-up from SDF (0.44 m) = EL 2.51 (max allowable WSEL).

2.3 *Structural design and analysis*

The proposed PKW was designed with guidance from the American Institute of Steel Construction's *Steel Design Guide 27 – Structural Stainless Steel*. Load factor and combinations were taken from USACE's *EM 1110-2-584 – Design of Hydraulic Steel Structures*. Three load cases were analyzed: Normal Pool (WSEL 2.01), the 100-yr frequency storm, and a dry reservoir to simulate construction conditions. Hydrostatic pressure from tailwater was conservatively neglected.

An analytical model of the PKW was built in STAAD.Pro V8i, a finite element modeling software, and appropriate loading, factors, and combinations were applied. Quadrilateral plate elements with a 7.6 cm target size were used to mesh the fabricated shapes; 316L stainless steel material properties and plate thicknesses were assigned. Anchorage locations were modeled as pinned connections while vertical compression-only springs along the base of the structure acted as the invert slab of the culvert. The structure was checked that maximum stresses and deflections did not exceed allowable values (Schnabel Engineering 2021).

Table 3. Results from STAAD.Pro model.

Result	Unit	Max Value	Utilization*
Principal Stress	MPa	57.30	36.9%
Von Mises Stress	MPa	54.61	35.2%
Shear Stress	MPa	2.48	1.6%
Membrane Stress	MPa	36.61	23.6%
Resultant Deflection	cm	0.493	97.0%

* Max modeled value as a percentage (%) of max allowable value.

The challenge in the structural design of the PKW was not with stress, but with deflection, which was found to control design. Under flood conditions, the sidewalls of the outlet keys pinch inwards, and the cantilevered portion arches upwards; this "accordion effect" can be seen in

Figure 4. Ideally, bracing would span across the outlet keys, but this would disrupt the flow path and potentially catch debris. Instead, additional material was added to the upstream base of each sidewall and cross braces were welded in between, below the keys. Two interior sidewalls were further extended on the downstream side along the base plate to provide additional stiffness.

The stainless steel epoxy anchors along the exterior sidewalls and base plate were designed in Hilti's PROFIS software using forces from the finite element model. Under hydrostatic loading, the entire structure was being pushed downstream, lifted vertically, and rotated about the center of anchorage. Concrete breakout of the existing culvert sidewalls from shear was the controlling failure mode (Schnabel Engineering 2021). A separate finite element model was also developed for the bridge culvert, to analyze the effects of the installed PKW loading on the existing structure. Load factors and combinations from the AASHTO *Load and Resistance Factor Design (LRFD) Bridge Design Specifications* and DelDOT's *Bridge Design Manual* were applied to concrete plate elements with reduced strength properties to account for potentially cracked and under-reinforced concrete. It was determined that the additional loading from the PKW increased stresses in mounting areas an insignificant amount when compared to structural capacity (Schnabel Engineering 2021).

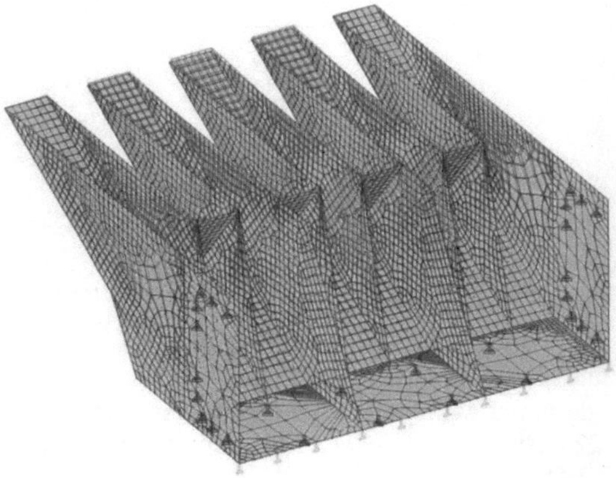

Figure 4. Maximum deflection (exaggerated by a factor of 12) from STAAD.Pro model.

3 FABRICATION AND INSTALLATION

3.1 *Fabrication*

The proposed PKW (as well as the slide gate, mounting pier and fishway gate) was fabricated entirely out of austenitic Grade 316L stainless steel to extend the lifespan of the structure in the brackish environment at the site. Epoxy-coated, higher grade carbon steels were also considered but the potential need for recoating offset the cost difference, and the lower yield strength of the stainless steel (172.4 MPa) was found to be sufficient with the addition of structural bracing.

Stainless steel's high thermal conductivity relative to carbon steel makes it more susceptible to warping, distorting and stressing of joints when welding, particularly for large structures or long weld lines (American Institute of Steel Construction 2013); thus, fabrication of the PKW required careful planning and expertise. Feasibility and access to weld locations were considered throughout the design process. Discussions with the fabricator, Whipps, Inc of Athol, Massachusetts (Whipps), who specializes in gates, valves and other hydraulic structures, were begun as soon as possible. Tight fabrication tolerances (1.59 mm maximum deviation from plane over 1.22 m) were required given the thin hydraulic margins at the site; the PKW crest needed to be level to function properly. Whipps used various techniques, including preheating, stitch welding, and temporary bracing, in addition to a large rigging and staging system, to successfully meet fabrication specifications. Heat treating was utilized after welding to reduce upward bowing of the central outlet keys.

3.2 *Installation*

During the design phase, Schnabel devoted substantial time and effort considering installation of the PKW. A conceptual installation plan was developed and provided in the contract drawings, and it was generally followed by the contractor during construction (Schnabel Engineering 2024). The structure, weighing approximately 6125 kg, was lifted with a John Deere 350G Excavator from SR-24, and lowered into the left bay of the culvert cell. A series of tiebacks and winches were used to pull the PKW downstream into position against temporary guide brackets, which would also act as grout seals. The crest was leveled and the mounting surfaces squared and plumbed with set and jacking screws provided along the exterior sidewalls and base plate of the PKW. To limit movement, several anchor bolts were drilled and installed prior to grouting through designed injection ports. Temporary bracing and pick points welded to the PKW to stabilize the structure during installation were cut and specific areas were re-passivated. Installation of the PKW began on the evening of March 25, 2023, and the structure successfully passed a leakage test on the morning of March 28, 2023: a total duration of less than 3 days.

Figure 5. Water discharging over the installed PKW.

4 IMPORTANCE OF BIM

During the concept phase, a digital 3-dimensional model of the PKW was built in Autodesk's Revit software, a building information modeling (BIM) program, along with the rest of existing and proposed conditions for the site. This proved vital to the success of the project for several reasons. Throughout the design process, numerous iterations of the PKW were developed and analyzed. Geometric ratios were adjusted to improve hydraulic efficiency. Sidewall shapes were altered, bracing was added, and various anchorage layouts were considered to provide adequate structural capacity while reducing costs. The configuration of the PKW changed after input from fabricators regarding access to welds and set up within the fabrication shop. Altering a digital model of the PKW within Revit allowed for a single source of truth for the latest iteration's configuration. Revised geometry could be exported from Revit and imported directly into the CFD model. 3-dimensional coordinates for the various corners and edges of the structure could be pulled and used to more quickly generate plate mesh for finite element analysis. Precise volumes and weights of individual structural components were calculated automatically. It saved valuable design time and reduced the risk of geometric inaccuracies, conflicts, and deviations across the various analyses.

The use of BIM also allowed for the generation of more sophisticated deliverables to the client, fabricator, and contractor. Revit geometry was used to create renderings of the site so the client could more easily understand the unique design. Isometric views were provided on the construction drawings and in construction meetings with the contractor to convey conceptual installation plans. The Revit model was shared in virtual meetings with the fabricator to discuss potential issues and changes to the proposed structures during the submittal process. Geometry was even exported to an augmented reality program so the fabricator could virtually "walk" around a full-scale model of the PKW to confirm feasibility of fabrication early on in design. BIM provided new and useful methods of visualization for the entire project team, from proof of concept through construction.

5 CONCLUSIONS

The PKW at Burton Pond Dam appears to be functioning as intended since its installation. A water level reading station at the site shows the PKW is passively managing reservoir levels similar to modeled results for storms below or around the 1-year frequency storm. Organic and other floating debris has been observed collecting around the upstream keys, but the structure seems to be "self-flushing" once sufficient head is available for discharge. No other maintenance or operational issues have been relayed to Schnabel.

While additional time and data collection is needed to confirm the hydraulic and structural performance of the PKW under more strenuous conditions, the design appears to be a feasible solution to a unique, but not uncommon, set of problems. Low-head environments often involve existing infrastructure and restrictions on space and construction. The stainless steel PKW at Burton Pond provides high-efficiency passive discharge in a small footprint. It was able to be installed quickly, without adversely affecting the existing structure, and without deep foundation elements. The cantilevered shape of the PKW could presumably be adapted to fit different geometric constraints and remain hydraulically and structurally feasible. The fabrication process took more time, foresight, and coordination when compared with more traditional structures, but the tight tolerances were able to be achieved by an experienced and adaptive fabricator. Schnabel believes that a similar design could be applied to a variety of low-head dams in need of rehabilitation, particularly those in coastal, tidal, or marshy areas.

REFERENCES

American Institute of Steel Construction. 2013. *Steel Design Guide 27 – Structural Stainless Steel.*

Anderson, R.M. 2011. Piano Key Weir Head Discharge Relationships. *All Graduate Theses and Dissertations, Utah State University.*

Khassaf et al. 2016. Hydraulic Behavior Of Piano Key Weir Type B Under Free Flow Conditions. *International Journal of Scientific & Technology Research* 5(3): 158–163.

Pennoni Associates, Inc. 2018. *Bridge 3-709, Burton Pond Dam Spillway, Design Level Inspection Report.*

Ribeiro, M.L, Pfister, M., Schleiss, A.J, Boillat, J. 2012. Hydraulic design of A-Type Piano Key Weirs. *Journal of Hydraulic Research* 50 (4): 400–408

Schnabel Engineering. 2015. *Hydrologic and Hydraulic Analysis, and Incremental Flooding Analysis, Burton Pond Dam.*

Schnabel Engineering. 2021. *PS&E Design Report, Burton Pond Dam Upgrading Design.*

Schnabel Engineering. July 2024. *Construction Plans for Burton Pond Dam Improvements – Record Drawing Set.*

Torre-Gomez, G., Erpicum, S., Pugliese, F., Giugni, M. 2022. Hydraulic Behavior Assessment of Type A and Type B Piano Key Weirs from Experimental and Numerical Results. *Environmental Sciences Proceedings* 21(1): 84.

United States Army Corp of Engineers. 2014. Design of Hydraulic Steel Structures. *Engineering Manual 1110-2-2105*, Department of the Army, Washington, D.C.

Labyrinth and Piano Key Weirs IV – Erpicum, Crookston & Paxson (Eds)

Piano key weir with orifice sluice

Dau Xuan Thuy
Power engineering consulting company 2 (PECC2), Ho Chi Minh city, Vietnam

ABSTRACT: In recent years, many Piano Key Weirs (PKW) have been built around the world. In Vietnam there are more than 18 PKWs presently in operation. However, during the operation, some experience has been drawn on the design and application conditions. This paper is about the Dak Mi 3 reservoir with a PKW that operated well for more than 4 years until an exceptional flood occurred following typhoons in October 2020, fully filling the reservoir with mud and sand. The result is the PKW became ineffective and the hydropower plant unable to operate. In order to dredge the reservoir and prevent mud deposits again, a part of the PKW was replaced by two deep orifice sluices. Using orifice sluices in combination with PKW was an effective solution for sand flush of reservoir in catchment areas sensitive to landslides and under climate changes.

1 PIANO KEY WEIR SPILLWAY AT DAK MI 3 HYDROPOWER PROJECT

Dak Mi 3 dam, located at 15°19′8"N 107°49′32"E, was built in 2017. It is a medium-large storage reservoir, and the dam height is about 40 meters. The initial spillway was designed as a free ogee spillway with a design nappe depth of 8.72m. Subsequently, to enhance investment efficiency, the project was optimized by applying a PKW spillway to reduce the nappe depth, thereby increasing the reservoir capacity and the power generation head of the plant while maintaining the same reservoir inundated area.

When the PKW was applied, the reservoir water level of the project was raised by 4 meters, resulting in an increase in reservoir capacity from 3.4 hm^3 to 5.0 hm^3 and an increase in plant capacity from 56MW to 63MW, thereby enhancing the investment efficiency of the project. Specifically, the project's annual revenue increased by approximately $520,000 USD.

Figure 1 is the graph of discharge capacity of the Dak Mi 3 PKW spillway, and Table 1 presents the comparative characteristics between the two alternatives: traditional Creager spillway and PKW spillway.

The Dak Mi 3 spillway is a Type B PKW with upstream overhang, and the calculated wall height is H=7.5m. The total spillway length of the PKW is W=144.0 m, and the dam height is P= 40.0m. The PKW discharge capacity is Q= 6670 m^3/s. The plan view and longitudinal section of the dam are shown in Figures 2 and 3, Figure 3 shows the PKW under construction and Figure 4 is the PKW in operation.

2 SEDIMENT FLOOD INCIDENT AT DAK MI 3 RESERVOIR

The Dak Mi 3 hydropower plant was put in operation in 2017. During the operation, the project's PKW spillway met the normal flood discharge capacity as initially designed.

Based on sediment flow monitoring data of the river basin, the reservoir annual inflow and the reservoir volume, amount of sediment deposited in the reservoir is approximately

DOI: 10.1201/9781003655541-19

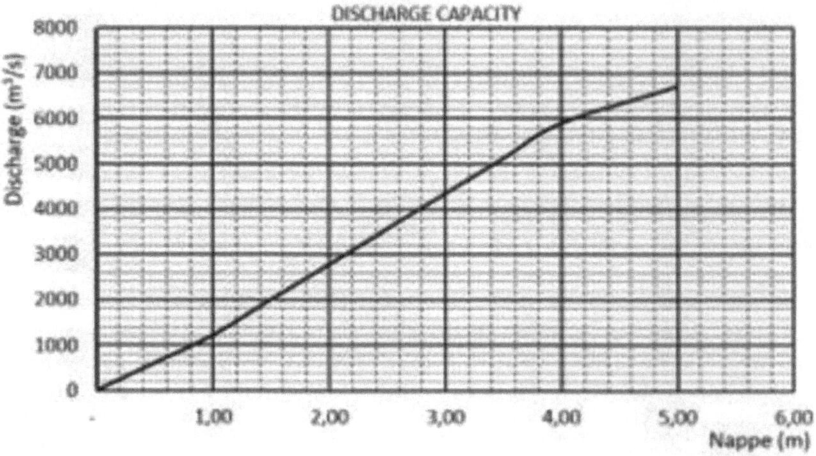

Figure 1. Discharge capacity of Dak Mi 3 PKW.

Table 1. Characteristics of creager spillway and PKW spillway.

Parameter	Creager spillway	PKW spillway
Spillway section	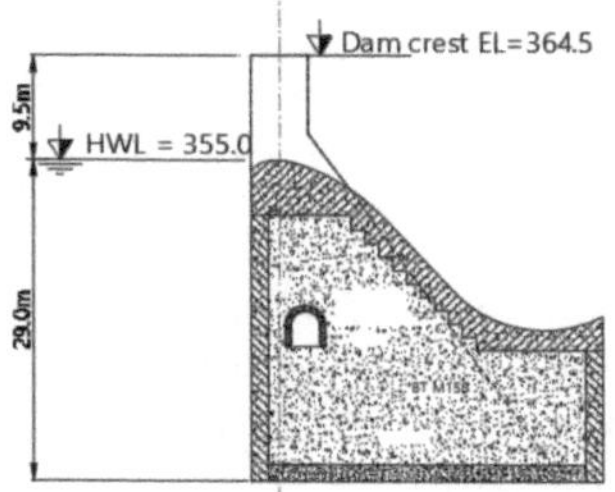	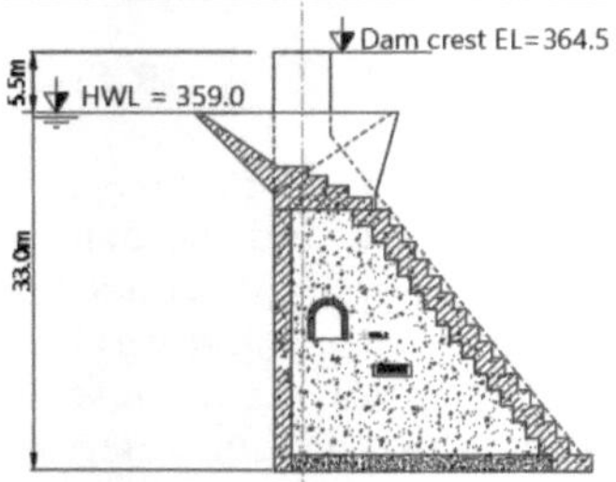
Dam crest	364.5m	364.5m
High water level HWL	355.0m	359.0m
Total Storage Volume	3.4 hm^3	5.0 hm^3
Active storage Volume	1.35 hm^3	2.30 hm^3
Dead storage Volume	2.05 hm^3	2.7 hm^3
Spillway construction cost	USD 9,760,000	USD 8,120,000
Powerhouse capacity	56MW	63MW
Annual output	197 GWh	210 GWh
Yearly income	7,880,000 USD/year	8,400,000 USD/year
Annual additional profit		520,000 USD/year

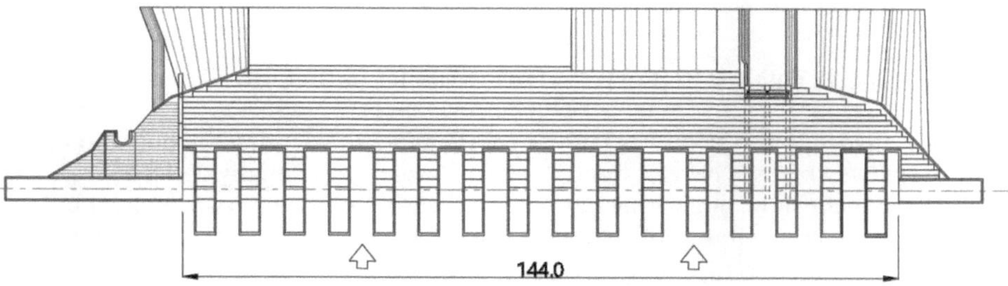

Figure 2. Dak Mi 3 PKW spillway, plan view.

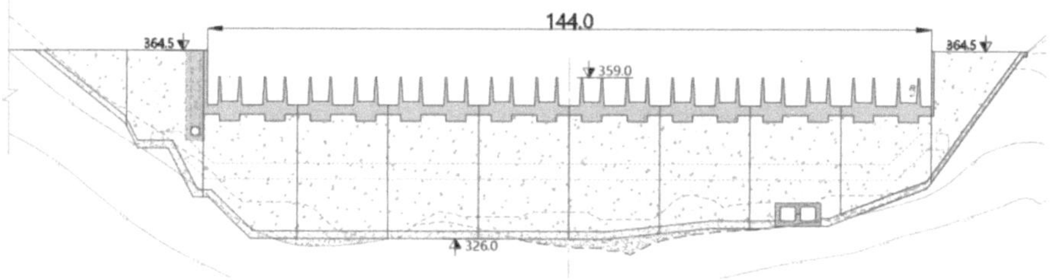

Figure 3. Dak Mi 3 PKW spillway, longitudinal section.

Figure 4. Dak Mi 3 PKW spillway under construction, view from upstream.

Figure 5. Dak Mi 3 PKW spillway in operation.

37,000 m^3/year (using Brune's method of Trap Efficiency). Meanwhile, the Dak Mi 3 reservoir is classified as a medium-large reservoir, with a total storage capacity of 5 hm^3 and a dead storage capacity of 2.7 hm^3. Therefore, the reservoir can operate for about 50 years before sediment deposition affects its useful capacity.

However, in October 2020, after a prolonged period of rain followed by an unusual large flood with a peak flow rate of approximately Q= 7800 m^3/s, the Dak Mi 3 reservoir was filled up with sediment, causing the plant to cease operations as the intake gates and headrace tunnels were clogged with sediment.

When the reservoir was filled with sediment, the PKW spillway no longer functioned effectively as a PKW spillway but operated as a linear spillway, reducing its flood discharge capacity by about 60%. At that time, the inlet keys of the PKW spillway were filled with sediment, therefore, the spillway operated nearly as a traditional free overflow spillway. With a spillway width of 144m, and its discharge capacity was estimated approximately using the following formula:

$$Q = C.B\sqrt{2g}H_o^{3/2} = 0.4 * 144 * (2 * 9.81)^{0.5*}(5.0)^{1.5} = 2852\text{m}^{3/\text{s}}, \quad (1)$$

Thus, the discharge capacity of the spillway was reduced by nearly 60%, from 6670 m^3/s to 2852 m^3/s. As a result, during the flood of October 2020 with a peak flood flow rate of about 7800 m^3/s, the Dak Mi 3 dam experienced overtopping with water levels 2-3m higher than the dam crest. Some photos of the Dak Mi 3 reservoir and PKW spillway after being filled with sediment are shown in Figure 6, Figure 7, Figure 8, and Figure 9 below:

Figure 6. Dak mi 3 reservoir filled up with sand after flood.

Figure 7. Dak mi 3 reservoir filled up with sand after flood.

Figure 8. Dak Mi 3 reservoir after flood (view from upstream of the dam).

Figure 9. Dak Mi 3 reservoir after flood (view from downstream of the dam).

3 PIANO KEY WEIRS COMBINED WITH DEEP SLUICE GATES TO ENHANCE FLOOD DISCHARGE CAPACITY AND SEDIMENT FLUSHING

In order to restore the operation of the hydroelectric plant, it is necessary to immediately conduct the dredging of sediment in the reservoir to restore useful storage capacity. However, dredging millions of cubic meters of sediment is extremely difficult, a costly task and not a long-term solution as similar incidents could occur in the future.

Therefore, a 28m long section (including 2 sections) of the PKW spillway had to be dismantled and replaced with 2 deep sluice gates to increase flood discharge capacity and effectively discharge sediment.

The two deep sluice gates have dimensions of 9.5m x 12m (Width x Height) and the bottom elevation of the sluice gates is equivalent to the riverbed elevation to ensure effective sediment discharge.

When replacing a section of the PKW spillway with deep sluice gates, the flood discharge capacity of the spillway increased by about 35% and the PKW spillway always ensures flood discharge capacity due to not being buried by sediment. The combined discharge capacity of the PKW spillway and deep sluice gates is shown in the diagram of Figure 10 below.

The layout drawings of the PKW spillway combined with deep sluice gates are shown in Figures 11, 12, and 13. Figure 14 is PKW under rehabilitation, and Figure 15 is the PKW after rehabilitation. The process of dredging sediment to restore the reservoir and upgrading the PKW spillway lasted nearly one year. By August 2021, the Dak Mi 3 hydropower plant had returned to normal operation. The total discharge capacity of spillway after rehabilitation is Q= 8777 m^3/s.

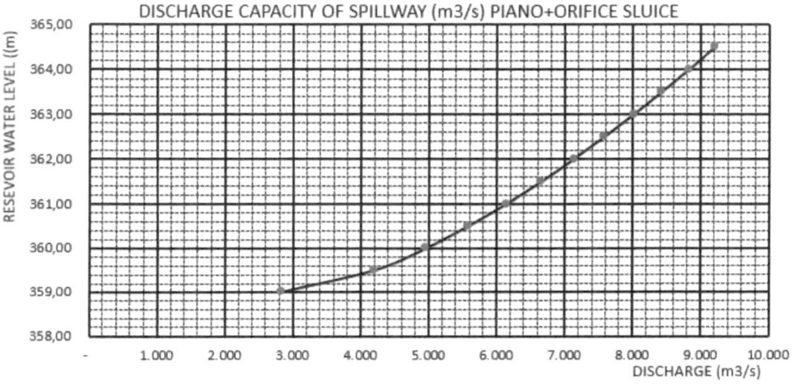

Figure 10. Total discharge capacity of PKW spillway & orifice sluice.

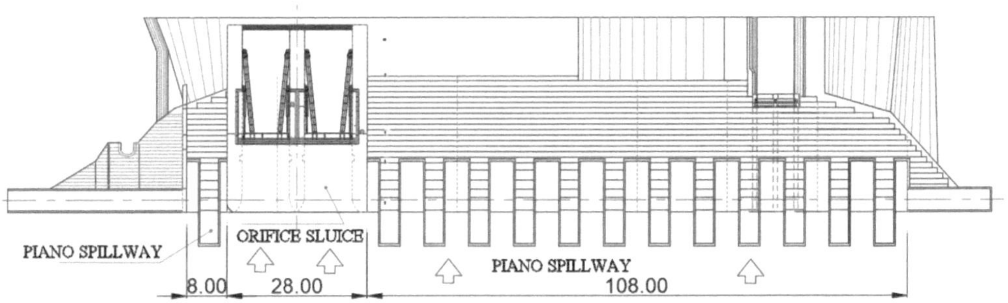

Figure 11. Dak Mi 3 PKW spillway with orifice sluice, plan view.

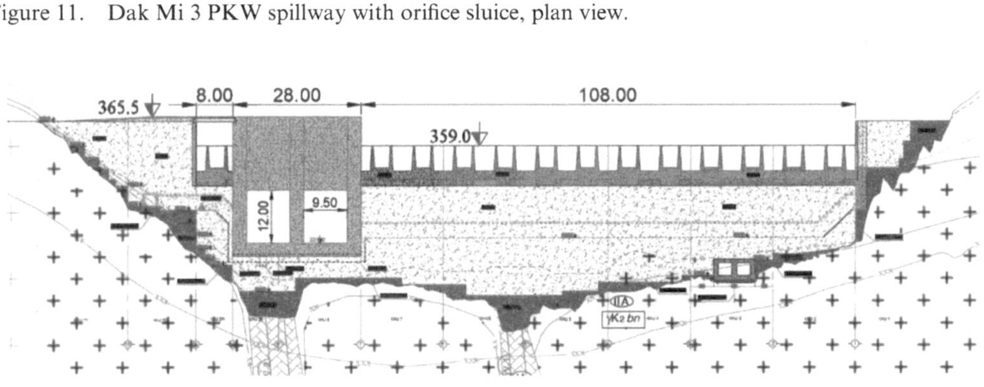

Figure 12. Dak Mi 3 PKW spillway with orifice sluice, longitudinal section.

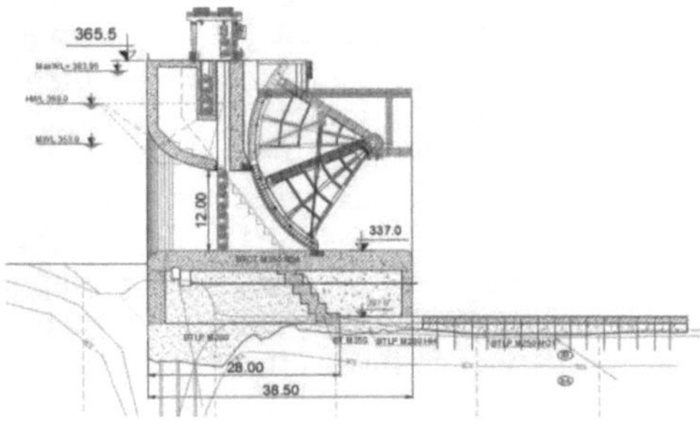

Figure 13. Dak Mi 3 PKW, orifice sluice section.

Figure 14. Dak Mi 3 PKW with orifice sluice, view from upstream.

Figure 15. Dak Mi 3 PKW with orifice sluice – view from downstream.

4 ASSESSMENT OF THE CAUSES LEADING TO RESERVOIR SILTATION

Since the Dak Mi 3 reservoir began operation in 2017, no abnormal sedimentation phenomena were observed during its more than three years of operation, and the spillway continued to function normally. However, in 2020, an unusually large and prolonged rainstorm occurred in central Vietnam. From September 15, 2020 to October 30, 2020, four consecutive storms hit the East Sea, including storms named NOUL (September 15-20), which caused approximately 324mm of rainfall in the basin; LINFA (October 10-11), which caused up to 1880mm of rainfall in the basin; Nanga (October 12-16), which caused about 400mm of rainfall in the basin; and finally, Molave (October 18-30), which caused up to about 2000mm of rainfall in the basin.

After a prolonged period of rain, the surface soil on the slopes of the basin became saturated with water and sensitive to landslides. Subsequently, a heavy rainstorm caused mass landslides on the hillsides of the basin, and a large amount of soil and rock was swept into the river, creating a sediment flood that filled the Dak Mi 3 reservoir. At this time, the Dak Mi 3 reservoir was the uppermost reservoir in the basin, so all the sediment was retained in the Dak Mi 3 reservoir until it was filled.

After the rainstorm, it was observed that the hillsides in the basin had experienced mass landslides, and soil and rock carried by the flood buried many fields and streams in the basin. Some photos of the landslides in the basin are shown in Figures 16 and 17 below.

Figure 16. Landslide at river side.

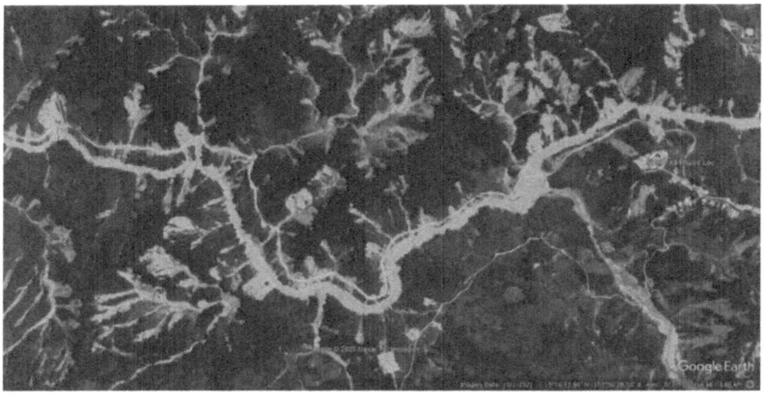

Figure 17. Mass hillside landslides in the catchment area (google earth image 2021).

5 CONCLUSION AND RECOMMENDATIONS

Abnormal heavy rains causing mass landslides in the river basin, creating sediment floods during major floods, are the reasons for the sediment filling the reservoir. The sediment filling the inlet cells of the PKW diminished its effectiveness making it functions only as a traditional straight weir and reducing the flood discharge capacity by about 60% compared to the original PKW.

The topographical and geological characteristics of the river basin, including steep hillside terrain and unstable surface soil layers, are factors leading to the risk of mass landslides during prolonged rains when the soil becomes saturated.

Currently, most reservoirs using PKW have medium and small storage capacities, making them sensitive to siltation.

The location of reservoir on a river cascade is also a factor to consider because if the reservoir is at the most upstream of a river cascade, it will retain the most sediment.

Therefore, when applying the PKW, it is necessary to conduct a comprehensive assessment of the topographical and geological conditions of the catchment area, the scale of the reservoir, and the order of the reservoir on the river cascade. If the PKW is applied for reservoirs with features sensitive to sediment deposits, it is necessary to arrange additional orifice sluices to both increase flood discharge capacity and combining sediment flushing to ensure that sediment can be discharged in the event of mud flooding in the basin, preventing sediment from filling the reservoir.

For efficient sand flushing, the size of sluice orifices gates should be large enough and should be installed at bottom of dam. So that it is costly because it needs a large force to open and close the gates, and the operation and maintenance is also complex.

Below is a suggestion for a decision-making flowchart for assessing whether sluice orifices for sand flushing cum flood discharge is needed when applying free PKW spillway in a reservoir design for reference in a preliminary study of project.

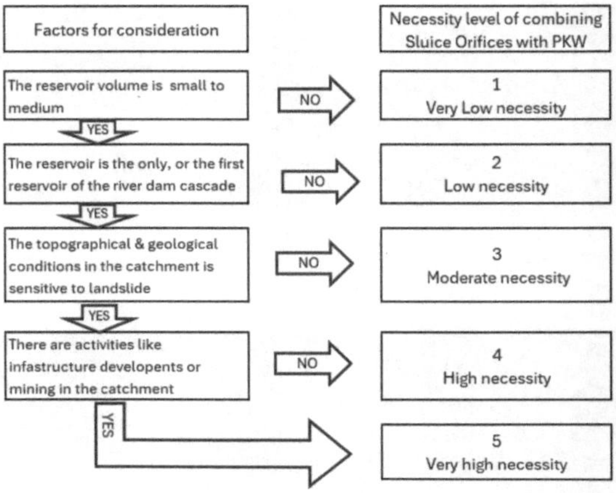

Figure 18. Decision making flowchart for orifice sluice installation.

REFERENCES

Brune, G.M. (1953) Trap Efficiency of Reservoirs. Transactions, American Geophysical Union, 34, 407–418

EASIA JSC. 2009. Technical design of Dak Mi 3 HPP.

Erpicum. S. et al. 2017. "Hydraulics of Piano Key Weirs: A Review". Labyrinth and Piano Key Weirs III - PKW 2017, CRC Press/Balkema.

Ho Ta Khanh, M, T. Chi Hien & N. Thanh Hai. 2011. "Main results of the P.K weir model tests in Vietnam (2004 to 2010)". Labyrinth and Piano Key Weirs-PKW 2011. CRC Press/Balkema.

Ho Ta Khanh, M, Dinh Sy Quat & D. Xuan Thuy. 2011. "P.K weirs under design and construction in Vietnam (2010)". Labyrinth and Piano Key Weirs-PKW 2011. CRC Press/Balkema.

Ho Ta Khanh. 2017. "History and Development of Piano Key Weirs in Vietnam (2004–2016)". Labyrinth and Piano Key Weirs III - PKW 2017, CRC Press/Balkema.

Kixelep P.G. et al. 1974. "Hydraulic Calculation Handbook".

Laugier F. 2007. "Design and construction of the first Piano Key Weir (PKW) spillway at the Goulours dam", International Journal of Hydropower and Dams 14 (5), 94–101

Lempérière, F & Ouamane, A. 2003. "The Piano Keys Weir: a new cost-effective solution for spillway". The International Journal on Hydropower & Dams, Issue Four.

PECC2. 2012. Revised Technical design of Dak Mi 3 HPP.

PECC2. 2021. Basic design (Revised) of Dak Mi 3, Spillway Upgrading.

PECI. 2021. Technical design (Revised) of Dak Mi 3, Spillway Upgrading.

Reynolds, P.2011. "PK Weirs in Vietnam". International Waterpower & Dam Construction. 26 January 2011.

Tullis, B., Young N. & Crookston, B.. 2017. "Physical modeling size-scale effects for labyrinth weirs with half-round crests". Labyrinth and Piano Key Weirs III - PKW 2017, CRC Press/Balkema.

Vietnamese Standard 12262:2018 – Hydraulic Structures – Spillway – Hydraulic Calculation of PKW.

II.2 Design

Labyrinth and Piano Key Weirs IV – Erpicum, Crookston & Paxson (Eds)

Design of a new arced Labyrinth spillway for Dog River Dam under adverse approach conditions

Brad Kirksey
Freese and Nichols, Inc. El Paso, Texas, USA

Rojin Tuladhar
Freese and Nichols, Inc. Austin, Texas, USA

Gregg Hudock
Freese and Nichols, Inc. Atlanta, Georgia, USA

ABSTRACT: The Douglasville-Douglas County Water and Sewer Authority (DDCWSA) in Georgia is raising the Dog River Dam a second time to increase raw water supply storage to meet anticipated future demands. The dam's first raise — by 10 ft (3.05 m) — increased the reservoir's normal storage capacity to 5,830-acre-ft (7.19 × 10^6 m^3). As the county continued to grow and develop, DDCWSA planned a second dam raise — by 35 ft (10.67 m) — to more than triple the reservoir's normal storage capacity to nearly 20,000-acre-ft (2.47 × 10^7 m^3). But raising the dam that much on a reservoir with steep abutments — and to meet a higher hazard classification — required an entirely new approach to designing the replacement spillway. The Freese and Nichols (FNI) team designed a nonconventional 39-cycle labyrinth weir with an arced shape, using a combination of hybrid 2D/3D computational fluid dynamics modeling and physical hydraulic modeling. The arced labyrinth weir is estimated to cost $8,200,000 less to build than the next cheapest alternative and can be completed six months faster. Using the hybrid CFD method also saved time and expense during the design.

1 BACKGROUND INFORMATION

1.1 *Project history*

The Douglasville-Douglas County Water and Sewer Authority provides water, sanitary sewer and stormwater services to residents, businesses, industries and government customers in an area west of downtown Atlanta with a growing population of more than 148,000. The Authority constructed Dog River Dam (GA State ID 048-048-04519; National ID GA05334) in the early 1990s to impound a new raw water supply reservoir on the Dog River. The Reservoir supplements the water supply at the Bear Creek Reservoir to meet the increasing water supply demands of growing Douglas County, Georgia. When completed in 1992, the dam impounded 3,700 acre-ft (4.56 × 10^6 m^3) of surface water from a 78-square-mile (202 km^2) watershed and was permitted as a Category II (low hazard), Large Dam by the State of Georgia Environmental Protection Division. The original earthen embankment dam was approximately 50 ft (15.24 m) tall from the dam crest to the Dog River channel. The crest of the dam, at elevation 760 ft (231.65 m), had a length of approximately 600 ft (182.88 m). The service spillway was a 240-ft-wide (73.15 m) reinforced concrete chute spillway with an 8-cycle labyrinth weir that maintained the 217-acre (0.878 km^2) reservoir at an elevation of 750 ft (228.6 m). The service spillway was constructed on the main dam embankment. The project also included a saddle dike embankment and a pump station on the left (east) side. The pump station delivers raw water to the Bear Creek Water Treatment Plant.

DOI: 10.1201/9781003655541-20

The Project had an initial withdrawal permit of 6.0 MGD (Million Gallons per Day) (0.263 m^3/s) in 1992 that was modified in 2002 to increase the withdrawal to 23.0 MGD (1.01 m^3/s). The DDCWSA secured a Section 404 permit from the United States Army Corps of Engineers (USACE) in 2004 to raise the normal pool elevation of the reservoir from 750 to 760 ft (228.6 -231.65 m) for additional reservoir storage capacity. The project included raising the dam crest elevation by 10 ft (3.05 m) by constructing a reinforced concrete parapet wall along the embankment crest. The chute spillway sidewalls and labyrinth weir walls associated with the existing service spillway were also raised to accommodate the modified normal pool elevation. With its completion in 2009, the raised dam impounds 5,800 acre-ft (7.19×10^6 m^3) of surface water and provides a 256-acre (1.04 km^2) reservoir. Collectively, this modification is known as the "First Raise" of the Dog River Dam, which represents the existing conditions at the Project site at the time of this Report.

1.2 *Project description*

The DDCWSA proposes to expand the Dog River Reservoir again to increase the raw water storage to meet increased water supply demands and address recent changes in the precipitation and hydrology models for the State of Georgia. The expansion includes three projects, as shown in Figure 1.

The new dam raise project will raise the normal pool of the Dog River Reservoir from elevation 760.0 ft to 795.3 ft (231.65 to 242.4 m) and reflects an additional reservoir storage of approximately 14,635.0 acre-ft (1.81×10^7 m^3). The top of the main and east dams will be raised, and a new embankment saddle dike will be added between the Dog River and Basket Creek basins on the west side to accommodate the higher reservoir level. The existing labyrinth spillway and principal spillway will be abandoned in place, and a new service spillway and low-level outlet will be constructed to accommodate the new reservoir level, required inflows/outflows per Georgia Safe Dams Program (GA SDP), and current environmental releases.

The main dam will consist of homogeneous embankment fill with selected zones of lower hydraulic conductivity soils and will include soil cement on the upstream slope and downstream toe for erosion protection. The embankment will be 110 ft (33.53 m) tall at its highest point and will have a top elevation of 815.0 ft (248.41 m). The upstream slope will be 2:5H:1V, and the downstream slope will be 3H:1V with terraces for surface water control. A grout curtain will be constructed through the foundation rock along the full length of the east and main dams. A cutoff wall will be constructed from elevation 800.0 ft (243.84 m) down through the embankment into the top of the grout curtain. The proposed east dam embankment will be constructed near the existing pump station and will have a cross-section configuration similar to the main dam including a grout curtain and cutoff wall.

A new saddle dike, referred to as the west dam embankment, will be required to close a low area near the west end of the reservoir. It will also be an earthfill embankment with a top elevation of 815.0 ft (248.41 m), similar to the other embankments, but will not have a grout curtain and cutoff wall.

The service spillway will consist of a 39-cycle, arc-shaped labyrinth weir that converges into a 200-ft-wide (60.96 m) concrete chute on the left abutment of the main dam. The chute is sloped at 15% for approximately 220 ft (67.06 m) and steepens to 25% (4H:1V) for approximately 140 ft (42.67 m) before discharging into a 140-ft-long (42.67 m), modified United States Bureau of Reclamation (USBR) Type III stilling basin. Flow exits the stilling basin into a soil cement and rock riprap-lined trapezoidal channel that connects with Dog River. The service spillway also includes Roller-Compacted Concrete (RCC) abutments on the left and right sides and a bridge over the chute including a center bridge pier. The service spillway also includes provisions for low-flow releases through the left abutment, and an emergency drain through the center pier of the bridge from an inlet upstream of the labyrinth weir. Releases for both the low flow and emergency drawdown are made manually by vertical lift sluice gates.

Other major components of this Project include a new raw water pump station, a new state highway bridge, new access roads, geotechnical instrumentation, and demolition of numerous existing facilities.

1.3 *Design constraints*

The DDCWSA is raising the Dog River Dam and Reservoir a second time to increase raw water supply storage for anticipated demands and address changes in the precipitation and hydrology models for the State of Georgia. The key design criteria for the proposed reservoir and dam(s) include:

- Normal Pool Water Surface Elevation 795.3 ft (242.4 m)
- 100 Year Flood Water Surface Elevation 797.8 ft+/- (243.17 m)
- Compliance with Safe Dams Program (GA SDP) and Metropolitan River Protection Act (MRPA)
- Design to Category I (high hazard) dam criteria due to its importance as critical infrastructure for water supply benefits, which includes passing the flood from the 100% Probable Maximum Precipitation (PMP)
- Dam Subclassification as a "very-large" dam due to its anticipated height exceeding 100 ft (30.48 m) and maximum storage volume of approximately 32,000 acre-ft (3.94×10^7 m^3)
- Meeting all requirements of the USACE 404 Permit, which includes limiting impacts to Waters of the U.S. downstream of the dam to less than 7.82 acres (31646.42 m^2) for wetlands and less than 1,396 linear feet (425.5 m) for streams

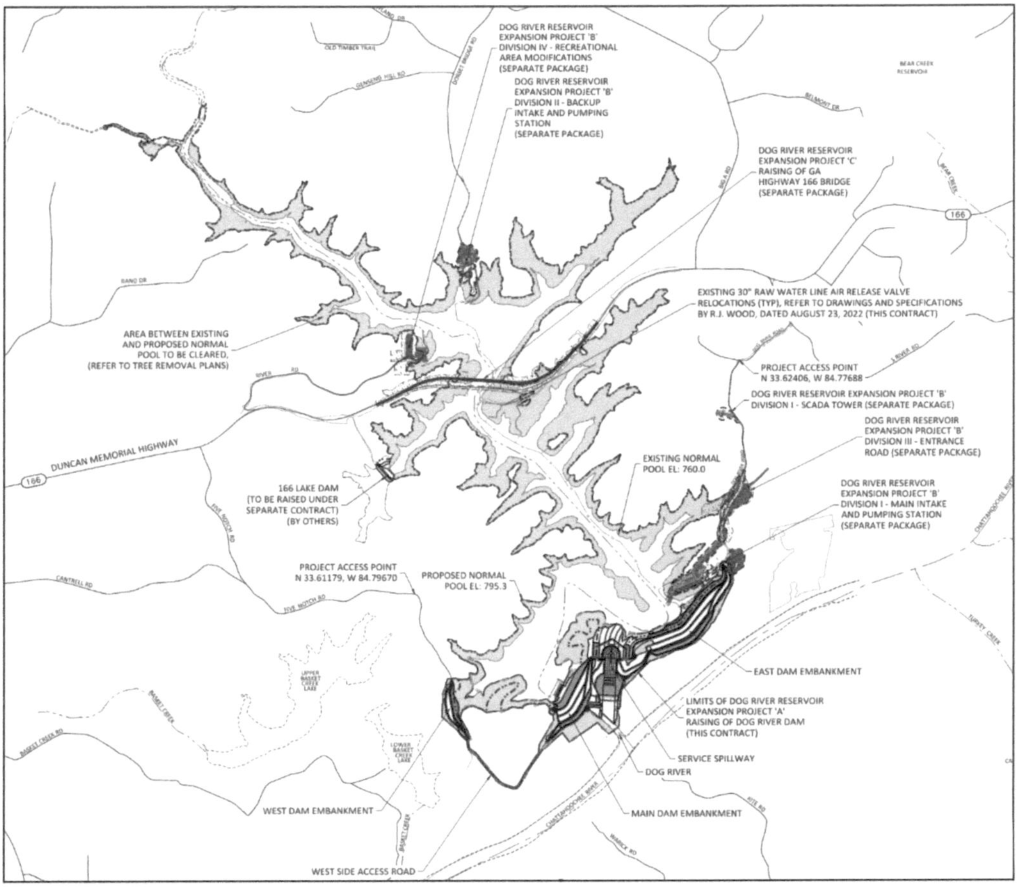

Figure 1. Overview of the second raise of dog river reservoir project.

2 ARCED CONFIGURATION EXPLAINED

The dam's location and original spillway configuration posed challenges for meeting the new needs, and an arced labyrinth provided to be the optimal new design. The following considerations factored into the plan.

2.1 *Care of water during construction*

The Dog River Reservoir must continue to serve as raw water storage for the duration of construction. In general, the DDCWSA does not wish to compromise dam safety during construction. As such, a prescriptive construction sequence and care of water plan is established to limit risks to the DDCWSA and define the division of responsibility between the DDCWSA and the construction contractor. FNI developed the construction sequence based on the following list of goals and assumptions:

- Deliberately lowering the reservoir pool by more than 2 ft (0.61 m) from the current normal water surface elevation (WSE) (Elev. 760 ft; 231.65 m) is not acceptable.
- Normal conditions during construction are defined as the typical state of the project site in which inflows to the reservoir, flows along Dog River and the Chattahoochee River (downstream of project site) are relatively low and steady. This condition is likely to exist most of the time during construction. Normal inflows are less than 350 cfs (9.91 m^3/s).
- Routine flooding during construction is defined as the state of the project site when usual flooding occurs due to increased inflows to the reservoir and/or increased flows along Dog River and/or the Chattahoochee River. This condition is likely to occur periodically during construction. Routine flooding inflows are between 350 cfs (9.91 m^3/s) and 6,000 cfs (170 m^3/s)
- Infrequent flooding during construction is defined as the state of the project site when significant flooding occurs due to increased inflows to the reservoir and/or increased flows along Dog River and/or the Chattahoochee River. This condition is not likely to occur during construction, but it is possible. Infrequent flooding inflows are greater than 6,000 cfs (170 m^3/s).

In general, to abide by the care of water plan, the construction contractor will be required to manage all reservoir inflows, resulting reservoir levels, and tailwater levels during normal and routine flooding conditions. Flood insurance will be required to cover the risk of infrequent flooding.

2.2 *Construction sequencing*

The construction is divided into three phases. Phase 2 is most relevant to the arced labyrinth spillway. During Phase 2, the new main dam embankment will be constructed immediately downstream of the existing service spillway. Thus, flows through the existing spillway must be prevented to limit the risk of eroding and breaching the new main embankment. To prevent flows over the existing spillway, the project requires a diversion channel that can be used for passing normal flows and routine flooding. Passing 6000 cfs (170 m^3/s)without the reservoir level exceeding elevation 760.0 (231.65 m) requires a 200-ft-wide (60.96 m) excavated open channel with a crest elevation of 754.0 (229.79 m) or a 100-ft-wide (30.48 m) channel with a crest elevation of approximately 751.0 (228.94 m). The only logical location for this Phase 2 diversion channel is between the main embankment and east embankment. Other locations are complicated by either the steep topography or environmental impacts.

The result of these project and site constraints is that significant excavation will be required to construct the diversion channel, and the diversion channel must be protected against erosion. The excavated material will be incorporated into the homogenous embankments, so that effort is not duplicative. Rather than making the diversion channel temporary (i.e. providing nominal erosion protection and then backfilling the channel at the completion of the project),

FNI determined that it was best to turn the majority of the diversion channel into the permanent spillway chute and incorporate it into the proposed service spillway.

Due to the depth and width of the diversion channel, the spillway chute would have sufficient depth to pass the Probable Maximum Flood (PMF) storm (approximately 195,000 cfs (5,520 m³/s)), though at a considerable flow depth. However, due to the relatively high local tailwater, dissipating the energy from a deep, fast flowing chute would be possible with a properly designed stilling basin. Therefore, the only remaining decision was how to maximize the flow through the chute while holding the normal water surface elevation at 795.3 ft (242.4 m). Theoretically, this could be accomplished either with a series of large gates, or with a non-linear weir. DDCWSA opted against a controlled spillway structure with large gates due to the risk of operation and continued operation and maintenance costs, so a non-linear weir was chosen. Through a series of iterations, the resulting arced or horseshoe-shaped weir was designed. Iterations began with a straight labyrinth weir but was later modified to have a three-sided rectangular footprint due to stability requirements. The latter design evolved into an arced labyrinth as it was hydraulically more efficient.

The labyrinth consists of 39 cycles, each of which measures approximately 30 ft (9.14 m) long and approximately 15 ft (4.57 m) wide (the arc results in slightly varied dimensions). The labyrinth walls are 17.3 ft tall (5.27 m) with an 18-inch (0.46 m) half round weir crest. The downstream face of the walls is battered. The labyrinth apron is 24 ft (7.32 m) above the spillway chute and sits on either competent rock or a RCC foundation at different parts of the spillway. The RCC foundation is primarily used through the center of the arc where the diversion channel is located during construction and must be closed following the completion of Phase 2. As a result, the approach depth for the labyrinth varies significantly from a minimum of 17.3 ft (5.27 m) on the left and right ends of the weir to 41.3 ft (12.59 m) at the centerline of the spillway. Figure 2 shows a view of the arced labyrinth weir from the physical model study.

3 LABYRINTH DESIGN AND MODELING

To evaluate the performance of the arced labyrinth, FNI utilized several types of computational modeling as well as a scaled physical hydraulic model.

Figure 2. View of the arced labyrinth physical model.

3.1 *Design criteria*

The proposed Dog River Dam will be classified as a Category I structure. As defined by the Georgia Safe Dams Act, a Category I dam is one where improper operation, or failure would result in a probable loss of human life. GA SDP provides a secondary classification (subclassification) based on the dam size. Dams are subclassified as "Very Large" when they are more than 100 ft tall or impound more than 50,000 acre-ft (6.17 x 10^7 m^3) at the top of dam. Following the second dam raise, Dog River Dam will impound approximately 32,000 acre-ft (3.94 x 10^7 m^3) at the top of dam. The proposed dam will be more than 100 ft (30.48 m) tall when measured from the base of the Dog River valley to the top of dam. The second dam raise is being designed as a Category I, Very Large structure and is required to pass 100% of the PMF which results from the Probable Maximum Precipitation (PMP) on the receiving watershed.

3.2 *First iteration CFD model*

The arced labyrinth rating curve was initially developed according to Crookston & Tullis (2012). However, it quickly became apparent that the published rating curve tools were not going to accurately calculate the rating curve for this spillway due to the approach conditions, small apex angles, rapid convergence of flow into the chute, and submergence. Thus, FNI developed the rating curve based on the results of Computational Fluid Dynamics (CFD) modeling. CFD modeling simulates fluid flow to analyze and predict its behavior by solving the Navier-Stokes equations. This is completed by discretizing the physical domain into a grid of connected cells.

We used the commercially available CFD software Flow-3D by Flow Science, Inc., which employs a cuboidal grid with the Volume of Fluid (VOF) method to track the water surface and apply boundary conditions on the surface as it moves through the computational grid. The model used a cell size of 1 foot (0.3048 m), used the Renormalization Group (RNG) turbulence model, used a water surface elevation and an outflow boundary condition to introduce and exit fluid, respectively, in the model. The cell size was deemed sufficient to model the half round labyrinth weir crest. The model run time was approximately 24 hours, models ran until the results converged:

- Multiple configurations of the labyrinth weir were evaluated that included labyrinth wall geometry, approach channel geometry, and apron contractions as variables.
- The results were analyzed to determine the most hydraulically efficient configuration.
- The CFD model also was utilized to optimize the shape and impacts of the bridge pier on the spillway flow capacity.
- Accounting for the bridge pier, approach conditions, and the apron contraction, the rating curve developed using available literature was adjusted in accordance with the CFD model results and subsequently the physical model results.
- The approach channel conditions in the initial model were assumed to be symmetrical about the spillway centerline.
- Figure 3 shows an isometric view of the results from the initial CFD Model. Table 1 indicates the PMF inflow to the reservoir as well as the outflow through the proposed service spillway as determined by FNI's initial CFD Rating Curve.

3.3 *Second iteration 2D/3D (Hybrid) CFD model*

As part of the hydraulic peer review of the project, FNI became aware of several approach conditions on the spillway that would seem to have consequential effects on the spillway hydraulics and the discharge rating curve. In particular, a peninsula within the Dog River Reservoir approximately 1,000 ft (304.8 m) upstream of the spillway that would be under the proposed normal pool elevation was hypothesized to affect the spillway hydraulics. Flow-3D's multi-block approach supports the integration of 2D and 3D models within a single simulation, capturing both large-scale and detailed flow patterns. The 2D model solves the standard shallow water momentum equations to model flows over large domains with minimal vertical velocity components.

Figure 3. Isometric view of the initial CFD model of the labyrinth spillway.

Table 1. Summary of discharges for varying load conditions on labyrinth spillway.

	Inflow		Outflow	
Storm Event	(cfs)	(m^3/s)	(cfs)	(m^3/s)
10% of PMF	20,683	586	19,475	551
20% of PMF	41,366	1,172	38,950	1,103
30% of PMF	62,049	1,758	58,425	1,654
40% of PMF	82,732	2,343	77,900	2,206
50% of PMF	103,415	2,929	97,375	2,757
60% of PMF	124,098	3,515	116,850	3,309
70% of PMF	144,781	4,101	136,325	3,860
80% of PMF	165,464	4,687	155,800	4,412
90% of PMF	186,147	5,273	175,275	4,963
100% of PMF	206,830	5,858	194,750	5,514

In this case, the 2D model simulates the reservoir, where vertical flow is negligible, while the 3D model simulates the arced labyrinth spillway, where vertical velocity, pressure changes, and detailed flow behavior are crucial. This hybrid approach effectively simulates large areas while accurately capturing the spillway's more complex flow dynamics and behavior. The 2D domain was modeled using mesh blocks of cell size 16 ft (4.87 m), 8 ft (2.44 m) and 4 ft (1.22 m). A terrain was developed that included the reservoir and peninsula as shown in Figure 4.

To verify, validate, and compare this approach, a 2D HEC-RAS (Hydrologic Engineering Center - River Analysis System) model was also developed. Shallow water equations (SWE) were used to solve for the flow patterns moving through the computational mesh. The HEC-RAS model shared the same reservoir domain extent as the CFD model. The proposed arced labyrinth spillway rating curve was used to identify the outflow boundary. We compared the results of the hybrid CFD model with the HEC-RAS model.

- The hybrid 2D/3D CFD model effectively identified potential impacts of the lake's orientation and the peninsula upstream of the spillway.
- It revealed that the peninsula split the reservoir flow, creating unsymmetrical hydraulic characteristics as the flow approached the proposed spillway, resulting in a non-uniform flow distribution across the arced labyrinth weir.
- The model also showed that the approach angle combined with the curved left (eastern) embankment at the entrance caused a significant lowering of the water surface elevations with higher velocities at selected locations on the east side of the arced labyrinth spillway.

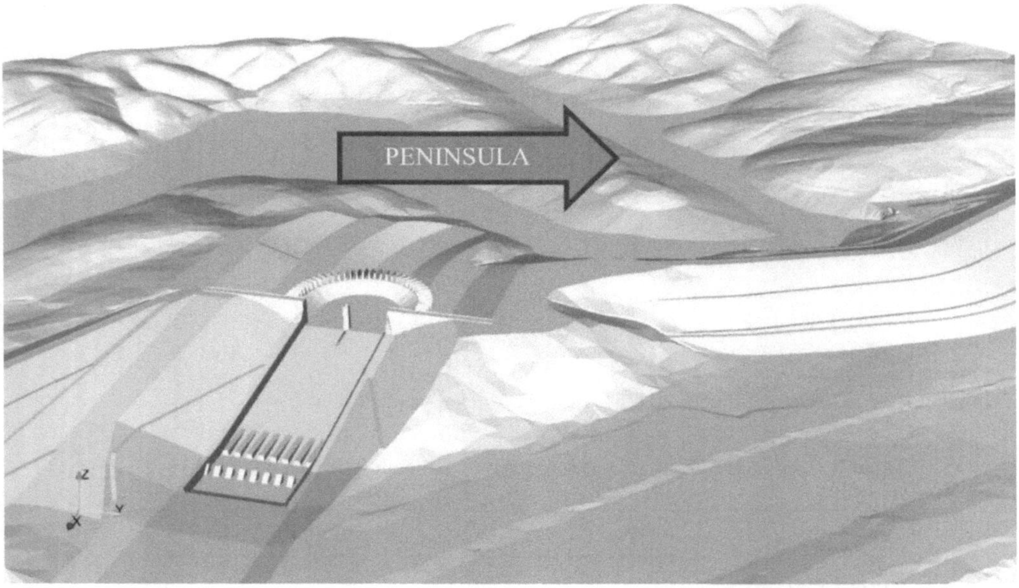

Figure 4. Terrain used in 2D/3D hybrid model.

- The 2D HEC-RAS model produced similar results as the CFD model.
- The multi-directional approach flow was also observed in the HEC-RAS model results.
- Effects on the approach flow caused by the lake orientation and the peninsula upstream of the spillway was notable.
- Hydraulic characteristics like the velocity distribution and water surface elevation along the arced labyrinth spillway was unsymmetrical.
- The higher velocities and the dip on the east side of the spillway was clearly visible.
- Model results are shown in Figure 5.

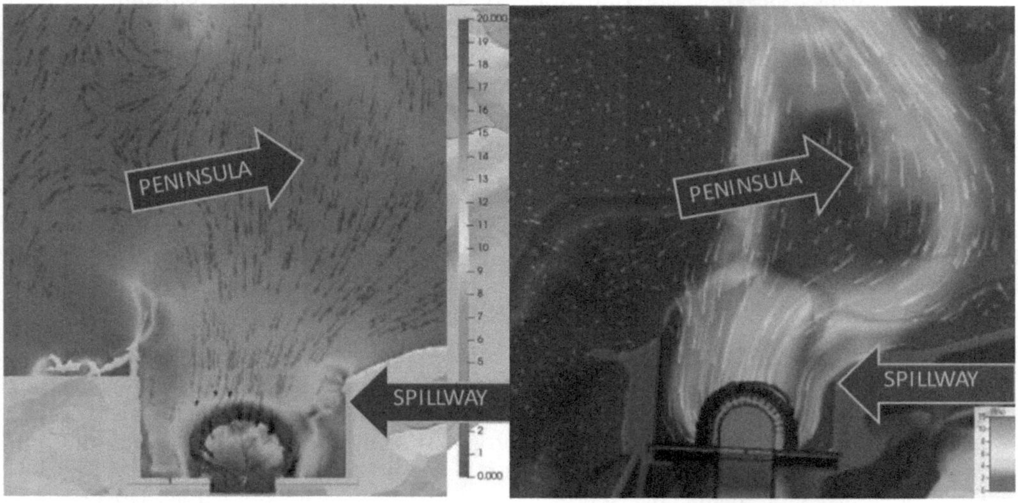

Figure 5. Model of spillway approach flow conditions, hybrid model (left) and HEC-RAS model (right).

3.4 *Phase 1 physical model*

The physical modeling efforts for the Second Raise of the Dog River Dam and Spillway were conducted in 2023 at Utah State University (USU) at the Utah Water Research Laboratory (UWRL) located in Logan, Utah, USA. Physical modeling included a large physical model (1:35 scale) based upon Froude similitude (Ettema 2000). All geometric and hydraulic information needed for scaling was provided by FNI. USU UWRL designed, built, and operated the scaled model.

The model included a portion of the reservoir, the proposed labyrinth crest, chute, stilling basin, and a portion of the downstream area. This model was intended to be a dynamic interactive resource to confirm hydraulic performance and test design options, reduce where permissible construction costs, and to provide detailed hydraulic information about the interaction between water, structure, and terrain. Materials like steel, acrylic, and concrete were majorly used for model construction.

The primary objective of Phase 1 of the Physical Model was to test the spillway design at the 60% design stage and identify any potential refinements or issues — specifically, how flows in the reservoir approached the arced labyrinth weir and how the hydraulic jump performed with basin walls being overtopped. As part of this effort, USU UWRL collected a preliminary data set to support the witness testing (a performance test observed by project stakeholders) with FNI.

An important component of Phase 1 was the calibration of the approaching velocity field in the reservoir as demonstrated by the hybrid CFD modeling of the spillway approach. Phase 1 also explored the downstream flow field and tailwater elevation and how each impacted basin performance. The arced labyrinth weir was observed for discharges ranging from the 100-year storm to the full PMF. It was observed that flows are not evenly distributed across the arced crest, with some bulking and vertical velocities at the left abutment. The reservoir elevations immediately adjacent to the labyrinth weir are also not equivalent, which reflects the approach conditions created in the reservoir. However, these less-than-optimal conditions did not present a concern in the physical model as the head-discharge rating curve was considered acceptable by FNI since good agreement was found between the CFD rating curve and hydrologic analyses by FNI and the experimental data obtained in the physical model.

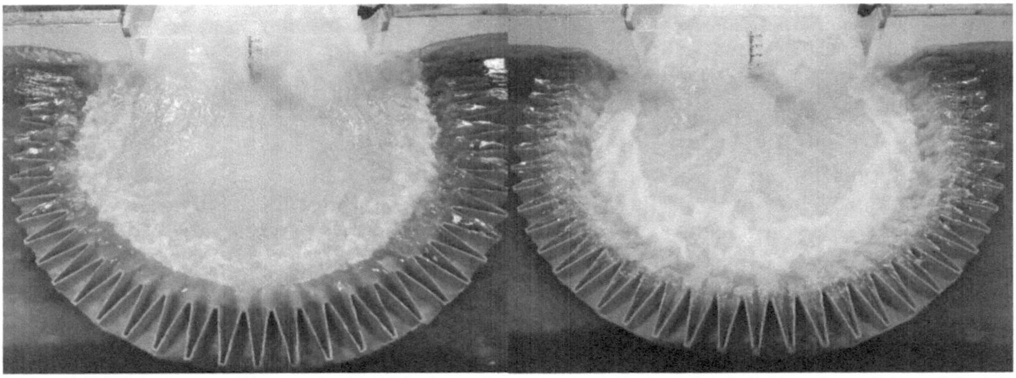

Figure 6. Arced Labyrinth weir at full PMF (left) and 50% PMF (right).

3.5 *Phase 2 physical model*

The primary objective of Phase 2 was to test the refined spillway design and obtain the final hydraulic data sets for design. The primary difference between the preliminary and final spillway design was a narrower excavated channel in the reservoir and relocating the stilling basin upstream, with a short steeper chute section. As part of this effort, USU UWRL held a second witness test with FNI and DDCWSA. During the witness test, the Second Raise of the Dog River Dam and Spillway Phase 2 testing and corresponding data collection by USU UWRL included the following:

- Reservoir elevation (headwater) vs discharge for 11 scenarios (50% PMF and ratios of the PMF) for the new reservoir geometry
- Chute average and maximum water surface profiles for the new chute geometry
- Test of a rounded upstream pier nose and a pier extension
- A modified Reclamation Type III stilling basin
- Velocity measurements in the headbox for the 0.5 PMF and Full PMF
- Detailed velocity measurements adjacent to the stilling basin
- Two PMF scenarios for tailwater elevation 744.9 ft (227.06 m) and elevation 738.5 ft (225.03 m)
- 25 piezometric pressure measurements at select locations in the model for 10% increments of the PMF, for 10 total discharges
- Dynamic pressure measurements and a Fast Fourier Transform (FFT) analysis at select locations
- Additional sensitivity of the basin performance to tailwater and any global direction of flows from the Chattahoochee River (simulated with tailbox vertical gate operations)
- Flow pattern observations

Velocity measurements were conducted using a 1D impeller probe (Global Water), 2D acoustic doppler velocimeter (ADV)(Sontek FlowTracker), 3D ADV (Nortek Vectrino), and a magnetic velocimeter (Marsh McBirney FLO-MATE). To address measurement challenges in aerated flows, probes less affected by bubbles were used – 3D ADV and FLO-MATE.

Pressure measurements were conducted using piezometers and Keller PR-35X pressure transducers. Piezometers provided static pressure readings with an accuracy of ±0.005 ft (model scale), equivalent to approximately 0.2 ft in prototype scale. The Keller PR-35X transducers, used for dynamic pressure measurements, had a total error band of ±0.1%.

4 LESSONS LEARNED

4.1 *Physical model head box design*

The USU UWRL designed the portion of the reservoir in the physical model to be contained in the model headbox. The headbox included the supply pipe with a diffuser located upstream of a baffle wall. The purpose of the baffle wall is twofold: 1) to transition from the turbulent nonuniform flows exiting the diffuser into more uniform quiescent flows, and 2) to be an adjustable component to help establish a flow field (velocity magnitude and direction) in the physical model to match the predicted flow field from the FNI CFD analysis. The baffle wall is used in combination with the terrain in the headbox to establish the upstream boundary condition for all tested discharges. To calibrate this boundary condition of the model, 2D velocity measurements were taken for the full and half PMF and compared to the CFD results; any needed adjustments were made to the baffle wall and the final velocity measurements are summarized in Figure 7. These results were discussed in detail with FNI so that the reservoir flow field was acceptable before proceeding with additional data collection. The discussion included some comments regarding the CFD model results and corresponding CFD setup; FNI approved this data as being appropriate and favored the physical model results when any differences between physical and CFD model velocity vectors occurred.

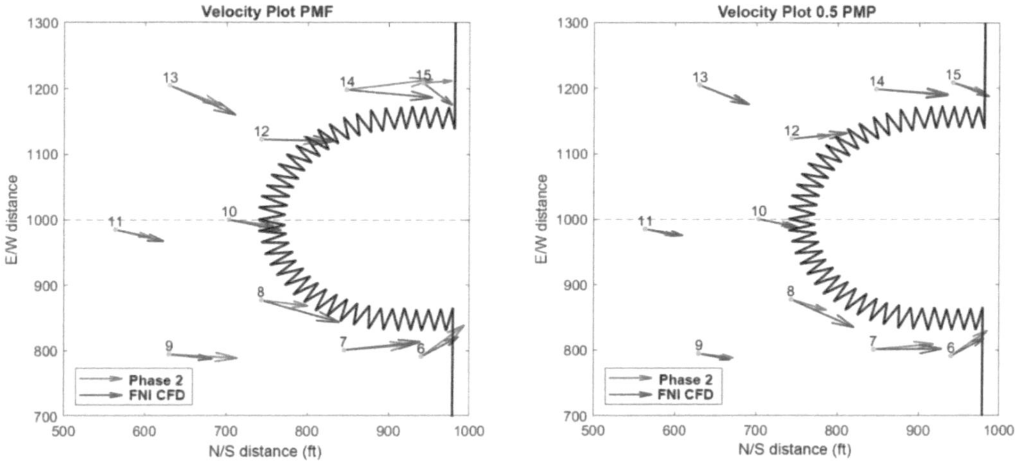

Figure 7. Phase 2 reservoir velocity magnitudes for 100% PMF (195,000 cfs; 5,520 m³/s) and 50% PMF USU UWRL physical model and phase 2 FNI CFD model.

4.2 *Modified approach channel configurations*

The upstream excavated channel in the reservoir was narrowed to consider reduced construction costs for the project. Velocity magnitudes were measured at locations noted in Figure 7 in 10% PMF increments (see Table 1). The arced labyrinth weir was observed for discharges ranging from the 10% PMF to the full PMF. It was observed that flows with the narrower reservoir channel continued to cause a nonuniform flow distribution across the arced labyrinth weir, with some bulking and vertical velocities remaining at the left and right abutments. The effects of the channel are more apparent for the head-discharge rating curve. Despite a less efficient labyrinth weir, the narrower channel was considered acceptable by FNI when balanced against the estimated savings to the project.

4.3 *Bridge pier and abutment affects*

Two components of the spillway that had outsized effects on the spillway performance were the abutment where the labyrinth weir adjoined the RCC non-overflow sections of the spillway and the central bridge pier, located in the center of the spillway chute. The physical model was invaluable for illustrating how these water structure interactions are likely to behave. For example, the left and right abutments of the spillway showed a significant surge of water against the RCC abutment where the flow is forced to make a ninety degree turn to flow over the labyrinth. However, due to the approach asymmetry discussed previously, the surge was significantly greater on the right side of the spillway than on the left. With the water levels on the right surging approximately 3 ft (0.91 m) above the "still" water reservoir level. Figure 8 shows the water surface at the left and right abutments during the 100% PMF simulation.

A modification to the bridge pier for the vehicular bridge was considered where a rounded bullnose was also added to the pier on the upstream face (see Figure 9). The bullnose had negative effects on chute hydraulics when compared to the unaltered pier geometry, but as is discussed below, there were still significant effects of the convergence of flow through the bridge opening.

4.4 *Reservoir elevation-discharge rating curve*

Each of the topics of discussion above had a nominal effect on the discharge rating curve of the spillway that wasn't fully realized until the completion of Phase 2 of the physical model

Figure 8. Reservoir-labyrinth abutments for full PMF. Left abutment (left) and right abutment (right).

Figure 9. Bridge pier rounded bullnose alteration.

study. Figure 10 shows the elevation discharge results of the final physical model configuration as compared to Phase 1 of the physical model and the CFD Modeling. As can be seen, the CFD model generated a reasonable first iteration of the rating curve. Available data points for the CFD Phase 2 model compared well with the physical model data. Minor inefficiencies that may be missed in a CFD model would be captured in a physical model so the physical model rating curve was selected for final design.

4.5 *Chute water surface profiles*

Immediately downstream of the arced labyrinth weir is an apron where flows converge and transition into the narrower spillway chute. This is a turbulent aerated region. The abrupt transition into the narrower chute resulted in shockwaves that were the full length of the chute and entered the stilling basin (see Figure 11).

The resulting conditions along the spillway chute resulted in a region of flow separation near the entrance and shockwaves impinging against the downstream walls, increasing and decreasing flow depths along spillway stationing, with some helical rotation observed via dye tracing. GA-SDP does not have specific criteria for training wall height, but the walls were designed to not overtop during the PMF events, with no freeboard. These flow dynamics were relevant to the energy dissipation provided by the stilling basin. Ultimately, the stilling basin was optimized for the conditions observed in the physical model study, but these shockwaves did have some effect on the necessary length of the basin, the height of the baffle blocks, and the height of the chute walls.

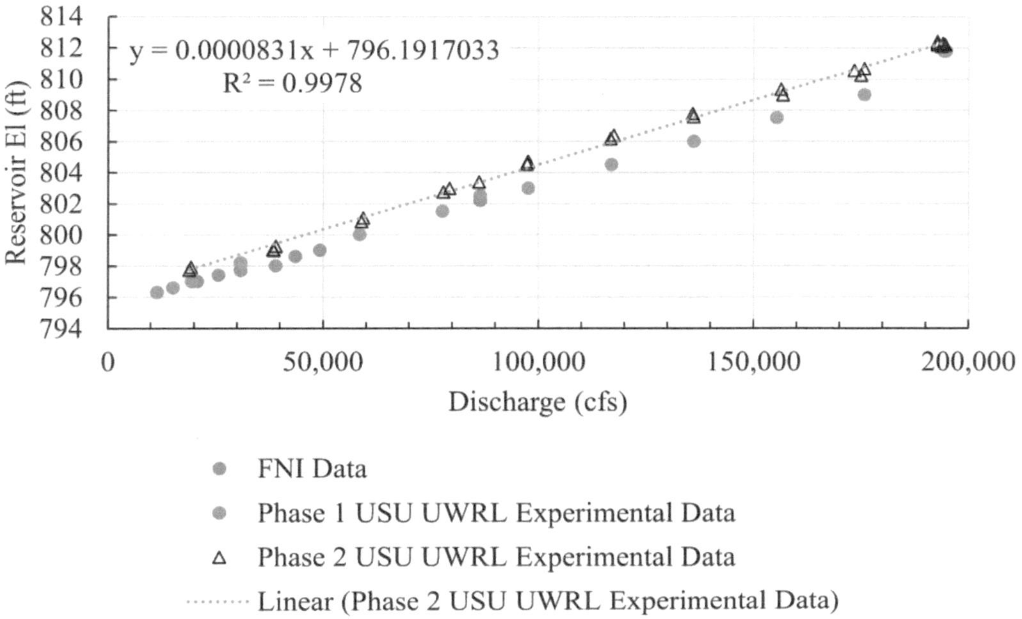

Figure 10. Phase 2 reservoir elevation-discharge rating curve.

Figure 11. Spillway chute for 50% PMF (left) and full PMF (right).

5 CONCLUSIONS

While the configuration of the Dog River Spillway is unique and somewhat complicated, it is a result of the physical setting in which it is being built and the design constraints and criteria established by the owner, regulators, and engineers involved with its design. It is a great example of what innovative thinking can create to solve the problems presented by dam rehabilitations and enlargements. When designing a spillway that is "outside the box," it is imperative to use the most up-to-date technologies and thoroughly study all calculations and models, revising as necessary, and checking, double-checking, and validating with other methods to get the most accurate results.

REFERENCES

Crookston, B. M., and B. P. Tullis. 2012a. "Arced labyrinth weirs." J. Hydraul. Eng. 138 (6): 555–562. https://doi.org/10.1061/(ASCE)HY.1943-7900.0000553.

Crookston, B. M., and B. P. Tullis. 2012b. "Discharge efficiency of reservoir-application-specific labyrinth weirs." J. Irrig. Drain. Eng. 138 (6): 564–568. https://doi.org/10.1061/(ASCE)IR.1943-4774.0000451.

Crookston, B. M., and B. P. Tullis. 2013a. "Hydraulic design and analysis of labyrinth weirs. I: Discharge relationships." J. Irrig. Drain. Eng. 139 (5): 363–370. https://doi.org/10.1061/(ASCE)IR.1943-4774.0000558.

Ettema, R. 2000. Hydraulic modeling, concepts and practice. Reston, VA: ASCE Press.

Falvey, H. T. 2003. Hydraulic design of labyrinth weirs. Reston, VA: ASCE Press.

Flow Science, Inc. 2024. FLOW-3D® Version 2023R1 User's Manual. Santa Fe, NM: https://www.flow3d.com

Tullis, B., B. M. Crookston, and N. Young. 2019. "Nonlinear weir free-flow hydraulics and size-scale effects." J. Hydraul. Eng. https://doi.org/10.1061/(ASCE)HY.1943-7900.0001661.

Tullis, B. P., J. Young, and M. Chandler. 2007. "Head-discharge relationships for submerged labyrinth weirs." J. Hydraul. Eng. 133 (3): 248–254. https://doi.org/10.1061/(ASCE)0733-9429(2007)133:3(248).

Tullis, B. P., J. Young, and M. Chandler. 2007. "Head-discharge relationships for submerged labyrinth weirs." J. Hydraul. Eng. 133 (3): 248–254. https://doi.org/10.1061/(ASCE)0733-9429(2007)133:3(248).

US Army Corps of Engineers, Hydrologic Engineering Center. HEC-RAS, "Hydraulic Reference Manual,". Davis, CA.

Labyrinth and Piano Key Weirs IV – Erpicum, Crookston & Paxson (Eds)
© 2026 The Author(s), ISBN 978-1-041-10527-5

Advances in arced labyrinth weir research and application at the Springton Spillway project

S.D. Thompson
Gannett Fleming, Salt Lake City, Utah, USA

B.P. Tullis and B.M. Crookston
Utah State University, Logan, Utah, USA

M.J. Marchisello
Aqua Pennsylvania, Philadelphia, Pennsylvania, USA

ABSTRACT: Arced labyrinth weirs, characterized by a curved cycle orientation, can enhance discharge efficiency compared to linear labyrinth weirs by taking advantage of the approaching flow field in reservoirs. This paper summarizes some of the Utah State University arced labyrinth weir hydraulic efficiency research findings, including sidewall and cycle arc angle influences, nappe aeration, and local submergence. Recent CFD modeling that replicates some of the experimental findings is also highlighted, demonstrating CFD utility for design. The Springton Spillway project, owned by Aqua Pennsylvania, leveraged this research when designing a two-stage, eight-cycle arced labyrinth weir to replace an aging ogee crest spillway in order to meet current flood safety standards. Drawing parallels to the Lake Isabella project, which integrated empirical data and physical modeling, the Springton spillway project exemplifies the application of research-based innovations in finding solutions to complex hydraulic challenges.

1 INTRODUCTION

Dam spillways typically use weirs as flow control structures to regulate reservoir flood discharges to the river system downstream. A trend in the dam safety industry over the past three decades has been to retrofit existing spillways with designs that increase flow capacity to accommodate larger probable maximum flood (PMF) estimates, more stringent inflow design flood (IDF) regulations, and reduce the risk of an overtopping dam failure. Non-linear weirs, such as labyrinth weirs, are commonly used to achieve this due to their ability to increase the weir length (L_c), and subsequently discharge, without increasing the existing channel width. Additionally, the increased hydraulic efficiency of the nonlinear weir, relative to the linear weir it replaces, often allows the reservoir normal pool to be increased without reducing dam and public safety. In some cases, passive-control non-linear weirs have also replaced or been added in parallel to gated spillway structures to reduce the risk of operational failures.

Typically, the cycles of labyrinth weirs are arranged linearly (i.e., a rectangular footprint or linear axis in plan). Initial research dates back over 80 years (Gentilini 1941) with numerous engineering and scientific studies, including Hay & Taylor (1970), Darvas (1971), and Tullis et al. (1995). The majority of published research has been conducted for channelized applications, including the design methodology developed by Crookston & Tullis (2013a) that has become the industry standard in the USA for labyrinth weir design. In more recent years, studies have explored an arced labyrinth cycle orientation (see Figure 1) to increase L_c and improve the cycle orientation to the converging approaching flow common in reservoir applications.

DOI: 10.1201/9781003655541-21

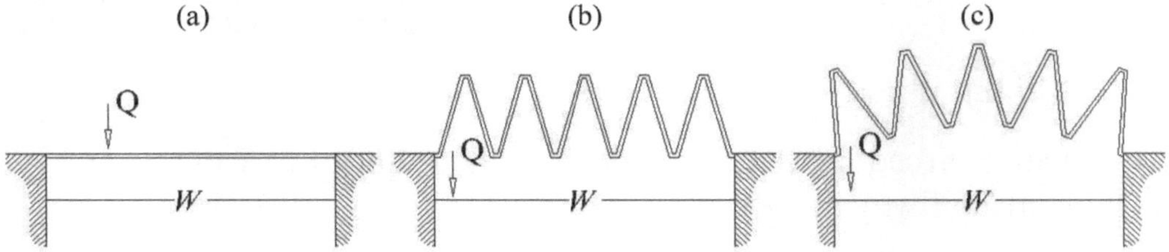

Figure 1. Geometric comparison between linear (a), traditional labyrinth (b), and arced labyrinth (c) weirs.

2 ARCED LABYRINTH WEIR LITERATURE

2.1 *Key geometric parameters*

In addition to the geometric parameters used to describe traditional labyrinth weirs, arcing the weir in plan introduces additional parameters that need to be considered in design. Crookston & Tullis (2012a) introduced a standard geometric nomenclature for arced labyrinth weirs (see Figure 2) that includes the cycle arc angle (θ), the weir arc angle (Θ), the downstream cycle sidewall angle (α), and the upstream cycle sidewall angle (α'). Both θ and α have been used to present hydraulic efficiency relationships, i.e., the relationship between the weir discharge coefficient (C_d) and the headwater ratio (total head [H] divided by weir height [P], H/P). Additionally, as the weir's radius of curvature (R) decreases, θ increases, and α' increases relative to α, which impacts C_d and local submergence effects.

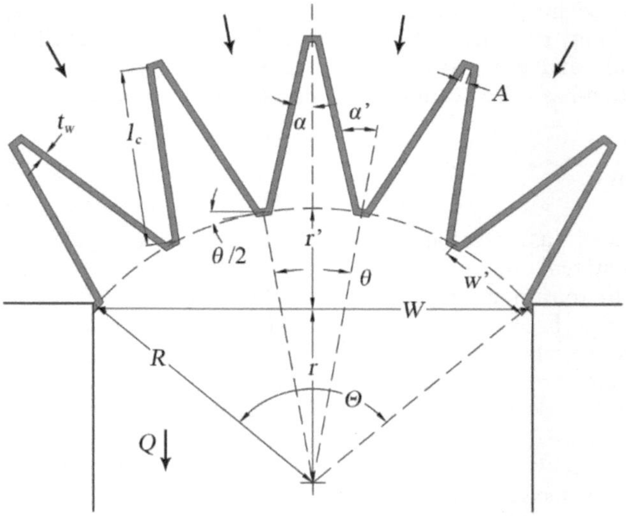

Figure 2. Arced labyrinth weir nomenclature (adapted from Crookston & Tullis 2012a).

2.2 *Experimental studies*

Experimental studies of arced labyrinth weirs started with Crookston in 2010 and continued with Christensen (2012) and Thompson (2019). Each study used an elevated reservoir headbox with flow entering through a diffuser and baffle wall from three sides, which allowed for a 180° flow to converge toward the weir and simulated ideal reservoir conditions. The weirs were installed on a slightly elevated, horizontal platform and the weir crests leveled to within ±0.8 mm. Discharges were measured with calibrated orifice plates and magnetic flow meters. The arced geometries evaluated by each researcher are listed in Table 1. All of the arced weirs were projecting into the reservoir, as shown in Figure 1c. For direct comparison with the arced weirs, Crookston & Tullis (2012b) and Christensen (2012) also evaluated non-arced labyrinth weir geometries for reservoir application, which included projecting, flush (square abutments), and rounded inlet (rounded abutment) geometries as depicted in see Figure 3a, b, and c, respectively.

Table 1. Summary of arced labyrinth weir geometries evaluated.

Researcher	α degrees	θ degrees	*Cycle number (N)*
Crookston	6	0, 10, 20, 30	5
	12	0, 10, 20, 30	5
Christensen	12	10	10
	12	10	7
	12	10, 20	5
	20	10	10
	20	0, 10, 20, 30	5
Thompson	16	10, 20, 30	5

Note: All model configurations featured: P=203.2 mm and w/P= 2.

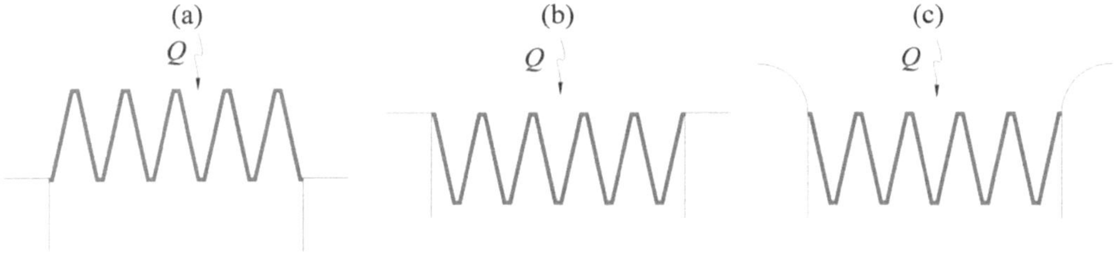

Figure 3. Traditional, linear labyrinth weir reservoir applications: projecting (a), flush (b), rounded inlet (c).

Hydraulic performance of each weir geometry was evaluated for H/P values ranging from approximately 0.05 to 0.8. The weir head-discharge equation (1) was used to relate the measured H and Q to C_d, and evaluate hydraulic efficiency. Due to minimal approach velocities in the reservoir, including at the piezometric head (h) measurement location, the measure h was assumed to equal H. This assumption was confirmed with the three-dimensional (3D) computational fluid dynamics (CFD) modeling conducted by Thompson (2019) which found that velocity head values at the h measurement location ranged from 0.1% to 0.25% of H. Therefore, C_d is computed as a function of h (i.e., $h \approx H$) and is influenced by weir geometry, approach flow field, nappe stability and aeration, and local submergence.

$$Q = \frac{2}{3} C_d L_c \sqrt{2g} H^{\frac{3}{2}} \tag{1}$$

where Q = discharge; C_d = weir discharge coefficient; L_c = weir crest length; g = acceleration of gravity, and H = total head measured relative to the weir crest.

Figure 4 provides the C_d versus H/P relationship for each arced labyrinth weir geometry [e.g., α and θ, (N=5)] evaluated. Geometric impacts, nappe behavior, and local submergence are discussed in Sections 2.2.1 and 2.2.2. In addition to these observations, the N=5 data were used to create curve-fit equations to facilitate designers' estimation of C_d at a given H/P;

moreover, Christensen found that varying N had a negligible impact on C_d for the heads and geometries evaluated. The curve-fit equation for weirs of α=6° is provided by (2), and (3) approximates C_d for α=12°, 16°, and 20°. The empirical coefficients, *a, b, c,* and *d* for each geometry are given on Table 2. Data in Figure 4 should still be referenced for the most accurate C_d information.

$$C_d = ab^{\frac{H}{P}}\left(\frac{H}{P}\right)^c + d \tag{2}$$

$$C_d = \frac{1}{a\left(\frac{H}{P}+b\right)^2 + c} + d\ln\left(\frac{H}{P}\right) \tag{3}$$

where *a, b, c, d* = are empirical coefficients.

Table 2. Empirical coefficients for equations (2) and (3).

θ	α		Empirical coefficients				
degrees	degrees	Equation	*a*	*b*	*c*	*d*	Correlation coefficient (R^2)
10	6	(2)	65.7156	-0.1048	2.0860	0.2005	0.997
	12	(3)	1.6633	0.2532	0.4867	0.3342	0.998
	16	(3)	1.2573	0.2784	0.4420	0.4128	0.995
	20	(3)	0.8774	0.5045	0.4539	0.2735	0.984
20	6	(2)	52.9460	-0.1108	1.9018	0.1784	0.998
	12	(3)	2.1060	0.1227	0.5972	0.3214	0.999
	16	(3)	1.43872	0.1636	0.6310	0.3100	0.996
	20	(3)	1.0847	0.3008	0.5319	0.3276	0.995
30	6	(2)	68.6077	-0.1088	1.7951	0.1623	0.996
	12	(3)	2.7769	0.0482	0.6341	0.3244	0.992
	16	(3)	2.1022	0.0410	0.7260	0.2751	0.998
	20	(3)	1.6654	0.1040	0.6576	0.3142	0.996

2.2.1 *Geometric influences*

As shown in Figure 4, all three studies observed an increasing trend in C_d at each α when θ increases from 10° to 20°. For θ=30°, however, the efficiency drops below the θ=10° and 20° geometries at higher heads ($H/P \gtrsim 0.3$). At lower heads ($0.3 \gtrsim H/P$), the increase in θ (i.e., θ=30° compared to θ=10° or 20°) better orients the labyrinth cycles to the approaching flow field. However, as H increases, the accelerating approach velocity immediately adjacent and upstream of the labyrinth cycles causes the streamlines to orient more toward the downstream channel and less toward the weir cycles. This behavior increases local submergence at the upstream cycle apexes and then moving downstream across the cycles with increasing H, ultimately shifting the control point downstream of the weir in the modeled configuration. In application, this control shift will depend on the downstream chute hydraulics. These factors decrease efficiency at higher values of H/P.

Christensen (2012) observed this decrease in efficiency by collecting velocity vector data on the α=20°: θ=30° arced labyrinth weir at H/P=0.3 and 0.6 and observing the shift in vector orientation for the two discharges evaluated. Thompson (2019) made a similar comparison using a CFD model (see Section 2.3). Out of the θ angles assessed and for each α, the θ=20° balances the benefits of improved discharge capacity at lower heads compared to θ=10° and at higher heads compared to θ=30°.

Additionally, arced weirs cause outflows to converge towards the center or origin of R. As θ increases, R decreases, causing this cycle outflows to converge closer to the weir. This creates

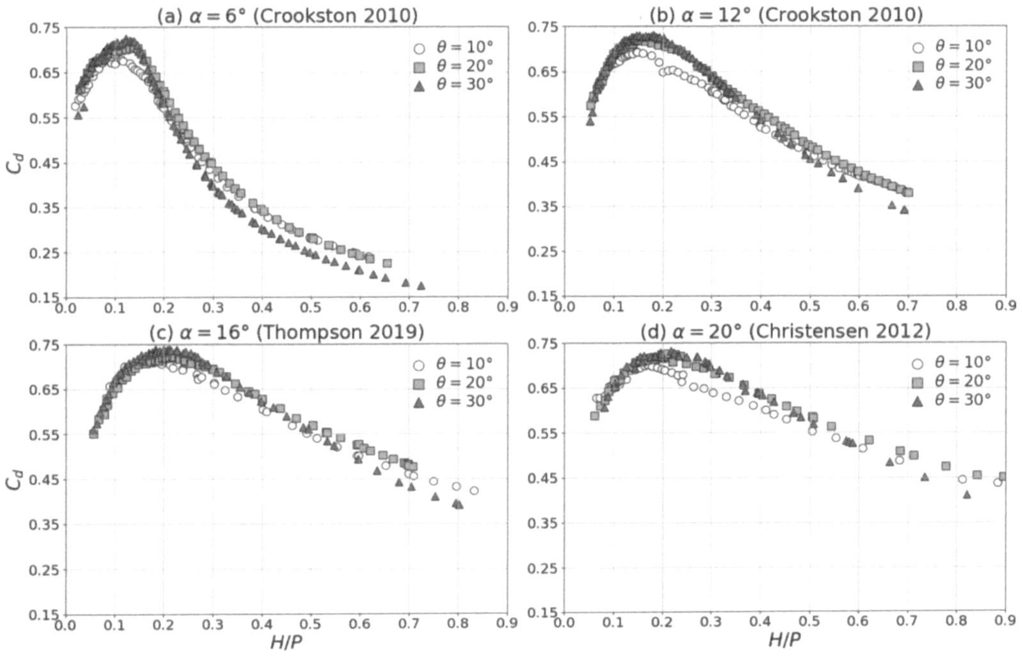

Figure 4. C_d versus H/P relationship for (a) α=6°, (b) 12°, (c) 16°, and (d) 20°, each with θ=10°, 20°, 30°.

a standing wave or mound downstream of the weir. In addition to the changes in the approaching flow field at high heads, this flow mounding can result in weir submergence and caused the θ=30° weirs to decrease in efficiency relative to θ=10° and 20° at $H/P \gtrsim 0.2$ to 0.5 (varies with α). Figure 5 highlights this behavior at H/P=0.3 and 0.6 for the α=16° weirs; for θ=10° and 20° the flow in the outlet channel is more aerated and shallower compared to the non-aerated outflow of the θ=30° weir. The lack of aeration for θ=30° is a result of the deeper tailwater caused by the converging outflow.

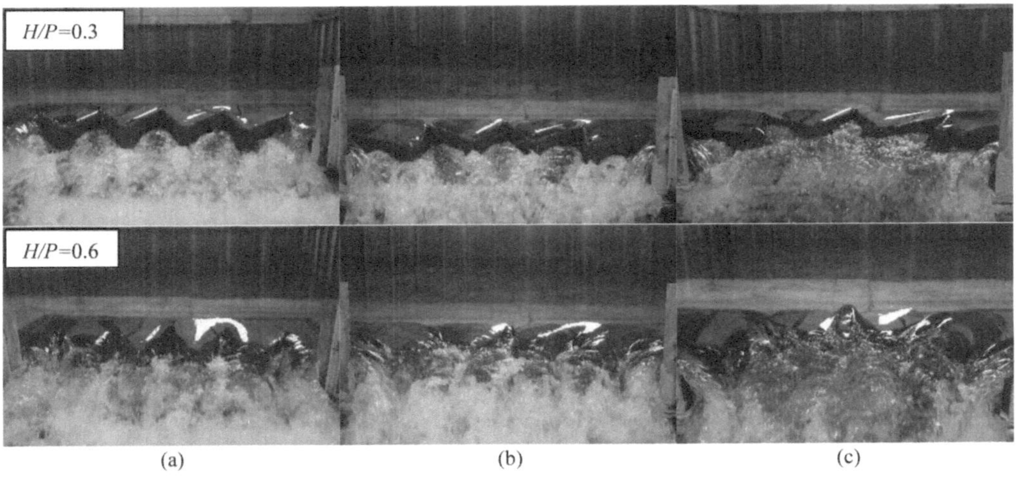

Figure 5. Comparison of tailwater submergence at H/P=0.3 and 0.6 for α=16°: θ=10°(a), 20°(b), and 30°(c).

2.2.2 *Nappe aeration and local submergence*

For arced and non-arced labyrinth weirs in reservoir applications, nappe aeration and local submergence behave uniquely compared to the in-channel traditional labyrinth weirs evaluated by Crookston (2010). Crookston describes a natural and abrupt shift in the nappe from clinging to aerated at $H/P=0.25$ for $\alpha=12°$ in-channel weirs (Crookston 2010). This abrupt shift from clinging to aerated was not observed in any of the reservoir geometries (arced and non-arced). For the arced weirs, only partial aeration (generally on the distal cycles) was achieved at heads between H/P of 0.15 and 0.3. For H/P above ~0.25 (varies with α) the nappe would become unstable and oscillate from aerated to drowned; air cavities traveling upstream from the aerated tailwater into the downstream cycles would become rapidly filled with water and forced downstream.

As α increases the nappe generally becomes more unstable with increasing H/P. Christensen (2012) observed an unsteady, partially aerated nappe for the $\alpha=20°$ geometries for $0.3 \leq H/P \leq 0.5$; for the $\alpha=12°$ weirs this instability region decreased to H/P between $0.25 \lesssim H/P \leq 0.34$. Thompson's (2019) $\alpha=16°$ weirs follow the trend with nappe instability occurring for $0.3 \leq H/P \leq 0.4$. This trend can be explained when considering the local submergence phenomenon of labyrinth weirs.

Local submergence occurs when the flow over an upstream weir apex exceeds the flow capacity immediately downstream, creating a localized backwater effect at the upstream apex. This behavior starts at the upstream apexes due to colliding nappes and the limited width of the upstream region of the outlet cycle. As head increases, the local submergence increases, lessoning the effective weir length, and eventually drowning out the nappe. At smaller α angles the outlet cycles have less flow capacity relative to larger α angles causing the nappe to drown sooner. This behavior is what causes the nappe instability to become more pronounced at higher H/P values as α increases. While local submergence and a drowned nappe decrease the weir's efficiency, as captured in the C_d versus H/P plots (Figure 4), it does stabilize the nappe and reduces the intensity of fluctuating pressures acting on the weir wall. Additionally, for arced labyrinth weirs the upstream cycle angles (α') is greater than the downstream cycle angle (α), and at lower heads, the larger α' angles reduce energy losses and improve efficiency compared to traditional weirs. The smaller outlet cycles relative to the inlet cycles cause arced labyrinth weirs to submerge at smaller heads relative to traditional labyrinth weirs; this is especially visible at $\theta=30°$.

2.3 *Numerical validations*

Initial numerical validations were conducted by Thompson (2019) using Flow-3D (solver version 11.2.4.01), a commercially available CFD software, which numerically solves the Reynolds-Averaged Navier Stokes (RANS) equations of fluid motion for a finite element (computational cell). The main objective of the CFD modeling was to compare the numerical C_d data with the physical models. To achieve this Thompson evaluated $\alpha = 16°$; $\theta=10°$ and 30° weirs at six discharges resulting in H/P values between 0.3 and 0.8.

The CFD modeling research efforts took advantage of the symmetrical laboratory reservoir setup and only modeled one half of the weir and reservoir, as shown in Figure 6, to optimize run times by reducing computational demand. The upstream (*y-min*) and right (*x-min*) boundary conditions were set to flow rates based on the laboratory C_d data, while the left (*x-max*) boundary condition was set to a symmetry or free-slip boundary. Additionally, the downstream (*y-max*) boundary was set to be a free outflow.

2.3.1 *CFD models and key parameters*

Flow-3D by Flow Science pioneered the volume-of-fluid (VOF) (Hirt and Nichols 1981) free surface tracking numerical model, which uses a ratio to calculate the fraction of a cell that is filled with the fluid, the surface interface area (A_x, A_y, A_z) is also calculated; these two parameters allow for water surface tracking with respect to time and space. The VOF method used by Thompson (2019) was Flow-3D's split-lagrangian advection model, which was coupled

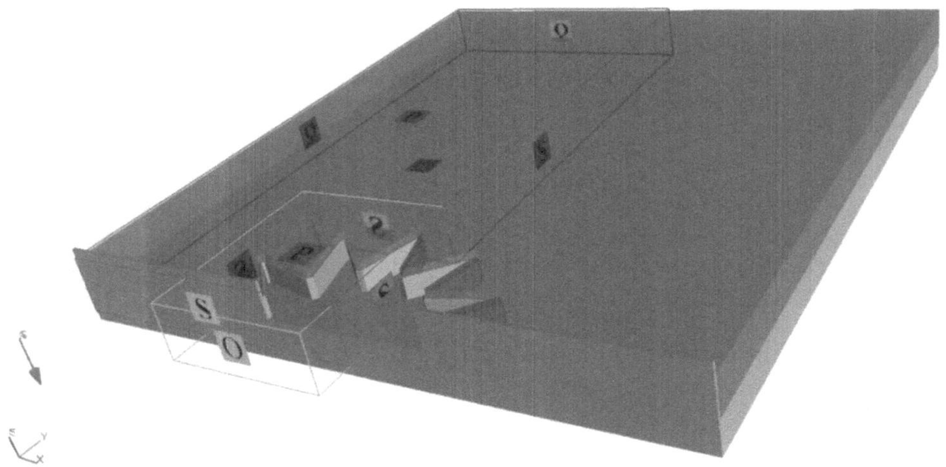

Figure 6. Flow-3D CFD model extents and boundary conditions.

with a second order momentum advection scheme and the k-ε Renormalized Group (RNG) turbulence model with a dynamically computed turbulent mixing length.

In 2023, Koldewyn tested some of these parameters in a controlled framework to understand their impact on predicting hydraulic performance (i.e., estimating C_d). Koldewyn (2023) found that a first-order momentum advection scheme computed C_d values that compared better to the experimental data at $H/P > 0.5$ by 2% to 10% when compared to the second order scheme. Additionally, the second order scheme resulted in more instability and significantly longer computation durations. Given the lack of accuracy at higher heads, Koldewyn (2023) states that the second order scheme might not be worth the added computational time and cost. Additionally, Koldwyn (2023) simulated various turbulence models, including the 1-Equation, k-ε, k-ε RNG, and k-ω models. The CFD models predicted C_d relatively consistently with both the k-ε models, and the k-ω model, indicating hydraulic efficiency may not be sensitive to the model selected. However, results from the 1-Equation model were scattered in nature with no clear pattern and are not recommended for use in modeling arced labyrinth weirs. Turbulence model selection between the k-ε, k-ε RNG, and the k-ω models becomes more important when other results such as negative weir pressure are needed (Koldwyn 2023).

Within Flow-3D, a hexahedral or cubic, non-conforming mesh is used with the use of mesh blocks to specify cell sizes for certain model regions. For the arced labyrinth models, two mesh blocks were used to define the mesh, a reservoir mesh block (RM) and a finer weir mesh block (WM), shown in Figure 6 by the green and orange meshblock outlines, respectively. Thompson (2019) evaluated the mesh independency using the grid convergence index (GCI) (Roache 2009) using the relative error in C_d between the physical and numerical data as the variable of interest. The computed GCI values were within 2% for nearly all discharges below H/P=0.6. At H/P values greater than 0.6 the solutions diverged from the physical model with GCI values between 3% and 5%. As a result of this phenomenon, one of the intermediate mesh sizes (WM2, 9.14 mm) between the most coarse (WM0, 12.2 mm, and WM3, 6.86 mm) best correlated with the laboratory experimental data.

Thompson (2019) hypothesized that this behavior was due to the large quantity of cells used to resolve the velocity profile at this location. Flow-3D version 11.2 used a log-law relationship to estimate the velocity profile near wall boundaries and just outside the viscous sub-layer; their documentation suggested sizing cells near wall boundaries such that the viscous sub-layer and log-law region of the boundary layer end within the first cell; otherwise, the near-wall velocity and shear stress computations deviate from the physical case (Flow Science 2019). Due to the high velocities and increased critical depths associated with $H/P \geq 0.6$, it is assumed that the finest mesh used multiple cells to resolve the viscous sub-layer, which can cause the numerical solution to deviate from physical observations. It is also possible that the second order momentum advection scheme simply overpredicts discharge efficiency at the higher heads due to the increased turbulence. As previously stated, Koldewyn (2023) found greater agreement at higher heads using a first order approximation.

2.3.2 *Model validation*

Figure 7 shows the results for both α=16° weirs simulated at the WM2 (9.14 mm) mesh compared to the physical data along with 5% error bounds, computed about (3). All results of the WM2 mesh had relative errors (error between numerical and [3]) in C_d of -2.5% to 2.8% for the α=16°: θ=10° weir and relative errors of -0.4% to 3.6% for θ=30°. Additionally, below H/P values of 0.6, the WM3 (6.86 mm) mesh produced accurate results with relative errors between -1.9% and 0.7% for θ=10° and 0.3% and 1.2% for θ=30°.

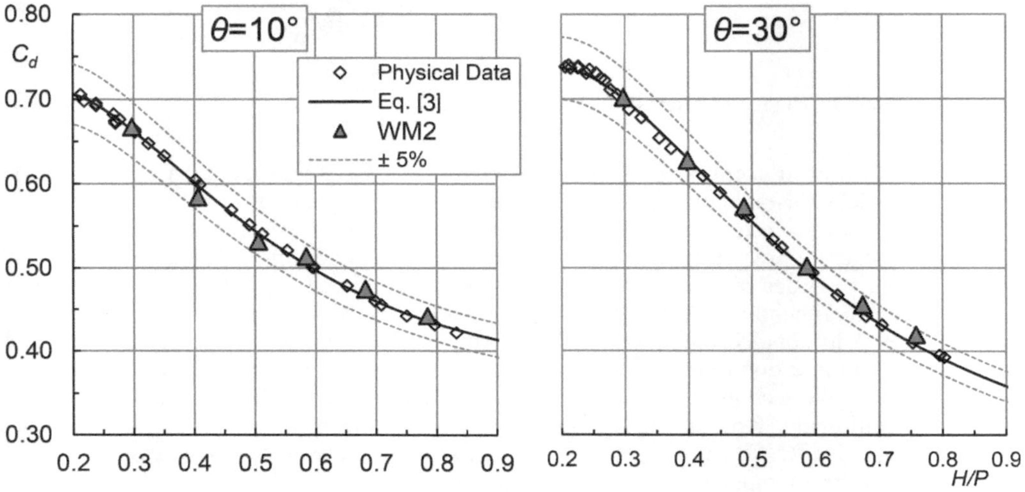

Figure 7. Comparison of Thompson's (2019) physical and numerical data at α=16°: θ=10° (left) and 30° (right).

In addition to comparing discharge efficiency with the physical data, the CFD models were used to compare nappe behavior and approach flow conditions. Both Thompson and Koldewyn found that nappe behaviors observed in the physical models were also predicted using CFD. As examples, Figure 8 shows a similar drowned nappe condition between the two models, and Figure 9 shows similar nappe aeration, turbulence, and tailwater conditions.

Additionally, Thompson (2019) used CFD to observe reservoir and outlet channel conditions to better understand hydraulic behavior. Approach streamlines between the θ=10° and 30° weirs are compared in Figure 10. As shown, streamlines are more uniformly approaching the weir crest for the θ=30° weir at H/P=0.4 indicating the weir's better alignment to the approaching flow field. However, at higher heads the orientation of the θ=30° cycles combined with the higher velocity flow and converging tailwater submerged the weir and the control section shifted from the weir crest to the outlet channel downstream; Thompson (2019) illustrated this shift in control by observing the localized Froude Number within the CFD models at H/P values of 0.4 and 0.8 (see Figure 11). This control shift validated the drop in hydraulic efficiency relative to the other θ geometries Thompson (2019) evaluated in the physical modeling.

2.3.3 *Design application insights*

The good correlation of CFD and experimental data suggest that CFD can be a useful tool for estimating hydraulic performance of arced labyrinth weirs; however, care must be taken to ensure

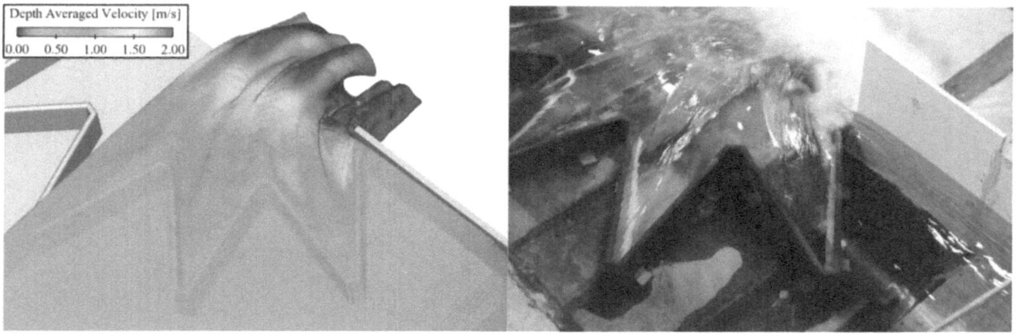

Figure 8. α=16°: θ=30° H/P=0.8 comparison between numerical and physical drowned nappe condition (Thompson 2019).

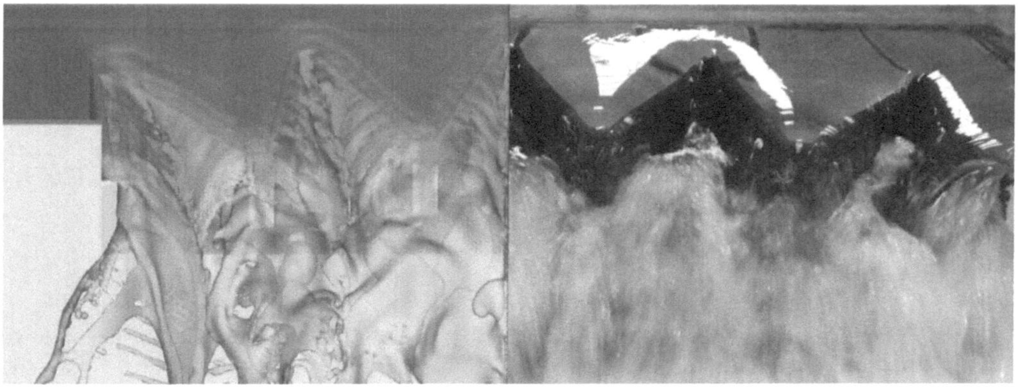

Figure 9. α=16°: θ=10° H/P=0.4 comparison between numerical and physical nappe and tailwater aeration (Thompson 2019).

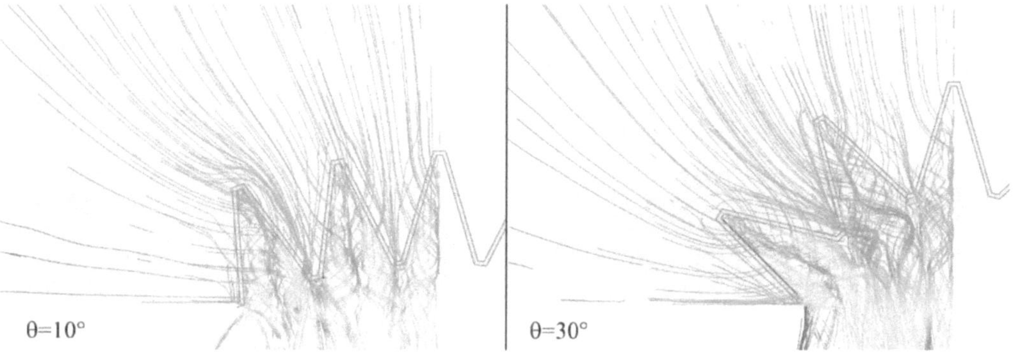

Figure 10. α=16°: θ=10° (left) and 30° (right) streamline comparison at H/P=0.4 (Thompson 2019).

the CFD is providing accurate results. Such care could include comparison to the experimental research paired with a mesh independency and parameter sensitivity analysis. Of these parameters, the momentum advection scheme is of important consideration (Koldewyn 2023); turbulence model selection is also important for design, but less so for estimating hydraulic efficiency. Therefore, sensitivity to turbulence model should be evaluated for other design criteria such as structural loading and/or energy dissipation.

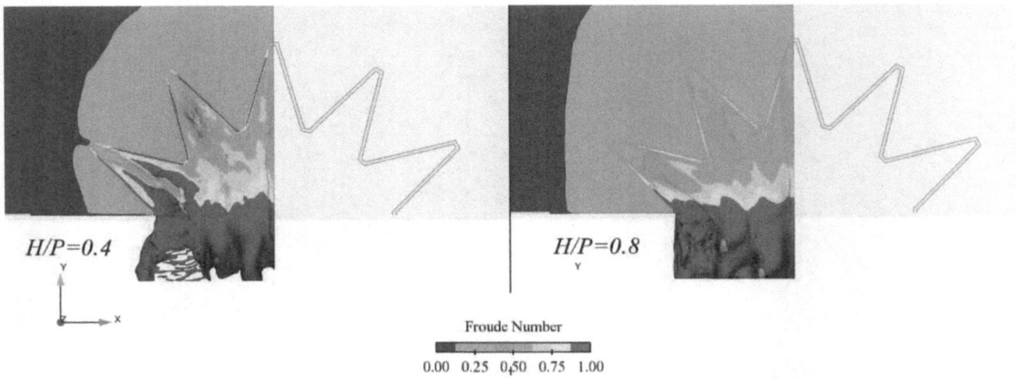

Figure 11. α=16°: θ=30° comparison of localized Froude Number for H/P=0.4 (left) and H/P=0.8 (right) (Thompson, 2019).

Physical modeling is also recommended for unique cases where more uncertainty is present due to a range of factors that may include, the upstream approach conditions, downstream submergence levels, a design outside the geometric bounds of the weirs evaluated, etc. The Springton Spillway Replacement Project provides a design example where CFD was used to augment the empirical C_d relationships derived by Thompson (2019).

3 SPRINGTON SPILLWAY PROJECT

3.1 *Background information and project drivers*

Springton Dam was constructed between 1929 and 1932 for water supply storage. The dam is currently owned and operated by Aqua Pennsylvania (AQUA PA) and located in Delaware County. The dam consists of an earthen embankment approximately 21.3 m high and 488 m long, protected with rockfill armoring on the entirety of both the upstream and downstream slopes.

The existing spillway consists of a 91.4 m long masonry arced ogee weir constructed of cyclopean concrete, keyed into rock, and capped with reinforced concrete. The spillway chute converges to a width of approximately 11 m under a highway bridge. The chute is lined with reinforced concrete and extends approximately 102 m downstream of the spillway crest where the spillway discharge then cascades approximately 15.2 m down an unprotected rock cliff face to the natural stream channel (see Figure 12).

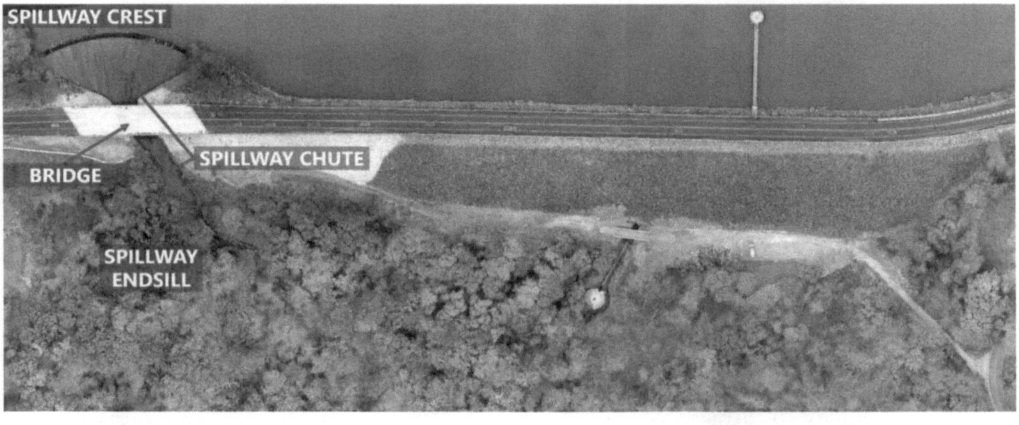

Figure 12. Aerial image of dam and spillway features.

According to Pennsylvania (PA) State Code, Springton Dam is classified by the PA Division of Dam Safety as a medium sized, high hazard structure. Therefore, the IDF for Springton Dam is the PMF. A 1978 U.S. Army Corps of Engineers Inspection Report identified insufficient spillway capacity as the dam's primary deficiency. Subsequent PA dam safety reviews designated the spillway as "seriously inadequate" and requested that AQUA PA address this deficiency. The peak reservoir inflow for the PMF has been established with concurrence by the Pennsylvania Department of Environmental Protection (PADEP) to be 1,510 m^3/s.

The existing narrow spillway chute constricts the flow as it passes under the highway bridge and limits the total capacity of the spillway to less than 388 m^3/s. Discharges exceeding 388 m^3/s would overtop the parapets on the top of the dam and potentially erode the embankment and cause the dam to fail. Embankment overtopping is predicted to occur during flood events exceeding 56% of the PMF. Inadequate spillway capacity remains the most significant dam safety issue at the dam. As such, the purpose of this spillway expansion project is to construct an enlarged spillway able to convey the dam's IDF.

In 2007, Gannett Fleming evaluated two alternatives for increasing the dam's spillway capacity to convey the PMF: (1) adding roller compacted concrete (RCC) overtopping protection to the embankment and (2) replacing the existing spillway with a new, enlarged spillway. Alternative 2 was selected as the most feasible option. The first step in implementing Alternative 2 was the replacement of the original 11 m span highway bridge with a new single, 44 m span bridge. Construction of the new bridge commenced in 2020 and finished in 2022.

With the new bridge constructed, the design of the new, enlarged spillway commenced. The new spillway design replaces the existing 102 m long arced ogee with an arced labyrinth weir, and the concrete spillway chute was enlarged along with adding modern drainage features, waterstops, and anchors to defend against seepage and uplift pressures. With the single span of the new highway bridge being 44 m, the spillway chute downstream of the arced labyrinth is required to converge to a width of approximately 27 m, allowing access on either side for maintenance. Initially this presented concerns for submerging the labyrinth weir and/or flows impinging on the lower chord of the bridge during the PMF. Through CFD modeling it was found that the weir would submerge prior to flows impinging on the bridge; however, the slope and convergence of the chute were designed such that submergence did not have a noticeable impact on the weir's hydraulic capacity. A rendering of the prototype is provided in Figure 13. Currently the hydraulic design of the spillway is complete, and the overall project is progressing towards final design.

3.2 *Empirical design*

The Springton Dam prototype weir geometry was primarily based on weir geometries studied by Crookston & Tullis (2012a, 2013a), and Thompson (2019) using data from the following labyrinth weir geometries they researched: α of 12°, 15°, and 16° and θ of 0° (linear) and 10°. The Springton Dam prototype consists of eight total cycles with α=14.9° and θ=8.5° with a slab or base at Elevation 57.0 m (based on the North American Vertical Datum of 1988 [NAVD88]) and a total weir length of 260.95 m. Three of the cycles have a crest elevation of 60.96 m, P=3.96 m, and L_1=97.72 m (Stage 1 length), which convey storm events up to the 1% (1:100) annual exceedance probability (AEP) event, without increasing this event's peak discharge or pool level compared to the existing spillway. I.e., Stage 1 of the labyrinth weir was designed to match the same peak reservoir level and discharge of the existing ogee spillway during the hypothetical 1% AEP event. The remaining five cycles have a crest elevation of 62.0 m, P=4.99 m, and L_2=163.23 m, to convey the PMF and are activated for all storm events exceeding the 1% AEP event. This configuration was selected based on site conditions (width and arc of existing ogee spillway), the U.S. Federal Emergency Management Agency's (FEMA) floodplain requirements, and based on the findings of multiple 3D CFD model and design iterations.

The U.S. Corps of Engineers were faced with a similar problem when designing Lake Isabella, in that their prototype needed to be of a geometry not evaluated experimentally in 2016, with α=16°. To overcome this, they used an interpolation method to develop C_d for their

Figure 13. Rendering of Prototype spillway at normal pool (top) and during the PMF (bottom).

geometry (Thompson 2016). The design team at Gannett Fleming used a similar approach for Springton Dam. To develop the discharge rating curve for this prototype geometry, the C_d values developed by Crookston & Tullis (2012a, 2013a), and Thompson (2019) for the aforementioned geometries were used to triple interpolate based on both the α=14.9° and θ=8.5° configuration of the Springton Dam prototype. The first interpolation occurred between the laboratory data for the α=12° and 16° weirs with θ=10° to develop a relationship for α=14.9° and θ=10°. The second interpolation occurred between α=12° and α=15° (Crookston & Tullis 2013a) both with θ=0° (non-arced). Finally, the third interpolation used the α=14.9° and θ=10° and θ=0° interpolated datasets to obtain the final data for the cycle orientation of α=14.9° and θ=8.5°. A visual of the relationship between the different laboratory weir datasets used for design and the final prototype is given in Figure 14, which plots the C_d versus H/P for each. The α=14.9°: θ=0° is not shown due to its similarity with the α=15° data.

Using the interpolated C_d and H/P relationship, a composite discharge rating curve for the proposed arced labyrinth spillway was developed; this rating curve, is shown alongside the existing spillway's rating curve in Figure 15; the proposed discharge rating curve was validated using Flow-3D.

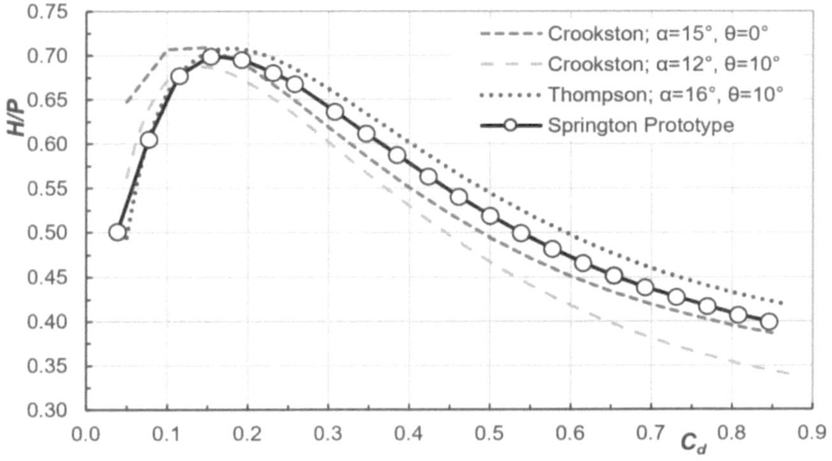

Figure 14. Comparison of C_d versus H/P curves used for design.

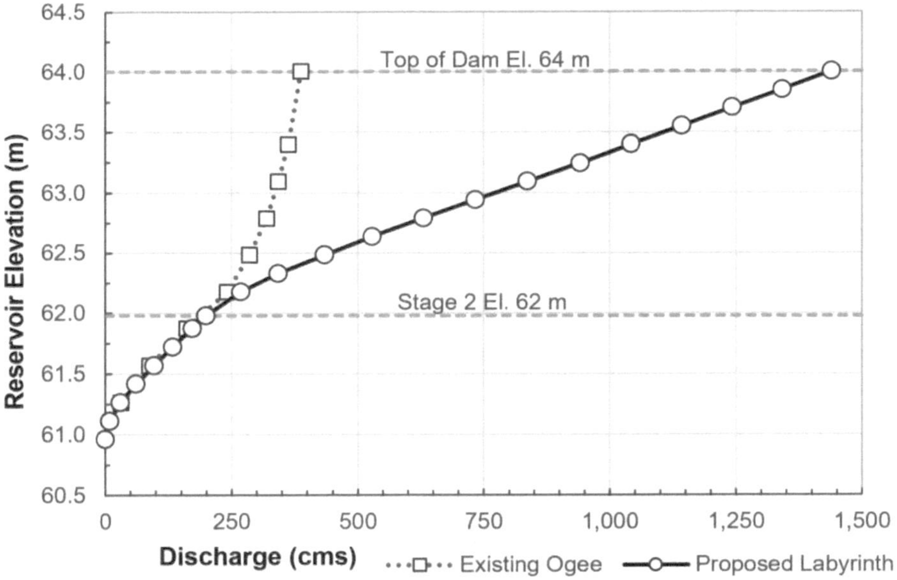

Figure 15. Prototype arced labyrinth discharge rating curve compared to existing curve.

3.3 *Numerical confirmation*

In order to validate the arced labyrinth weir prototype discharge rating curve (Figure 15) and to design the spillway chute, a 3D CFD model was developed using the Flow-3D HYDRO (version 2022R2) software. Four discharge scenarios were evaluated between the PMF and the 1% AEP event, the design events for the entire weir and Stage 1, respectively.

3.3.1 *Numerical model setup*

Initially CFD was used to model various weir prototypes in a similar reservoir geometry to those evaluated in literature. Once suitable performance of the model was obtained (within 5% agreement), Survey, publicly available digital elevation models (DEMs), and 3D design data were used to develop the reservoir, spillway chute, and downstream model geometry. The roughness for concrete within the model was set to 1.5 mm. Similar to the research by Thompson (2019) a split-Lagrangian VOF method was employed paired with a second order monotonicity preserving momentum advection approximation algorithm. A second order approximation is often recommended for complex spillway chute flows and for modeling air entrainment. The immersed boundary method was used for near-wall momentum flux calculations. The timestep was set to dynamically compute based on numerical convergence and model stability. Final results were reported using at least 30 seconds of steady state time-averaged values.

The air entrainment physics model was enabled to understand the effects of flow bulking in the chute and to size the heights of the training walls. Air entrainment can be difficult to model numerically without detailed calibration data. Numerically modeling air entrainment is sensitive to the rate at which air is entrained and the assumed air bubble size. Due to lack of entrained air calibration data for the proposed spillway, the Flow-3D HYDRO default variables were generally selected. In essence, the lack of calibration data translates to a lack of justification for selecting different variables due to the uniqueness of the site's spillway and hydraulics. Flow bulking and buoyancy of the air bubbles were also modeled. The air bubble diameter was set to be automatically computed using an initial diameter of 1 millimeter (default). Only one parameter was adjusted from the default value: the turbulent diffusion multiplier, which was adjusted from 0 to 1. After observing the initial models, it was seen that air was accumulating at the free surface and not diffusing throughout the water column. A multiplier of 1 was selected, which allows the air to diffuse at the rate of the fluid's turbulent diffusion of momentum. Ultimately air entrainment caused minimal flow bulking and was not a major driver in sizing downstream training walls, this agreed with the observations of Thompson's (2019) physical models where increasing tailwater drowned the nappe and decreased its aeration.

Three distinct mesh configurations were applied to this model, each used to assess mesh independence using the probable maximum flood (PMF) scenario. Initially, a coarse mesh was used, which was then refined twice to a medium and fine mesh. In the upper parts of the reservoir and in downstream areas (denoted as Mesh block 1 in Figure 16) 121.9-, 91.4-, and 61-cm cells were used while parts of the reservoir located closer to the spillway and the chute (Mesh block 2 in Figure 16) used a 61-, 45.7-, and 30.5-cm cell size for the coarse, medium, and fine mesh resolutions, respectively. The weir and spillway itself (Mesh block 3 in Figure 16) were modeled using 30.5-, 22.9-, and 15.2-cm cells for the coarse, medium, and fine meshes, respectively. The entire fine mesh model consists of 21 million cells.

Gannett Fleming used total weir discharge during the PMF event to compute the GCI of each mesh configuration, and found relative errors between the meshes of 0.8% to 1.4%, which resulted in GCI values of 1.3% and 1.4% between the coarse-medium, and medium-fine refinements, respectively. In addition to evaluating the sensitivity to the mesh and selecting appropriate numerical parameters, the location of boundary conditions had to be evaluated.

Various reservoir boundaries were evaluated with the size progressively getting smaller until a point where the boundary started to influence results; at which point the reservoir mesh-block bounds were increased to the previous, size for use in production modeling. Sensitivity to the downstream boundary was not as crucial due to a large vertical drop and plunge pool downstream of the chute termination.

3.3.2 *Numerical results*

The four CFD-modeled scenarios were used to validate the arced labyrinth weir's design discharge rating curve shown in Figure 15. This discharge rating curve is repeated in Figure 17 alongside the CFD datapoints. A comparison between the CFD datapoints and the empirical data points used to derive the rating curve are also given in Table 3. For Scenarios 2 through 4, the CFD model overpredicts spillway discharge by 5% to 8%; for Scenario 1, the PMF, the CFD model underpredicts discharge by 2.4%.

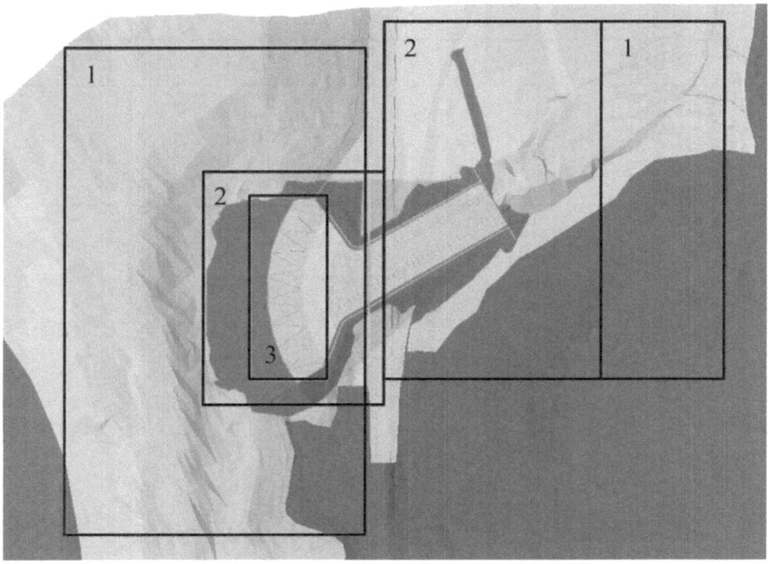

Figure 16. Springton Dam Prototype CFD model extents and meshblock size and location.

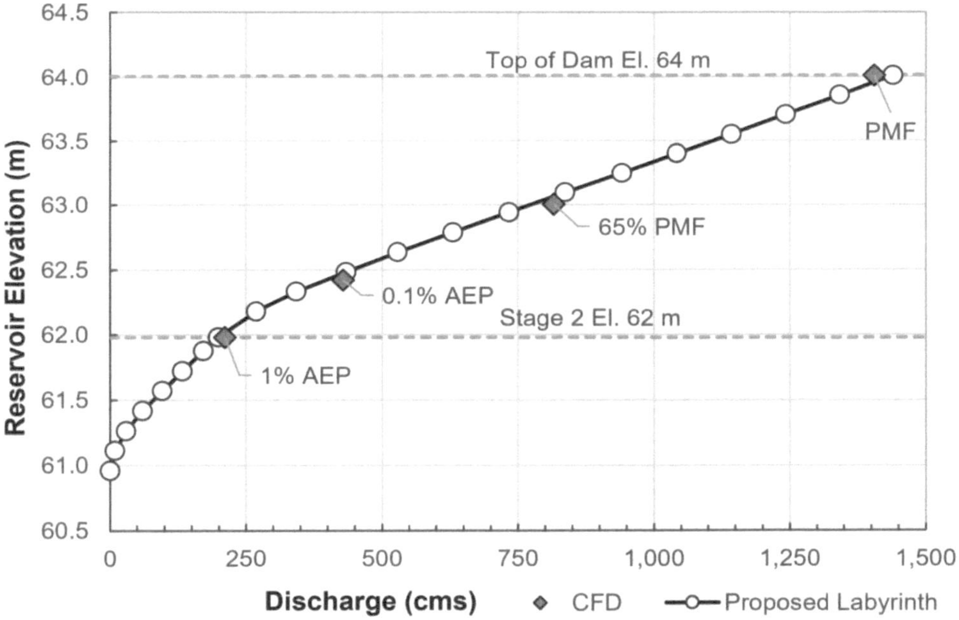

Figure 17. Prototype discharge rating curve and CFD validation points.

At the lower flow events (i.e., lower heads on the weir) the model tends to overpredict the flow capacity compared to the empirical discharge rating curve. A reason for this could be due to the nearly ideal approach conditions of Stage 1; the higher cycles of Stage 2 train the flow uniformly towards the crest of Stage 1 as shown in Figure 18, showing the approach conditions for Scenario 3, the 1% AEP event. In contrast, the laboratory data used to derive the design discharge rating curve experienced decreased efficiency at the distal cycles due to the approach conditions.

Table 3. Empirical and CFD discharge data comparison.

Model Scenario	Pool El. (m)	Empirical Q (m^3/s)	CFD Model Q (m^3/s)	% Difference
1	64.0	1,439.6	1,405.6	-2.4%
2	63.0	775.0	816.4	5.3%
3	62.4	397.6	428.4	7.8%
4	62.0	199.1	211.0	6.0%

Figure 18. 1% AEP CFD approach velocity streamlines.

Figure 19. PMF CFD approach velocity streamlines.

For the PMF Scenario (see Figure 19), the decrease in discharge capacity, according to the CFD model, can be attributed to model uncertainty, partial submergence of the weir due to the converging chute downstream, and the differences between the prototype geometry and the laboratory models used to develop the design discharge rating curve, which required interpolation. Such differences include differences in α and θ, *a* two-stage weir, irregular approach conditions, and different abutment geometries. With the potential reduction in capacity at the PMF, a CFD-adjusted discharge rating curve, which reduced the design rating uniformly by 2.4%, was input into a reservoir routing model and used to route the PMF hydrograph through the reservoir. The results showed a 3 cm increase in the peak PMF reservoir level, but still below the dam's freeboard requirement.

4 SUMMARY AND CONCLUSIONS

The authors offer the following conclusions and considerations when hydraulically designing arced labyrinth weirs:

- Empirical equations relating Cd to H/P are provided for the arced labyrinth weirs researched in the hydraulics laboratory at Utah State University's Utah Water Research Laboratory. These empirical equations simplify the use of the research for design; however, care should be taken to validate the data's applicability to a prototype setting using CFD and/or physical modeling of site-specific conditions.
- Arced labyrinth weirs can improve the discharge efficiency over linear labyrinth weirs in reservoir application due to better alignment of the weir cycles to the approach flow. However, care must be taken at higher heads to not submerge the weir due to the increased tailwater levels caused by the converging flow patterns. The $\theta=20°$ balances the benefits at low and high heads compared to the other two θ geometries evaluated.
- The nappes of the weirs assessed were generally less stable than the in-channel models evaluated by (Crookston & Tullis 2013b), with nappe instability observed at a larger range of H/P as α increased
- In many cases when the mesh and numerical parameters are carefully calibrated, CFD models can replicate experimental data with reasonable accuracy, indicating its usefulness as a design tool for practitioners. This was exhibited both in the studies completed by Thompson (2019) and Koldewyn (2023) and Gannett Fleming's implementation of their research in the hydraulic design of the arced labyrinth prototype weir at Springton Dam.

ACKNOWLEDGEMENTS

The authors would like to acknowledge, Utah State University for the funding and support to complete the research on arced labyrinth weirs, Aqua Pennsylvania, for providing the opportunity to apply the research and present the design concept herein, and their families for their endless love and support.

REFERENCES

Christensen, N.A., 2012. Flow characteristics of arced labyrinth weirs. Utah State University, Logan, UT.

Crookston, B.M., 2010. *Labyrinth weirs*. Ph.D. thesis, UT State Univ., Logan, UT.

Crookston, B.M. & Tullis, B.P., 2012a. Arced Labyrinth Weirs. *J. Hydraul. Eng.*, 138 (6), 555–562.

Crookston, B.M. & Tullis B.P., 2012b. Discharge efficiency of reservoir-application-specific labyrinth weirs. *J. Hydraul. Eng.*, 138 (6), 564–568.

Crookston, B.M. & Tullis, B.P., 2013a. Hydraulic design and analysis of labyrinth weirs. I: Discharge relationships. *J. Irrig. Drain. Eng.*, 139 (5), 363–370.

Crookston, B.M. & Tullis, B.P., 2013b. Hydraulic design and analysis of labyrinth weirs. II: Nappe aeration, instability, and vibration. *J. Hydraul. Eng.*, 139 (3), 371–377.

Darvas, L., 1971. Discussion of 'Performance and design of labyrinth weirs', by Hay and Taylor. *J. Hydraul. Eng.*, 97 (80), 1246–1251.
Flow Science, 2021. FLOW-3D HYDRO Documentation.
Gentilini, B., 1941. Stramazzi con cresta a planta obliqua e a zig-zag. *L'Energia Electrica*, XVIII (X), 653–664.
Hay, N. & Taylor, G., 1970. Performance and design of labyrinth weirs. *J. Hydraul. Div.*, 96 (11), 2337–2357.
Hirt, C.W. & Nichols, B.D., 1981. Volume of fluid (VOF) method for the dynamics of free boundaries. *Journal of Computational Physics*, 39 (1), 201–225.
Koldewyn, S.J., 2023. Using computational fluid dynamics for predicting hydraulic performance of arced labyrinth weirs. Utah State University, Logan, UT.
Roache, P.J., 2009. Chapter 2: Terminology and basic methodology. *In*: *Verification and Validation of 3D Free-Surface Flow Models*. Reston, VA: American Society of Civil Engineers.
Thompson, S.D., 2019. Reservoir applications of arced labyrinth weirs. Utah State University, Logan, UT.
Tullis, P.J., Amanian, N., & Waldron, D., 1995. Design of labyrinth spillways. *J. Hydraul. Eng.*, 121 (3), 247–255.

Labyrinth and Piano Key Weirs IV – Erpicum, Crookston & Paxson (Eds)
© 2026 The Author(s), ISBN 978-1-041-10527-5

Design approach for labyrinth weir: Case study of La Laye dam

M. Roy, S. Derrien, J. Schaguene and G. Dautois
ARTELIA, Grenoble, France

ABSTRACT: This paper presents ARTELIA's gradual approach for the design and construction of labyrinth weirs, from the preliminary concept design to the commissioning. It is illustrated with the case of La Laye dam, where an additional lateral spillway with a labyrinth weir has been developed to increase the existing discharge capacity. The design process involved preliminary hydraulic calculations, 3D numerical modeling and a physical scale model. The paper elaborates on the technical solutions and optimizations that were considered, to provide the Client with the most hydro-economical solution (best compromise for satisfactory hydraulic operation at minimum cost).

1 CONTEXT

1.1 *Short presentation of the dam*

The La Laye dam is a 30-meter-high earthfill dam with a clayish core located next to the city of Forcalquier in the South-East of France. It was constructed between 1962 and 1964 to supply water for irrigation (85% of the volume of the reservoir) and for drinking water. The dam is owned by the SIIRF (French public agency) and operated by the Société du Canal de Provence (SCP). A typical cross-section of the dam in its 2023 configuration is shown here below.

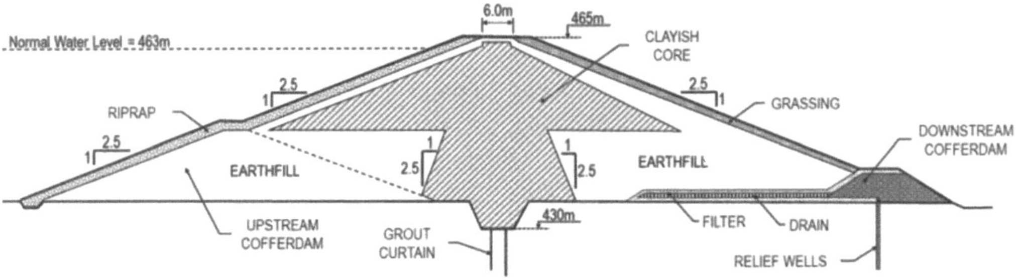

Figure 1. Typical cross-section of La Laye dam.

1.2 *Current spillway capacity*

Currently, the spillway capacity is mainly ensured by a morning glory spillway located in the reservoir on the left bank. The sill of the morning glory is set at the elevation 460 m asl. It is supplemented by a steel cylindric gate (diameter 7.5 m, height 3 m), which purpose is to regulate the reservoir water level at the Full Supply Level (FSL) of 463 m asl. The opening of the cylindric gate is thus automatically controlled to maintain the reservoir at the FSL.

DOI: 10.1201/9781003655541-22

The spillway discharge capacity can be increased using the bottom outlet, which is controlled by a 2.4 x 3.6 m roller gate located at the bottom of the intake tower, a few meters upstream the morning glory.

Both morning glory and bottom outlet discharge in the same horse-shoe shape tunnel (diameter 5.20m) which crosses the dam on the left abutment. The total discharge capacity is limited by the dimensions of the tunnel. When both gates are fully open, the maximum discharge capacity is about 400 m^3/s which was the design requirement considered at the construction of the dam. The tunnel then releases the flows to the downstream river through a concrete bucket provided with concrete blocks.

Owing to the regulation capacity of the cylindrical gate on the morning glory and of the roller gate of the bottom outlet, the Maximum Water Level (MWL) was initially considered equal to the Full Supply Level: MWL = FSL = 463 m asl.

Figure 2. Satellite picture of the dam and appurtenant structures.

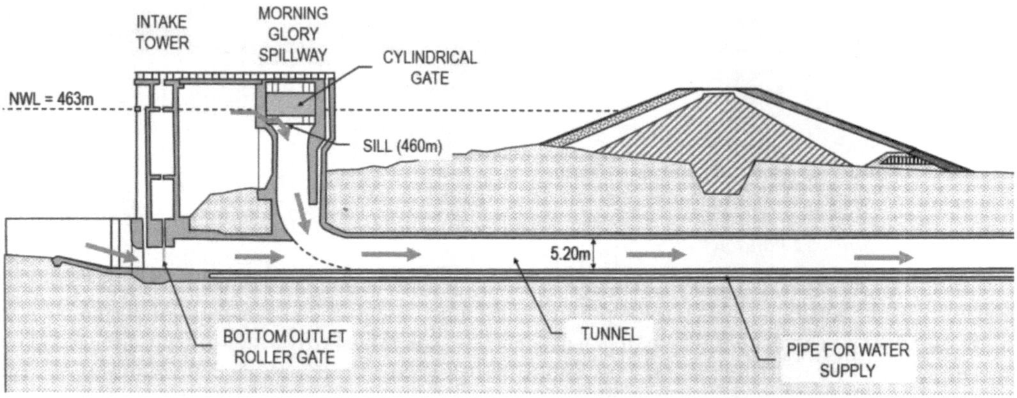

Figure 3. Upstream (left) and downstream (right) faces of La Laye dam.

1.3 *Flood issue*

The hydrological behavior of the catchment area is significantly influenced by the presence of karsts. Based on the SCHADEX method (2012), the hydrological study performed by ARTELIA (2019) advantageously considers a progressive saturation of the karsts for the most frequent floods and a probabilistic distribution of the rainfall. The methodology and results of this study were validated by comparison with a real flood event which occurred in Decembre 2019, after the completion of the study.

The main output of the hydrological study is the peak discharge of the "rare" flood (leading to Maximum Water Level) which is the 10,000-years flood (Q10000) for an embankment dam higher than 20m according to French regulations.

The peak discharge of this Q10000 has been assessed at 603 m^3/s. A safety factor of 15% corresponding to the confidence interval according to the method used has been additionally considered, leading to a final design value of 694 m^3/s. This value is 75% higher than the existing spillway capacity (400 m^3/s, see above). The spillway capacity must therefore be increased to fulfil the safety requirements, and the construction of an additional spillway is found to be necessary.

Pending the creation of an additional spillway the operating level has been lowered to the elevation 460 m asl by a prefectural decree in July 2015, leading to a significant loss in the storage capacity of the reservoir. ARTELIA was involved as Designer and Owner Engineer from the preliminary design to the commissioning of the structure.

2 CHOSEN SOLUTION

2.1 *Constraints*

In a region of dry climate where the water resource is at stake, the first constraint is to restore the Full Supply Level at the elevation 463 m asl.

The second main constraint is related to the space available to implement a large hydraulic structure. Although some previous studies mention the possibility of founding an additional spillway on the downstream face of the dam, this option was discarded for reliability and safety reasons. The right abutment of the dam is also discarded because it is steep and not accessible by any existing road. Eventually, the best option is deemed to be the left bank, even though it is already occupied by the existing hydraulic structures.

2.2 *Design criteria*

The previously mentioned constraints encouraged ARTELIA's experts to minimize the dimensions of the new spillway, which is thus designed to release the remaining discharge over the capacity of the existing structures. The design discharge is thus $694 - 400 \approx 300\ m^3/s$.

In addition, it is considered acceptable to increase the Maximum Water Level up to approx. 465 m asl, on condition that the stability of the dam is duly checked and that the top of the impervious core remains higher than this elevation. A parapet is also to be designed to provide sufficient freeboard against waves generated by wind or landslides in the reservoir.

2.3 *Description of the chosen solution*

The chosen solution to increase the spillway capacity is to create an additional free-surface lateral spillway on the left abutment of the dam. This structure is divided in several parts which are shortly described here after, from upstream to downstream:

- A labyrinth weir. The sill is set at the elevation of the Full Supply Level (463 m asl). The structure comprises 3 labyrinth cycles, for a total length of 30 meters. The dimensions of the weir are optimized to the available space between the crest of the dam and the existing morning glory. Since the weir cannot be placed on the dam, a lateral inlet has been designed. This type of structure was preferred to a Piano Key weir for its simplicity of construction and reduced cost, while meeting design criteria and site constraints.

- A collecting basin. This structure aims to collect and steady the flows coming from the labyrinth weir and redirect them towards the spillway chute downstream. The shape of the basin has been optimized to minimize the excavation works and ensure the best hydraulic performance (checked on a physical scale model)
- A spillway chute. In order to minimize the excavation works, the chute has been divided in an upstream part with low slope (0.5%) and a downstream part with steep slope (60%). An elbow is implemented between both parts in order to direct the flows in the downstream river. The downstream part is provided with concrete steps in order to dissipate a part of the energy.
- A concrete stilling basin. This structure aims to dissipate the flow energy in a controlled manner before releasing the discharge to the river. The retained design is a USBR type IV, with dissipation blocks.

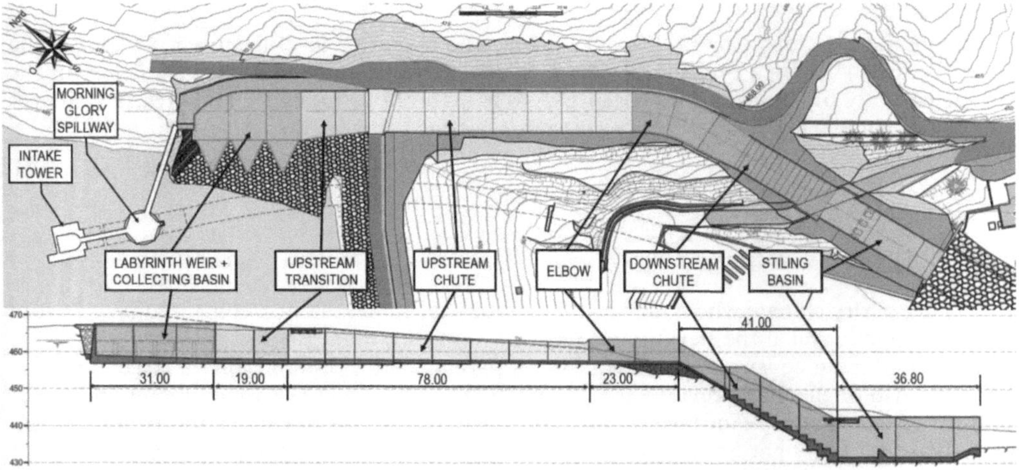

Figure 4. Plane view and longitudinal section of the additional spillway.

3 HYDRAULIC DESIGN OF THE LABYRINTH WEIR & COLLECTING BASIN

3.1 *Preliminary design based on analytical and empirical formulas – Basic Design Stage*

At Basic Design stage, the lateral spillway was first preliminary designed using analytical and empirical formulas from the literature.

The labyrinth weir was designed using the equation 5.2 in ICOLD Bulletin No. 172 (2016) considering a target discharge capacity of 300 m^3/s under a head of 2 mWC. After an iterative process, the main dimensions of the weir were chosen as follows:

• Number of cycles	3
• Opening angle of the cycles	50°
• Length of a cycle	9.94 m
• Width of a cycle	10.21 m
• Total width of the weir	30.53 m

The total width thus remains shorter than the 35 m available on site. It must be noted that to reach the same discharge capacity, a straight free-surface weir with a Creager hydraulic profile would require a total width of about 48.3 m i.e., 58% more than the labyrinth weir.

The collecting basin was calculated using the formula proposed by Ven Te Chow (1959), chapter 12.2, which is based on the momentum conservation between the upstream and downstream sections. Once again, the determination of the dimensions of the basin results from an iterative process aiming to optimize design considering the following constraints:

- The basin shall be deep enough to avoid the submergence of the weir (which would reduce its discharge capacity).
- It is preferable to limit the depth of the basin to avoid huge excavation works, especially at the vicinity of the dam axis, where such excavations would interfere with the dam body and the existing grout curtain.
- The shallower the basin is, the wider it must be. However, the steep natural slope of the left bank would raise excavation issues if the basin was too wide. In addition, it is deemed more appropriate to have approximately the same width for the collecting basin and the chute downstream, in order to avoid complex transition between two subsequent parts of the structure.

Eventually, the main dimensions of the collecting basin were chosen as follows:

- Bottom width 12 m
- Depth from FSL (d/s section) 5 m
- Bottom slope 0.01 m/m

3.2 *Numerical 3D model – Detailed Design stage*

Given the complex inlet configuration (especially due to the redirection at 90° of the flows downstream the weir, and to the presence of the intake and morning glory towers a few meters upstream), a 3D numerical model appeared necessary at Detailed Design stage to validate and refine the design. In addition, the numerical model aimed to (i) assess the discharge capacity law of the spillway, (ii) assess the flow conditions in the different considered flood scenarios, and (iii) optimize the design when possible.

The 3D numerical calculations were performed using the software OpenFOAM (Open Field Operation and Manipulation) CFD Toolbox. This calculation code solves the Reynolds equations that correspond to the turbulent equations of Navier-Stokes.

The model comprises a 3D meshed reproduction of the fluid domain, including a part of the reservoir, the existing bottom outlet and morning glory spillway, and the free-surface lateral spillway. The model extends in the riverbed downstream up to 60 meters from the stilling basin. The model eventually includes almost 3.3 million elements with typical dimension varying from 10cm to 3m. Several mesh configurations were tested to check that the calculation results are not influenced by the mesh size (nine versions tested with mesh size increase by 50% and reduction by 33%), and obtain a good compromise between accuracy and computation time.

The calibration of the model was checked owing to the existing hydraulic structures (bottom outlet and morning glory), which discharge capacities were determined at the time of the construction of the dam. The discrepancy between the theoretical maximum capacity (from initial design study) and the one calculated on the 3D model was found to be 3% only.

In general, the results of the numerical simulation were quite satisfactory and confirmed the preliminary design based on analytical and empirical formulas. The total discharge capacity was confirmed, with a slightly different distribution between the three structures (101 m^3/s through the bottom outlet, 307 m^3/s through the morning glory and 284 m^3/s through the lateral spillway) compared to the theoretical capacity. Only the flow conditions downstream the stilling basin must be confirmed, due to the interference with the flows coming from the existing spillway.

Three modifications to improve the hydraulic performances were tested and retained owing to the numerical model: (i) increase of the length of the transition between the collecting basin and the chute, (ii) increase the slope of the collecting basin and (iii) inclination of slab from

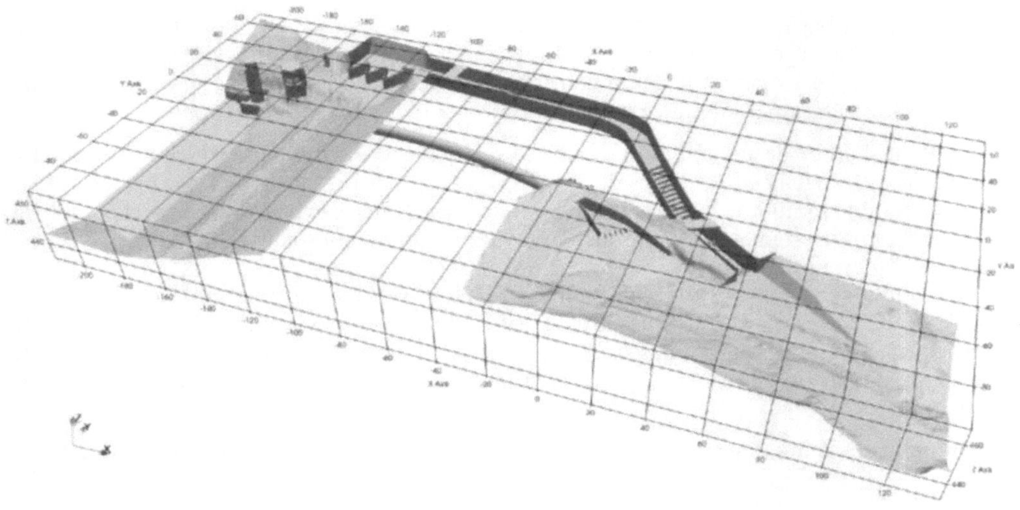

Figure 5. Extension of the 3D numerical model.

toe of labyrinth cycles to collecting basin. As a result, for a given discharge, the inlet conditions above the labyrinth cycles are better, increasing the overall discharge capacity and reducing the reservoir water level by 9 cm. Additionally, the velocity increase in the collecting basin is more progressive. See figure below.

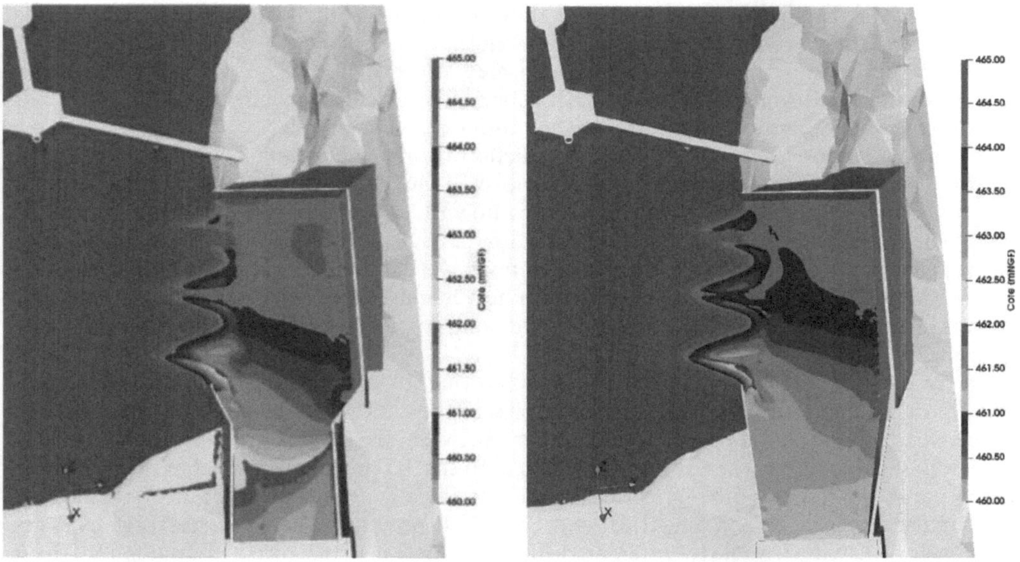

Figure 6. Water surface level in the numerical model. Left: before optimization. Right: after optimization.

3.3 *Physical scale model*

The physical scale model was built after the 3D numerical model between the Detailed Design stage and the Tender in order to (i) ultimately confirm the flow conditions and determine the discharge capacity law of the lateral spillway (ii) study the energy dissipation and especially check the potential interaction between the flows coming from the lateral spillway and those coming from the existing stilling basin and (iii) confirm the most hydro-economical design.

The extension of the physical scale model is approximately the same as for the numerical model (see Figure 5). The existing hydraulic structures (morning glory spillway and bottom outlet) are thus represented. The retained scale is 1/50, resulting in a total length of about 6 m and on a scaled design discharge of approx. 17 l/s (equivalent of 300 m^3/s in nature) for the new spillway and a total scaled design discharge of approx. 40 l/s (equivalent of 700 m^3/s in nature).

The reservoir, the dam and the downstream river are shaped using cement mortar applied between wood pieces representing the contour lines of the topography. The hydraulic structures such as the labyrinth weir, the morning glory and the bottom outlet are 3D printed.

The upstream boundary condition is the discharge implemented in the model. The discharge is controlled by a dedicated valve and precisely measured using a electromagnetic flowmeter. The downstream boundary condition is the water level in the downstream section, which is regulated owing to a flap valve. The downstream water level had been formerly assessed by a 1D HEC-RAS calculation.

The model is provided with several sensors in order to measure the elevation of the water surface and the velocity at relevant points. In addition, blue dye can be used to highlight specific flow conditions, especially the inlet conditions at the vicinity of the sills.

Figure 7. Pictures of the physical scale model.

The testing phase on the scale model was very useful to dissipate any uncertainty related to the hydraulic behavior of the lateral spillway and was of a great support to present the concept to the Client and to the authorities in charge of the design approval.

The comparison between the physical and numerical models shows very similar results in trend, although some differences can be noted (see figure below).

The discharge capacity curves (either from physical or numerical modelling) have the same general appearance, following a first linear trend up to around 200 m^3/s, followed by a transition between 200 and 250 m^3/s to the second steeper linear trend (above 250 m^3/s). The transition zone corresponds to the progressive drowning of the labyrinth weir, which leads to this curve inflection.

The difference between the curve from the physical model and the curve from the numerical model is about +5 to +10cm for the lowest discharge values up to the 100-year flood, and then reaches about +15cm for the higher discharge values. Even though difficult to explain, this higher level would have no effect on the dam's stability.

Furthermore, both discharge capacity curves (from numerical and physical models) are very close to the theoretical one up to a value of approx. 250 m^3/s. Beyond this value, the weir is submerged so the curves divert from the theoretical one.

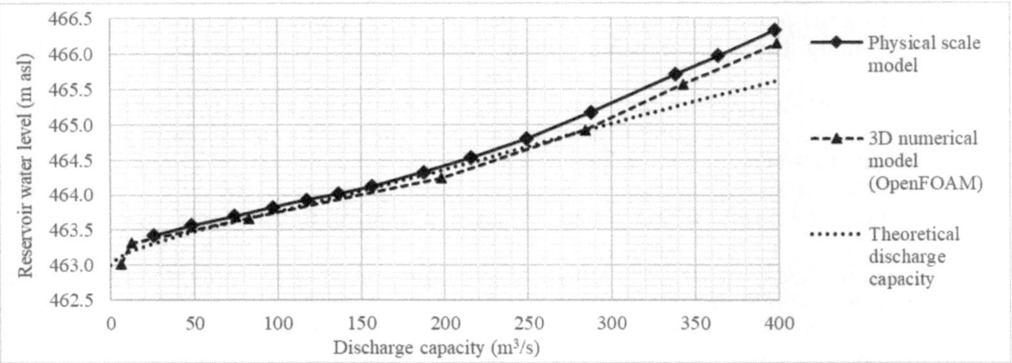

Figure 8. Discharge capacity curves of the new spillway.

As the new spillway is designed primarily for high head applications (i.e. the passage of extreme floods), scale effects are likely negligible, as indicated by Tullis (2017). With a dimensionless head ratio (upstream total head normalized by the weir height) of approx. 0.70 for the design discharge, scale effects are not expected to have led to a high error in the prediction of the flow behavior for the flood conditions the new spillway is planned to operate.

The flow conditions above the labyrinth weir are satisfactory and compliant with what was expected. For discharges lower than 280 m^3/s, the flow conditions are quite homogeneous from one cycle to another, and the sill remains under free fall conditions. Higher discharges (extreme situations) lead to the submergence of the weir, a detachment of the flow from the upstream and downstream walls and a higher level of turbulence.

The collecting basin fulfils its role to redirect the flow towards the downstream chute while dampening part of the turbulence coming from the weir. However, the upstream left corner of the basin is almost not involved in the conveyance of the flow (Figure 8.a). The shape of the stilling basin was therefore modified to fit the observed flow conditions. The corner of the basin was thus replaced by a bend and the left wall was straightened up to the vertical (Figure 8.b). Such modification greatly improved the flow conditions in the collecting basin, without any noticeable effect on the overall discharge capacity. In addition, it significantly decreases the volume of excavation and concrete works, allowing approx. 100 k€ savings on the total budget.

Figure 9. Geometry of the collecting basin, before optimization (a) and after optimization (b).

4 CONSTRUCTION

The construction of the weir and the collecting basin was performed from July 2024 to January 2025. No specific difficulty arose during the works, owing to the robustness and simplicity of the designed arrangement. The main figures of the construction are summarized in the table below.

The completion of the construction is expected in July 2025.

Table 1. Main figures of the weir and collecting basin construction.

	Duration of the works	Concrete volume
	days	m^3
Excavations	27	
Labyrinth weir	45	550
Collecting basin	65	1050

13.08.2024 - Excavations

15.10.2024 – Construction of the weir

11.12.2024 – Construction of the collecting basin

13.03.2025 – Works completed

Figure 10. Excavation and concrete works.

5 CONCLUSIONS

The progressive approach used by ARTELIA for the hydraulic design of the lateral spillway of La Laye dam was found adapted to the different stages of the design process. Each step allowed to confirm and optimize the previous, resulting in a robust and economical design.

Numerical – and even more– physical 3D modelling is deemed essential in such complex configuration with 3D behavior and several structures discharging at the same time. The efficiency of the labyrinth weir in constraint geometrical conditions was validated, despite the presence of the other hydraulic structures at the close vicinity.

REFERENCES

ARTELIA. 2019. *Barrage de la Laye. Construction d'un évacuateur de crues et travaux annexes. Etude complémentaire: analyse hydrologique au droit du barrage de la Laye.*

Chow, V.T. 1959. *Open-Channel Hydraulics*. New-York: McGraw-Hill.

ICOLD. 2016. *Bulletin No. 172: Technical advancements in spillway design – Progress and innovations from 1985 to 2015*. Paris: ICOLD.

OpenFOAM. http://www.openfoam.org/.

Paquet, E. & Lang, M. & Carré, J.C. 2012. *Schadex method for extreme flood estimation: overview, applications and perspectives*. 24th ICOLD Congress. Kyoto.

Tullis, B.P. 2018. *Size-Scale Effects of Labyrinth Weir Hydraulics*. 7th IAHR International Symposium on Hydraulic Structures. Aachen.

Labyrinth and Piano Key Weirs IV – Erpicum, Crookston & Paxson (Eds)
 ISBN 978-1-041-10527-5

Value added 3D modeling – Lower Crow Dam spillway modification and rehabilitation project

J. Bartels
Stantec, Chicago, Illinois, USA

C. Bernedo Sanchez
Stantec, Denver, Colorado, USA

ABSTRACT: The Confederated Salish and Kootenai Tribes (CSKT) have been operating Lower Crow Dam under restricted reservoir levels since 2008 due to significant structural and spillway capacity deficiencies that have increased risk to dam safety. Stantec was contracted by the CSKT to evaluate rehabilitation options and provide final design for the new spillway. The focus of this paper is to describe the design methodology and spillway selection process with an emphasis on the hydraulic modeling completed to support the final spillway arrangement. The design approach included the evaluation of various spillway types including gated, Obermeyer, fuse gate, ogee, labyrinth, and piano key weirs. Multiple models were developed through the project including one-dimensional computer models for preliminary design purposes and three-dimensional Computational Fluid Dynamics models for alternative evaluation and to confirm the final hydraulic performance of the spillway structure, inlet channel and chute.

1 INTRODUCTION

Stantec was contracted by the Confederated Salish and Kootenai Tribes (CSKT) to evaluate rehabilitation and modification options for the existing Lower Crow Reservoir Spillway, an irrigation supply source to Lower Crow Creek and the Flathead River located within the Flathead Indian Reservation. The dam is located on Lower Crow Creek approximately 11km southwest of Ronan, Montana. Lower Crow Reservoir releases water into Lower Crow Creek that flows approximately 8km until its confluence with the Flathead River.

Several alternatives were considered by CSKT, BIA and the Stantec design team for the spillway rehabilitation, from initial evaluation of standard spillway types including an ogee, labyrinth, and piano key to later consideration of more complex gated, Obermeyer and fuse gate type systems discussed during value engineering (VE) sessions. A target inflow design flood (IDF) of roughly 425m3s-1 was used for the evaluations.

The following sections of this paper focus on the design approach adopted for the project and knowledge gained when evaluating the spillway rehabilitation alternatives particularly those related to the hydraulic analyses, the alternative evaluation and decision-making process, and the importance of having verifiable hydraulic results to aid the civil and structural analysis of the final design.

DOI: 10.1201/9781003655541-23

2 PRELIMINARY DESIGN

Discharge capacity for the new Lower Crow Dam spillway were developed for several standard spillway types including an ogee, labyrinth, and piano key weir (PKW) using available empirical data and guidance references. The proposed geometric parameters and estimated discharge efficiency for each spillway type considered a target inflow design flood (IDF) discharge of roughly 425m3s-1 based on a 0.7m rise in the dam crest elevation from El. 878.3m to El. 879.0m and an incremental flood hazard assessment (IFHA) following the methods described in USBR (1988). Results of the IFHA with the proposed dam raise confirmed a target weir crest at El. 876.9m with the reservoir at a proposed normal high-water level (NHWL) of EL. 878.1m.

Validation of the empirically based results for use in the spillway and chute design was conducted using three-dimensional (3-D) Computational Fluid Dynamics (CFD) numerical modeling. CFD modelling was used to identify the need for large-scale or major modifications to the physical layout of the project before final design and construction was undertaken. This analysis also provided a mechanism to allow for iterative discussions among project stakeholders during the design process, an important outcome that commonly leads to improvements to preliminary layouts and a higher overall confidence level in the final design.

The 3-D CFD modelling of the spillways was constructed using the Flow-3D HYDRO computer program developed by Flow Science, Inc. Flow-3D is a well-known, commercially available CFD modelling package that uses structured, rectangular gridding based upon the Fractional Area Volume Obstacle Representation (FAVOR) and Volume-of-Fluid (VOF) methods originally developed by Hirt & Nichols (1981) and Hirt & Sicilian (1985), respectively.

The CFD models were run under a free surface, one-fluid flow condition utilizing the implicit renormalized group theory (RNG) κ-ε turbulence model, the generalized minimum residual implicit pressure-velocity solver with the dynamically adjusted pressure convergence tolerance setting, and a second-order momentum advection option, all default recommendations and settings provided by Flow Science for most hydraulic applications. Surface roughness heights applied to the concrete-lined spillway structure, inlet channel and chute and existing reservoir bathymetry were assumed at 3mm and 350mm, respectively within the models. The models also assumed "clean" water all existing and proposed spillway components and appurtenances free of debris and/or sedimentation. Note that air entrainment, conveyance, and the bulking of the fluid through use of the air entrainment, drift-flux and first-order variable density options available within Flow-3D HYDRO were not considered for this phase of model testing.

The limits and dimensions of the preliminary CFD models are provided in Figure 1 and matched the prototype geometrical and hydraulic boundary conditions tabulated within each of the subsequent sections of this report. The boundary conditions included a user-specified pressure boundary or water surface elevation (WSEL) applied 36.6m upstream of each weir crest and an outflow or free discharge boundary located between 18.3m and 36.6m downstream along an artificial spillway chute. These distances were finalized during initial model testing and were shown to provide sufficient approach and departure lengths for uniform flow profiles to be established ahead of and downstream of the weir crests.

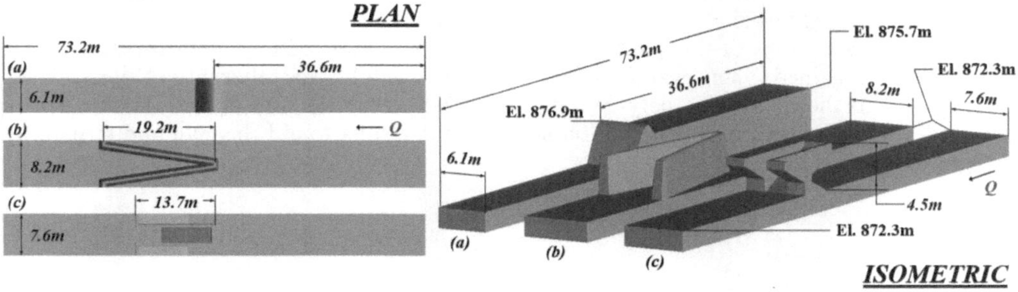

Figure 1. Preliminary design CFD model renderings – (a) Ogee, (b) Labyrinth, (c) PKW.

Each CFD model for this phase of design was composed of three meshes of varying resolution. The final mesh sizes for each spillway type ranged between 0.08m and 0.24m based on results of a mesh sensitivity analysis resulting in a total cell count of approximately 26.8 million cells. Figure 2 provides an example of the preliminary CFD simulation results for a test case with the reservoir set slightly above the NHWL at El. 878.3m after quasi-steady state conditions with respect to total system volume, surface area and kinetic energy were established.

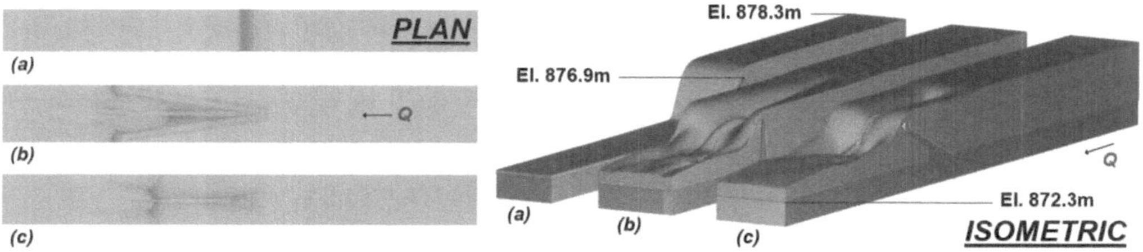

Figure 2. Preliminary design CFD model results – (a) Ogee, (b) Labyrinth, (c) PKW.

2.1 *Ogee configuration*

The preliminary sizing of the ogee arrangement was based on the guidance from the USBR (1987) shown in Figure 3. The main design parameters for the crest arrangement are provided in Table 1 and includes the total developed length (L) needed to pass the peak IDF flow through the dam at the NHWL. Note that the design head (H_o) was set to 70% of the maximum head to maximize discharge capacity for the full range of heads per USBR (1987).

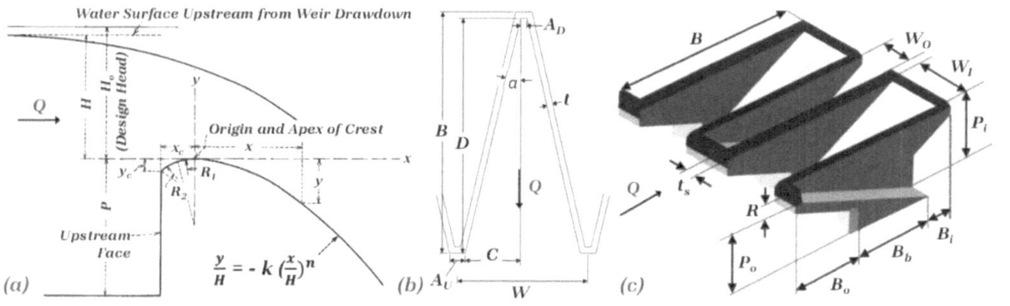

Figure 3. General Notation – (a) Ogee (USBR 1987), (b) Labyrinth (Crookston & Tullis 2013), (c) A-Type PKW (Ribeiro et al. 2012).

Table 1. Ogee design parameters.

H_o (m)	P (m)	K	n	x_c (m)	y_c (m)	R_1 (m)	R_2 (m)	L (m)
0.85	1.22	0.21	1.85	0.23	0.09	0.46	0.18	146.30

The discharge efficiency for the proposed ogee crest was based on Equation 1 from USBR (1987) assuming free flow conditions and omitting pier and abutment effects with a discharge coefficient (C_d) calculated between 1.71 and 2.38 depending on the upstream head.

$$Q = C_d L H^{3/2} \tag{1}$$

2.2 *Labyrinth configuration*

The preliminary sizing of the labyrinth weir arrangement was based on the guidance from Crookston & Tullis (2013) detailed in Figure 3. The main design parameters tested under the preliminary design is provided in Table 2 and includes the total number of cycles and total developed length needed to pass the peak IDF flow through the dam at the NHWL.

Table 2. Labyrinth design parameters.

H (m)	P (m)	t (m)	W (m)	A_D (m)	A_U (m)	B (m)	C (m)	D (m)	α (°)	N ()	L (m)
1.22	4.57	0.61	8.23	0.70	1.83	19.26	2.86	18.65	8.7	6	49.38

The discharge efficiency for the labyrinth weir was based on Equation 2 from Crookston & Tullis (2013) assuming free flow conditions and the discharge rating equation and coefficients presented in Equation 3 and Table 3, respectively.

$$Q = 2/3\ C_d L\sqrt{2g}H^{3/2} \tag{2}$$

$$C_{d=}a(H/P)^{b(H/P)^c} + d \tag{3}$$

Table 3. Labyrinth discharge rating coefficients.

a	b	c	d	Validity
0.02165	-3.31224	0.40690	0.23693	$0.05 < H/P < 0.90$

2.3 *Piano key configuration*

The preliminary sizing of the A-Type PKW configuration was based on guidance from Ribeiro et al. (2012) shown in Figure 3. The main design parameters tested under the preliminary design is provided in Table 4 and includes the total number of keys and total developed length needed to pass the peak IDF flow through the dam at the NHWL.

Table 4. PKW design parameters.

H (m)	P_o (m)	P_i (m)	R (m)	B_o (m)	B_b (m)	B_i (m)	B (m)	W_o (m)	W_i (m)	t_s (m)	N ()	L (m)
1.22	4.57	4.57	0.61	4.57	4.57	4.57	13.72	2.74	3.66	0.61	7	53.34

The discharge efficiency for the proposed A-Type PKW was based on Equation 2 assuming free flow conditions and discharge rating equation and coefficients presented in Equation 4 and Table 5, respectively from Cicero et al. (2013).

$$C_{(d=)}a_0 + a_1(H/P) + a_2(H/P)^2 + a_3(H/P)^3 + a_4(H/P)^4 \tag{4}$$

Table 5. Type A PKW discharge rating coefficients.

a_0	a_1	a_2	a_3	a_4	Validity
1.63	0.59	-11.56	21.72	-12.46	$0.10 < H/P < 0.71$

2.4 *Preliminary design results*

The preliminary spillway arrangements and discharge comparison between the empirically- and CFD-based methods are provided in Table 6 and Figure 4 for a range of reservoir levels after the spillways were scaled up to pass the peak IDF flow of $425m^3s^{-1}$ through the dam at the NHWL (El. 878.1m).

The estimated discharges calculated by the empirical and CFD methods were closely matched for all three spillway arrangements considered with values falling within the ±5% allowance adopted by Stantec for the project. This allowance was included to account for reductions in hydraulic efficiency in the final design related to side wall and potential abutment effects and the likelihood for partial to full submergence of the weir structure created by elevated water surface profiles within the downstream spillway inlet channel and chute.

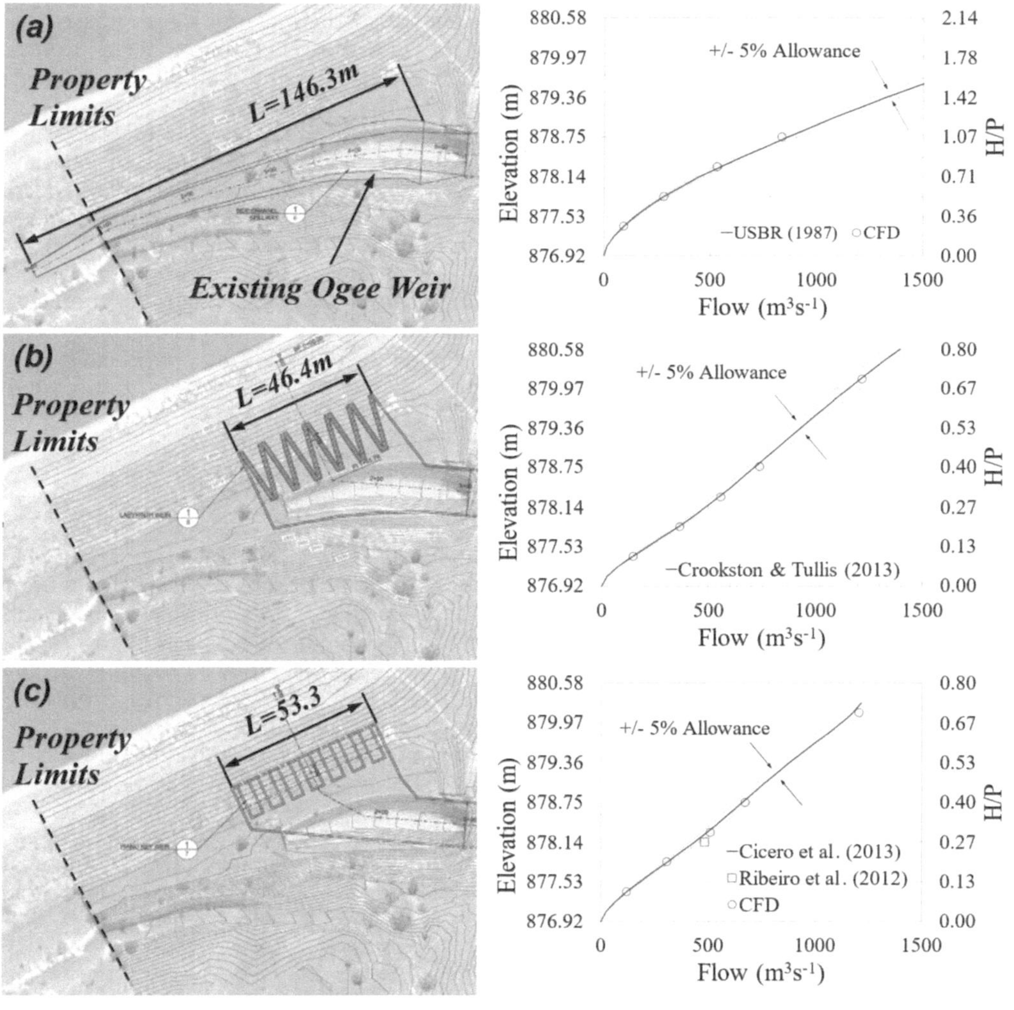

Figure 4. Preliminary design layouts and discharge estimates – (a) Ogee, (b) Labyrinth, (c) PKW.

3 SPILLWAY ALTERNATIVE SECTION

Selection of the preferred spillway alternative for the Lower Crow Dam was made following a series of VE workshops conducted with the CSKT and BIA in August 2015 and

Table 6. Preliminary design discharge results.

WSEL (m)	Ogee $Q_{EMPIRICAL}$ (m^3s^{-1})	Q_{CFD} (m^3s^{-1})	Δ (%)	Labyrinth $Q_{EMPIRICAL}$ (m^3s^{-1})	Q_{CFD} (m^3s^{-1})	Δ (%)	PKW $Q_{EMPIRICAL}$ (m^3s^{-1})	Q_{CFD} (m^3s^{-1})	Δ (%)
880.1	1,994.1	1,969.3	1.2	1,208.6	1,216.5	0.7	1,175.7	1,208.1	2.8
878.7	861.3	835.5	3.0	717.3	740.4	3.2	659.9	675.6	2.4
878.3	544.6	533.1	2.1	547.5	559.6	2.2	493.5	511.2	3.6
877.8	279.9	281.8	0.7	355.6	364.9	2.6	308.8	307.5	0.4
877.4	90.8	94.0	3.5	143.7	147.6	2.7	120.9	121.3	0.3

September 2016. A total of five spillway alternatives were ultimately considered for the final design that included the three standard types discussed earlier as well as more complex gated, Obermeyer and fuse gate type configurations. Evaluation of each alternative was based on a breakdown of relative construction and replacement costs and other non-monetary factors including hydraulic performance, overall size/layout, aesthetics, operation, and maintenance (O&M), safety, reliability, and potential for vandalism.

Results of the initial VE session conducted in 2015 resulted in the removal of the gated and Obermeyer configurations from consideration due to long-term O&M and vandalism concerns due to the remoteness of the site, while the fuse gate option was removed due to the added replacement costs, loss of reservoir volume and potential damage to downstream spillway components following an activation. Discussions during the 2016 workshop focused on the three remaining spillway configurations contained herein with the PKW arrangement ultimately being selected, despite its more complex construction requirements, due to its aesthetics and a comparative cost estimate that was slightly less than the labyrinth arrangement. The ogee configuration was removed in 2016 from consideration by CSKT due to schedule and permitting concerns related to the required extension of the existing spillway footprint beyond the existing property limits in order to pass the peak IDF flow at the NHWL.

4 FINAL DESIGN

Following its selection as the preferred alternative following the VE sessions, the preliminary A-Type PKW layout was further developed to optimize the final location, size and orientation of the spillway structure and spillway inlet channel located immediately upstream of the chute. Initial layouts of these features were assessed using an iterative analysis of the downstream spillway system based on standard one-dimensional (1-D) computer programs including the U.S. Army Corps of Engineers' HEC-RAS and USBR's WS77 computer programs. These 1-D models were quick, efficient and valuable tools that helped the Stantec design team balance CSKT's desire to reduce the overall size and cost of the facility against the impacts to hydraulic performance those reductions have on head loss and submergence levels at the PKW keys during normal spillway operation.

Following this initial optimization exercise, CFD modeling of the chosen spillway location and inlet spillway channel design was conducted to confirm the overall hydraulic performance of the rehabilitated facility including its ability to safely pass the peak IDF flow at the NHWL. The model setup for this final phase of CFD model testing, including the general model setup, boundary conditions and mesh resolution, mirrored the preliminary design described earlier but also included air entrainment, conveyance and the bulking of the fluid through use of the air entrainment, drift-flux and first-order variable density options available within Flow-3D HYDRO. The entrainment rate coefficient, initial average air bubble diameter, minimum volume fraction of air and the escape rate coefficient utilized in these settings were set at the defaults of 0.50, 12.7mm, 0.1 and 1.0, respectively recommended by Flow Science due to the lack of prototype and/or scaled physical hydraulic model test observations to calibrate to.

The layout of the final Type A PKW spillway arrangement is provided in Figure 5 and shows the weir structure being set back further upstream and into the embankment along with a slightly wider chute and a deeper and more tapered inlet channel when compared to the layout from the preliminary design phase.

Graphical results of the CFD analysis for the final arrangement is presented in Figure 6. Results of the CFD modeling for the final phase of design showed gradual submergence of the upstream PKW keys as the reservoir level reached the NHWL (i.e. El. 878.1m) and full submergence occurring when levels approached the new dam crest elevation (i.e. El. 879.1m) due to the increase in head loss through the spillway inlet channel and chute. As shown in the accompanying figure, the final configuration achieved the 425m^3s^{-1} IDF flow requirement for the project at the NHWL with the CFD-based rating curve taking into consideration the hydraulic impacts the reservoir bathymetry, spillway design and orientation, approach/departure conditions, turbulence, and head loss within the spillway inlet channel have on the flow through the facility.

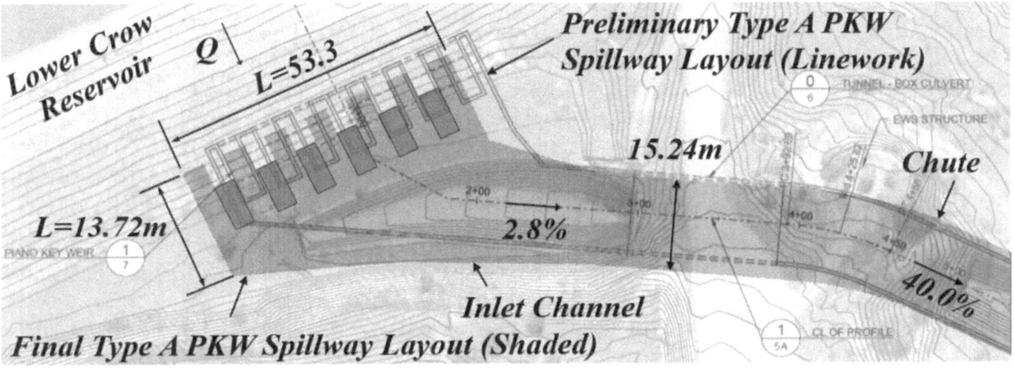

Figure 5. Final design PKW layout.

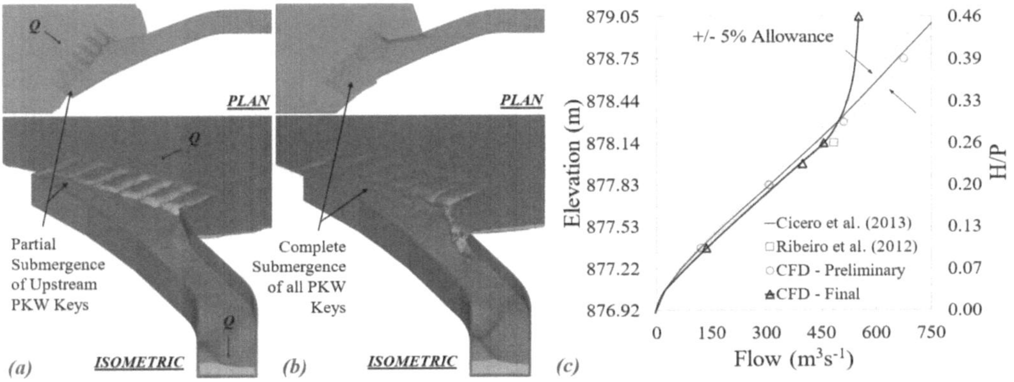

Figure 6. Final design CFD model results – (a) El. 878.14m (NHWL), (b) El. 879.05m (New Dam Crest Elevation), (c) Discharge Estimates.

5 SPILLWAY CONSTRUCTION

The construction of the new spillway was conducted between April 2020 and January 2022 and consisted of the replacement of the existing side-channel spillway tunnel and chute with an A-Type PWK spillway structure, spillway inlet channel, and chute crossing beneath Crow Dam Road. Photographs during and after construction of the spillway are provided in Figure 7.

Figure 7. Spillway Construction Imagery – (Left) March 2021 – Base Slab Installation; (Center) August 2021 – PKW Wall Placement; (Right) April 2022 – PKW and Spillway Inlet Channel After Construction.

6 CONCLUSIONS

Crow Dam, under a reservoir level and storage restriction, issued by the BIA and USBR since 2008 required the evaluation of rehabilitation options and design for a new spillway. The selected final spillway configuration has the capacity to pass the $425m^3s^{-1}$ peak IDF flow through the dam with the reservoir at the NHWL.

Based on the analyses and hydraulic modeling performed for the Lower Crow Dam spillway rehabilitation project presented herein, the following conclusions were drawn and applicable to future spillway projects:

- The phased modeling approach adopted for this project coupled with active communication and the VE workshops proved to be an efficient means to make informed decisions and maintain stakeholder engagement throughout the design process.
- The use of CFD modeling provided additional value to the design of the new spillway features when coupled with more traditional 1-D models by allowing Stantec to optimize layouts and minimize hydraulic impacts related to the approach conditions and the potential for submergence due to head loss within the downstream inlet channel and chute.
- The estimated discharges calculated by the empirical and CFD methods were closely matched for all three spillway arrangements considered under this study with values falling within the ±5% allowance adopted by Stantec for the project.
- Having verifiable hydraulic results is an important and invaluable component in the hydraulic, civil, and structural designs of spillway projects as it allows designers to have a high degree of confidence to make small- to large-scale modifications to preliminary layouts before final design and construction are undertaken, thereby reducing overall risk.

REFERENCES

Cicero, G.M. & Delisle, J.R., 2013. Discharge Characteristics of Piano Key Weirs Under Submerged Flow. *Labyrinth and Piano Key Weirs II – PKW 2013*. Paris: CRC Press/Balkema: 101–109.

Crookston, B.M. & Tullis, B.P. 2013. Hydraulic Design and Analysis of Labyrinth Weirs, I: Discharge Relationships. *Journal of Irrigation and Drainage Engineering, Volume 139, Issue 5*: 363–370.

Hirt, C.W. & Nichols, B.D. 1981. Volume of fluid (VOF) Method for the Dynamics of Free Boundaries. *Journal of Computational Physics, 39*: 201–225.

Hirt, C.W. & Sicilian, J.M. 1985. A Porosity Technique for the Definition of Obstacles in Rectangular Cell Meshes. *Proceedings of the 4th International Conference on Numerical Ship Hydrodynamics.* Washington, D.C.: Office of Naval Research.

Ribeiro, M. L., Pfister, M., Schleiss, A.J. & Boillat, J. 2012. Hydraulic Design of A-Type Piano Key Weirs. *Journal of Hydraulic Research, 50:4*: 400–408.

U.S. Department of the Interior, Bureau of Reclamation. 1987. *Design of Small Dams, 3rd Ed.* Washington D.C.: U.S. Government Printing Office.

U.S. Department of the Interior, Bureau of Reclamation. 1988. *Downstream Hazard Classification Guidelines. ACER Technical Memorandum No. 11.* Denver, CO: U.S. Dept. of the Interior, Bureau of Reclamation.

II.3 Construction

Nachtigal dam: A 4917 m long labyrinth spillway

D. Magnan and B. Blancher
EDF Hydro – Centre d'Ingénierie Hydraulique, La Motte Servolex, France

A. Chapuis and T. Guillemot
ISL Ingénierie, Paris, France

S. Erpicum
UEE–HECE – Université de Liège, Liège, Belgium

ABSTRACT: The Nachtigal Dam, part of the Nachtigal Amont Hydropower Project in Cameroon, features a 4917 meter-long labyrinth spillway, designed to safely manage floodwaters. This spillway was optimized for hydraulic safety, cost efficiency, and environmental considerations. Physical scale modeling (at a relatively rare 1:12.5 scale) and detailed design adjustments ensured robust performance under various flow conditions. Despite construction challenges for such a large-scale nonlinear structure, the project achieved reliable spillway operations, demonstrating the suitability of labyrinth weirs for large-scale hydraulic infrastructure.

1 GENERAL PROJECT OVERVIEW

1.1 *Nachtigal hydropower project*

The Nachtigal dam is an integral part of the Nachtigal Amont Hydropower Project in Cameroon. It dams the Sanaga River and diverts its waters into a headrace channel to feed a 420 MW hydropower plant located 3.3 km downstream.

The project is located 70 km northeast of Yaoundé. It will meet nearly 30% of Cameroon's electricity demand. The hydroelectric project is being carried out by Nachtigal Hydro Power Company (NHPC), a special purpose vehicle (SPV) formed by EDF, IFC, the government of Cameroon, STOA, and Africa 50. The company is responsible for the design, financing, construction, and operation for 35 years.

The civil engineering works of the project (including the dam) have been constructed since 2018 under an Engineering, Procurement, and Construction contract by the Camerounaise de Construction du barrage de Nachtigal (CCN) consortium composed of three civil contractors: NGE (France), SGTM (Morocco), and BESIX (Belgium). EDF–CIH has been supporting NHPC since 2013 with project design and construction supervision, while the engineering joint-venture Tractebel (leader)/ISL has been assisting CCN as Engineer for the execution since the tender launched in 2016.

The hydropower project is composed mainly of:

- An approximately 1.5 km long and 14 meters high Roller-Compacted-Concrete (RCC) dam forming a small reservoir of around 27 hm^3,
- A 3.3 km canal with a capacity up to 980 m^3/s,
- A hydroelectric plant, which includes 7 turbines, each generating 60 MW.

DOI: 10.1201/9781003655541-24

The dam was impounded in the summer of 2023, and the general commissioning of the project is scheduled for spring 2025.

1.2 *Nachtigal dam spillway*

The hydrological design criteria adopted for the Nachtigal dam were to be able to discharge the Design Flood (DF) (Q_{5000} = 7,180 m³/s) below the maximum water level (MWL) set at 1.00 m above the Full Supply Level (FSL) (513.5 m NGC) and the Safety Flood (Q_{10000} = 7,500 m³/s or Q_{5000} with the largest gate blocked) below the Extreme Water Level (EWL) set at 1.20 m above the FSL (Q_x means flood return period of x years).

Accordingly, the dam incorporates a gated spillway (2 segment gates L:17 * H:10.7 m²) and a labyrinth spillway with a capacity up to 6,401 m³/s under 1.16 m water head above the crest spillway set at 513,60 m NGC (+10 cm above the FSL).

The labyrinth spillway is atypical in two ways:

- It is located on the crest (7.68 m thick) of a low-height RCC dam with a steep stepped downstream face (0.5H/1V).

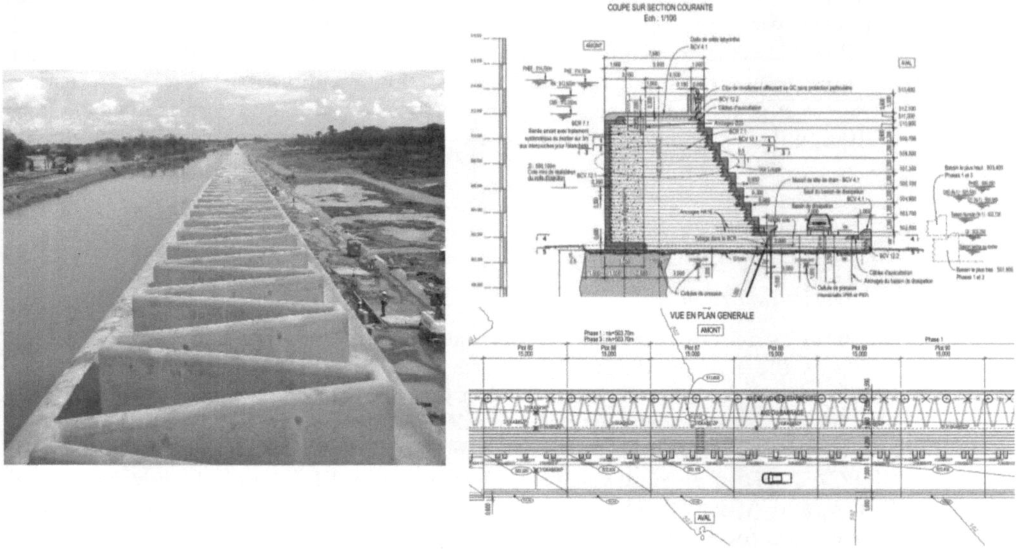

Figure 1. Nachtigal dam crest and cross section.

- It is made up of 463 cycles, creating a total overtopping section of W = 1,389 m/L_c = 4,917 m long, where L_c corresponds to the developed spillway length. Table 1 summarizes the geometric parameters of the cycles based on the nomenclature proposed by Crookston, 2013a.

Table 1. Nachtigal labyrinth cycles parameters according to (Crookston, 2013a).

w	3 m	A	0.25 m
α	12.5°*	D	0.65 m
l_c	4.86 m	t_w	0.25 m
B	5.00 m	w/P	2
P	1.50 m	l_c/w	1,62
H_T/P	< 0.61	H_T/P	< 0.74
MWL		EWL	

* exactly: 12.47°

2 PRELIMINARY DESIGN

2.1 *General presentation of the dam site*

The dam axis spans approximately 1,700 meters across the Sanaga River, which is divided into four branches. Environmental and Social (E&S) analyses have emphasized the importance of maintaining a continuous flow to the main branch to support biodiversity, as well as ensuring seasonal flows to the three secondary branches for biodiversity preservation and hydraulic safety downstream of the dam.

During the preliminary design phase of the project, several design options for flood spillways were evaluated to meet the sizing criteria, with the aim of developing the most suitable solution in terms of:

- Dam {hydraulic} safety (reliability & robustness),
- E&S consideration and seasonal flooding of the right bank branches
- Ease and cost of operation and maintenance,
- Cost and construction time.

Figure 2. The Sanaga river at Nachtigal dam site.

The reservoir's configuration, the dam's length, and the width of the river downstream of the dam made it possible to consider a long spillway capable of evacuating the maximum flood with a low specific flow rate and minimal water level elevation.

2.2 *Spillway alternatives*

2.2.1 *Assumptions*

To maintain gravel cleaning capacity during flood events, in front of the intake of the headrace canal, it was decided to consider the construction of at least two large segment gates (L:17 * H:10.7 m^2) at the natural level of the Sanaga River.

The additional evacuation capacity required can be obtained through various combinations of:

- Additional segment gates: The temporary diversion structure includes four additional concrete sluices designed to manage floodwater from work protection. Various solutions were considered, including the installation of up to four additional segment gates, each of the same dimensions as the two mandatory gates. These gates would equip all or part of the diversion channels, with the remaining channels being concreted to maintain the dam's continuity. The risk of obstruction by floating debris was deemed negligible due to the large size of the flow sections.

- Flap gates: The flap gates considered are elements of 30 ml by 1.50 m high,
- Standard ogee-crested weirs (Creager),
- Fuse gates: Fuse gates of H = 1.5 m high, and a first triggering for a flood with a return period of at least 100 years was considered,
- Labyrinth weirs, either traditional or PK-Weirs.

The PK-Weirs variant, as opposed to the traditional straight labyrinth weirs, was analyzed at the preliminary stage. However, given the length of the structure to be equipped and the significant construction time constraints tied to the project's financing model, this sub-variant was found to be less advantageous than the traditional labyrinth design, despite the potential flow coefficient improvements. Additionally, constructing a large series of PK Weirs introduced technical challenges (less easy-to-master technicality) and constraints that would be more difficult for contractors to manage, especially considering the spillway structure was a critical component of the project's timeline. As a result, the PK-Weir variant was excluded from the comparison during the optimization study

2.2.2 *Technical alternatives*

The table below summarizes the various alternatives considered for flood evacuation to meet the hydraulic flood evacuation criteria (mainly governed by MWL < FSL+1 m and EWL < FSL+1.6 m).

Table 2. Summary of possible flood evacuation systems.

Number of segment gates	Variant		
2 + 0	2.1	Flap Gates	L=650 m
	2.2	Labyrinth	L=1,389 m
	2.3	Labyrinth + Creager	L=1,120 m + L=220 m
2 + 1 = 3	3.1	Flap Gates	L=540 m
	3.2	Labyrinth	L=950 m
	3.3	Labyrinth + Creager	L=350 m + L=990 m
2 + 2 = 4	4.1	Flap Gates	L=420 m
	4.2	Fuse gates	L=440 m
	4.3	Labyrinth	L=670 m
	4.4	Creager	L=1,080 m
2 + 3 = 5	5.1	Flap Gates	L=310 m
	5.2	Labyrinth	L=490 m
	5.3	Creager	L=660 m
2 + 4 = 6	6.1	Flap Gates	L=190 m
	6.2	Labyrinth	L=310 m
	6.3	Creager	L=420 m

An expert appraisal was conducted among the different alternatives, which helped identify the most suitable options for further analysis, based on criteria previously cited, drawing on EDF's hydroelectric operational experience. The variants selected for further analysis (cost analysis) are as follows:

- Solution n°2.3: 2 segment gates, labyrinth over 1,120 m, and Creager over 220 m,
- Solution n°3.3: 3 segment gates, labyrinth over 350 m, and Creager over 990 m,
- Solution n°4.4: 4 segment gates and Creager over 1,080 m.

The cost comparison of the different solutions was conducted using simplified unit costs, intended solely for comparing the solutions against each other. These costs were relevant only for comparing the proposed variants (relative values, rather than absolute ones). Solution No. 2.3 emerged as the most economical solution, and also the most robust in the event of a malfunction of one of the sluice gates.

Table 3. Estimated cost per flood evacuation system.

Variant	2.3	3.3	4.4
Total ($)	24,819,910	30,657,392	37,495,480

2.2.3 *Final selection*

The proposed solution above involved combining a labyrinth weir with a Creager weir, which lacked significant hydraulic or structural benefits. It appeared more judicious, both from the overall economic perspective of the project and its execution, to equip the entire spillway with labyrinth weirs over a linear length of 1,389 m (Variant 2.2).

3 DETAILED DESIGN

3.1 *Physical scale model*

In order to accurately understand the behavior of the structure compared to the bibliographic analysis, a physical model was built at the Laboratory of Hydraulic Constructions – HECE of ULiège in 2019.

The Nachtigal spillway structure is a wide construction consisting of a large number of cycles. The discharge capacity of such a structure is minimally affected by edge effects, especially when operating heads are limited and the approaching flow is uniform. It was therefore possible to analyze it by studying a limited number of cells, as the labyrinth cycles have the same longitudinal profile, with no curvature in plan.

Due to the small height of the dam, a physical model at a relatively rare 1:12.5 scale was able to be built. This scale factor helped avoid the influence of scale effects (undesirable viscosity effects) on the discharge capacity of the weirs and the shape of the downstream jets, even at Q100 (61 cm real-world prototype – 4.8 cm scale model).

The model represents a dam block 15 meters wide topped with 5 labyrinth cycles and the dissipation basin. The block is placed in a rectangular-section channel 7.2 meters long, aligned with the transparent walls of the channel. The choice of 5 cells ensured proper representation of the jump position, which would be observed at the center of the model in the flow, minimally influenced by the edge effects of the channel (the relative importance of wall friction compared to energy dissipation in the jump).

Figure 3. Physical scale model 1/12.5.

The preliminary design featured a Quarter-Round (QR) crest shape. During the general project review before the detailed design phase, a possible overestimation (3%) of the gate discharge capacity was identified, likely resulting from the iterations and design changes inherent to such a complex project. This required taking precautions regarding the labyrinth discharge by finding a way to boost it to compensate. The first modification was therefore to directly incorporate a Half-Round (HR) crest shape. According to (Crookston, 2013a), a favorable impact of approximately +3% of the labyrinth discharge for the geometry of the Nachtigal labyrinth was expected between a hydraulic performance of HR and QR crest.

3.2 *First unexpected results*

The initial results indicated that the discharge capacity of the labyrinths was 6–7% lower than the values predicted in the literature for a HR crest shape and even 4–5% lower than those for an QR crest shape. This deficit leads to water levels 3–4% (equivalent to 4–5 cm) higher than the project's contractual design levels, leading to an overall discharge capacity deficit of 3–4% compared to the total flow required to be evacuated (gates + labyrinth).

The discharge coefficient values obtained from model measurements was so slightly lower than those calculated based on the bibliographic reference. This discrepancy has been attributed to the approach flow conditions of the Nachtigal spillway, which appear to be different from those of the scaled models used in the literature.

The Nachtigal labyrinth is placed at the top of a gravity dam with a sharp 90° transition between the vertical upstream face and the horizontal platform on which the labyrinth is installed. In contrast, the formulas in the literature were derived from tests conducted with a labyrinth spillway positioned directly on the bottom of a horizontal channel. The acceleration of the flow as it "climbs" onto the dam could be the cause of a slight decrease in the labyrinth's efficiency. A recirculation zone may thus form just downstream of the upstream face, potentially influencing the discharge capacity of the labyrinth.

As feedback, this discharge deficit due to approach flow conditions can be compared to the classical abatement coefficient used in ogee-crested weir type flow discharge, which accounts for approach depth ($p/H \approx 1.5$; $C_{abbat\_Creager} = 97.5\%$).

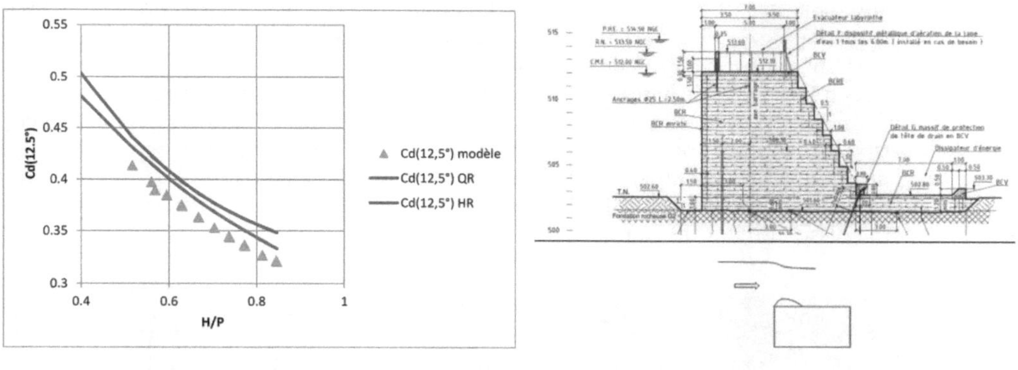

Figure 4. Unexpected results – Preliminary design – Analysis scheme.

Several mitigation measures have been considered:

- Integrate tolerance of contractual elevations: only one centimeter accepted;
- Decrease the labyrinth crest elevation which is at FSL + 10 cm: not possible for exploitation purposes;
- Improve the feeding of the labyrinths by relocating the structure upstream of the crest: this solution was not possible for dam stability concerns. Moreover, shifting the structure upstream does not completely alter the approach conditions for the downstream apex (the upstream angle remains steep), and the gain in discharge capacity was not straightforward;
- Wait for the discharge results of the gated spillways: The expected discharge capacity was already considered to be overestimated;
- Spillover on the closing structures of the temporary diversion passages: cannot compensate sufficiently for the lack of discharge;

– Improve the supply to the labyrinths through an upstream profiling of the thick crest of the dam: this solution was selected.

3.3 *Implemented mitigation measure*

Several forms of hydraulic profiled caps have been tested (trapezoidal shape, quarter-circle shape with a radius based on Miller's recommendations (Miller, 1994); upstream quadrant shapes of the Creager threshold, offset and cantilevered upstream in accordance with Hydraulic Design Criteria sheet 111–19).

The final shape implemented is presented next Figure (called GB.B1). Due to the distinctive rounded shape and the associated formwork challenges, which would require additional time during construction, the Contractor decided to prefabricate the components.

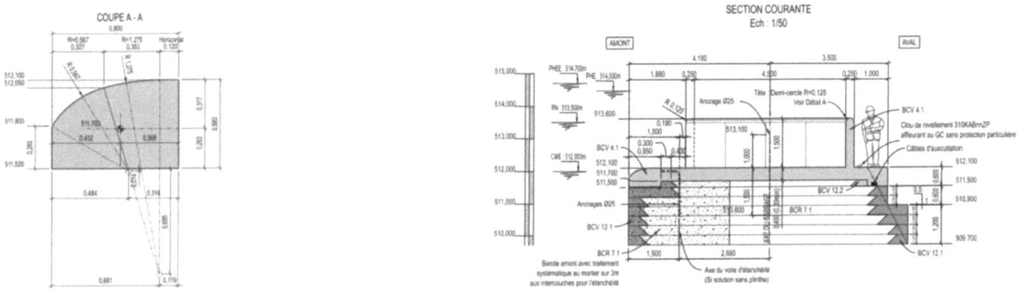

Figure 5. Detailed design and implemented hydraulic profiled caps.

3.4 *Discharge curve law*

The following curves present the specific discharge and the discharge coefficient of the final geometry of Nachtigal labyrinth.

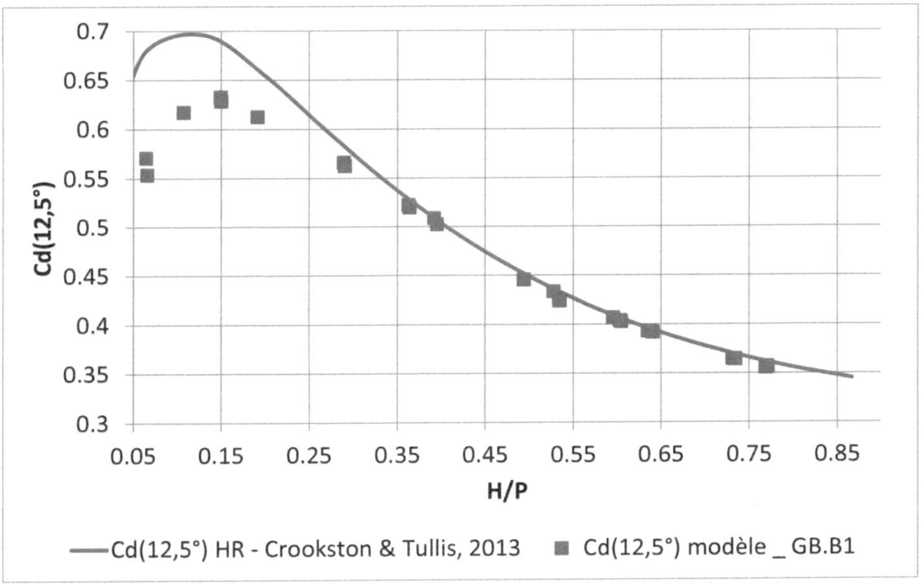

Figure 6. Discharge curve laws of Nachtigal labyrinth.

This figure shows a good alignment with the values from the literature (Crookston, 2013a) for a load ratio $H/P > 0.3$. The match is less accurate for smaller load ratios (upstream load on the model less than 3.5 cm; 45 cm in real-world prototype), but these are affected by scale effects that differ from those affecting the models of (Crookston, 2013a) (P = 305 mm versus P = 120 mm on Nachtigal model). It is common practice not to consider discharge values for these small water heads as reliable, as they are affected by surface tension. For water layers <0.45 m ($H/P < 0.3$) = 514.05 m NGC, it was so recommended to follow the theoretical discharge curves. In the range of $H/P > 0.3$, the average deviation from the theoretical curve Cd (12.5°) is 1.07%.

3.5 *Flow conditions on and downstream of the labyrinth*

As a reminder, the downstream steep face 0.5H:1V of the dam is composed of RCC steps (1,2 m high) and a stilling basin set at the dam foot about 8 m long with a small downstream control weir. The resulting observations of the flow conditions are:

- TYPE I Behavior: For the lowest flows ($q_{prototype} \leq 0.983$ m³/s/m), a calm and uniform flow is observed across the entire labyrinth. The turbulence at the top of the dam downstream of the labyrinth is low, and the nappe flow impact on the downstream chute occurs relatively uniformly across the width, first on the 1st step from the top and then shifting to the 5th step as the flow increases.
- TYPE II Behavior: For intermediate flows ($q_{prototype} < 2.188$ m³/s/m), the turbulence generated downstream of the labyrinth walls increases, resulting in a more disturbed water flow, less uniformly distributed across the width of the plot, impacting the chute between from the 4th to 7th steps from the top.
- TYPE III Behavior: For higher flows ($q_{prototype} \geq 2.399$ m³/s/m), the flow becomes highly turbulent and aerated downstream of the walls, as well as at the downstream end of the labyrinth. The water flow becomes even more disturbed and impacts the chute diffusely between the 4th and 7th steps, similar to the intermediate flow behavior.

With the exception of TYPE I behaviour, where the flow is of the "nappe flow" on the first steps, energy dissipation on the steps is limited. The water flow behaves more like a highly aerated free jet rather than a flow supported by the stepped chute conveyance face. In all cases, the impact occurs within the dissipation basin, with subcritical flow exiting the basin.

Even though the downstream face of the dam does not significantly contribute to flow energy dissipation, no improvement has been studied due to the limited energy to dissipate (low dam height, rapidly rising downstream water level, etc.)

3.6 *Water nappe aeration behavior and nappe instability.*

In the literature, two distinct unsteady phenomena are described for flow over a labyrinth spillway. One is the "flow surging" of the nappe, associated with its aeration downstream of the crest, while the other involves nappe instability creating vibrations (oscillations producing substantial sound vibrations with a frequency of less than 50 Hz). First, it is important to note that both these phenomena on labyrinth spillways are still not fully understood or documented and remain an ongoing research topic. It is also highly likely that these phenomena are significantly influenced by scale effects when studied using reduced-scale models (Lodomez 2019).

The water nappe aeration behavior is analyzed in (Crookston, 2013b) considering different crest shapes and angles of the weir walls. The water nappe downstream of a labyrinth spillway can fluctuate between an aerated flow and a partially aerated flow within a well-defined range of dimensionless head (H_T/P). Within this range, the size of the air pocket beneath the nappe varies both temporally and spatially along the spillway walls. Initially continuous and stable along the walls, the air pocket becomes non-uniform (dividing into smaller isolated pockets) and temporally unstable (with changes in pocket size and position over time). This instability of the air pocket is linked to pressure fluctuations on the downstream face of the spillway walls, which affect the trajectory of the nappe but do not impact the spillway's discharge capacity.

Figure 7. Flow conditions on and downstream of the labyrinth. *TYPE I:* $q_{prototype}$ = 0.983 m³/s/m (top) and *TYPE III:* $q_{prototype}$ = 4.624 m³/s/m (bottom).

In the case of the Nachtigal dam, the H_T/P values corresponding to the different project flow rates are as follows:

Table 4. H_T/P rate for several return period at Nachtigal dam.

Return period	Total flow rate	Labyrinth flow rate		H_T	H_T/P
Annual flood	3400 m³/s	1404 m³/s	1.01 m³/s/ml	0.28 m	0.19
Q_{10}	4400 m³/s	2359 m³/s	1.70 m³/s/ml	0.43 m	0.29
Q_{100}	5500 m³/s	3428 m³/s	2.47 m³/s/ml	0.61 m	0.41
Q_{1000}	6500 m³/s	4348 m³/s	3.13 m³/s/ml	0.78 m	0.52
Q_{5000}	7180 m³/s	5021 m³/s	3.61 m³/s/ml	0.90 m	0.61
Q_{10000}	7500 m³/s	5280 m³/s	3.80 m³/s/ml	0.96 m	0.64
$Q_{5\,000\,-\,N\text{-}1}$	7180 m³/s	6102 m³/s	4.39 m³/s/ml	1.11 m	0.74

Based on these HT/P values and the angle α = 12.5° of the labyrinth weir walls, the types of flow encountered range from aerated flow for an annual flood to drowned flow for the Q_{5000} Design Flood (DF). The flows are distributed as follows on the Crookston diagram (Crookston, 2013b):

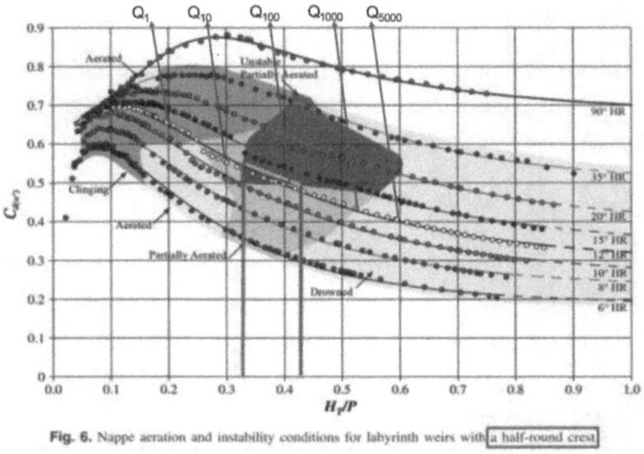

Figure 8. Flow return period distribution for Nachtigal labyrinth geometry on Crookston diagram (Crookston, 2013b).

The unstable partially aerated phenomenon reported in the literature occurs between H_T/P {0.33 and 0.42}, i.e., for flow rates varying between Q_{20} to Q_{100} (considering that the main hydropower plant (930 m^3/s) is shut down) and is postponed to Q_{100} to Q_{1000} when the hydropower plant is operational.

In scale model, the unstable partially aerated phenomenon was observed into the previous H_T/P range. Its frequency is variable but remains below 1 Hz. It does not affect the discharge or the overall flow conditions on or downstream of the labyrinth.

Figure 9. Unstable partially aerated phenomenon on scale model – $q_{prototype}$ = 2,4 m³/s/m – H_T/P=0,39 - Time between pictures 0.25 s.

The implementation of an artificial aeration system for the water nappe on such a long labyrinth is expensive to construct and introduces maintenance challenges, regardless of the system chosen. For simplicity in implementation, and given the low probability of unstable flows and their potentially minor impact on a reinforced concrete structure, it was decided by the all the actors in the project not to install an aerator on the spillway initially. However, a subsequent installation, based on feedback from the first few years of operation, remains a possibility. A triangular-section water nappe separator, 0.90 m in height, would then be installed at the downstream apex of the labyrinth, with the tip of the triangle facing against the flow direction.

4 CONSTRUCTION FEEDBACK

4.1 *Challenges related to structural aspects From ISL's perspective – CCN's Engineering for the dam execution*

Some details and challenges regarding the structural aspects:

- The wall thickness t_w was only 25 cm (larger than the recommended value P/8 (Crookston, 2013a) for construction purposes) and presenting both a calculation and construction challenge. Considering the very strict contractual watertightness classification (Class 2 according to Eurocode NF EN 1992–3), the mechanical loads, the seismic requirements with crest dam amplification, and Fluid-Structure Interaction (FSI), as well as the thermal loads (water, air temperature, and solar irradiation), the concrete shrinkage action of C40/50 concrete (XA2 concrete class according to Eurocode), and a concrete cover of 5 cm for reinforcement, the reinforcement was dense (220 kg/m^3) and required the use of a self-leveling concrete mix to avoid honeycombing.
- The waterstop at the RCC construction joint block, located every 15 meters, extends 5 cm above the FSL into the labyrinth wall, and 5 cm below the labyrinth crest level. A specific procedure for sealing the bulb of the waterstop joints has been developed by the contractor.
- The worksite access is limited to 0.9 m upstream and 1 m downstream which was quite challenging during the construction stage and commissioning inspections. This will remain a challenge for the annual topographical survey and will require specific small elevator platform.

4.2 *Challenges related to in-place casting of the labyrinth walls From EDF-CIH's perspective – site controller during the construction*

One of the main difficulties encountered on the construction site was determining the method for constructing the complex labyrinth weir on the crest of the RCC structure, considering the overall work schedule and available resources.

This topic had been deeply analyzed during the preliminary design phase by EDF–CIH. Although the initial feeling might suggest that the weir itself should be constructed using conventional vibrated and reinforced concrete due to its geometry and the hydrostatic forces to be tackled, several construction solutions were indeed considered possible:

- Option 1: Weir cast in place and embedded in a conventional reinforced concrete slab,
- Option 2: Weir made of prefabricated concrete elements, placed on a mortar bed, then concreting the upper slab in reinforced concrete,
- Option 3: Weir made of prefabricated concrete elements, placed on a mortar bed, directly anchored on an enriched RCC layer, then keyed together,
- Option 4: Weir made of pre-wall type elements, assembled, anchored in an enriched RCC layer, then filled with concrete.

The table below provides a qualitative comparison of the advantages and disadvantages of each solution:

Option 1 emerged as the simplest to implement and the most robust. Option 2, involving prefabricated elements, remained feasible if there was significant pressure on the project's execution deadlines. Options 3 and 4 were deemed unsuitable in terms of stability, sealing, and durability.

In the context of an EPC contract, the implementation solution remains the responsibility of the contractor. After careful analysis of Options 1 and 2 proposed by the contractor's method, CCN decided to cast the labyrinth weir walls in place (Option 1), rather than using prefabricated elements (only V-elements) as initially planned by the contractor method. Both solutions were fully developed until formwork, reinforcement drawings and construction methodology. The final decision was decided only several weeks before the first labyrinth construction. This decision was driven by the Contractor's goal to simplify and gain better control over the execution methods

Table 5. EDF–CIH construction analysis of labyrinth.

Option	1 – Labyrinth and upper slab in cast-in-place CVC	2 – Prefabricated labyrinth then upper slab in CVC	3 – Prefabricated labyrinth anchored in enriched RCC	4 – Labyrinth in pre-walls anchored in enriched RCC
Technical simplicity of execution	++	-	–	–
Required handling means	++	–	–	- (blocks lighter than option 3)
Interface sealing	++	+	–	-
Simplicity of weir alignment	++	–	–	-
Execution speed	-	+	+	+
Durability	++	+	–	-
Material cost	-	-	++ (no CVC slab)	++ (no CVC slab)
Main advantages	Simple, robust, usual construction techniques, basic handling and construction means, durability, sealing	Execution time, durability, sealing	No CVC slab, execution time	No CVC slab, execution time
Main disadvantages	CVC volume, execution time	CVC volume, heavy handling, difficult alignment of elements	Sophisticated techniques, heavy handling, difficult alignment of elements, sealing, durability	Sophisticated techniques, heavy handling, difficult alignment of elements, sealing, durability

compared to the prefabricated option (including transportation and lifting complications of high weight and fragile structure), eliminate the risk of leaks through concrete joints (for keying), and optimize the schedule by reducing the need for surface repairs caused by the presence of lifting hooks.

Initially, the in-place casting was carried out in two lifts using wooden formwork. Once the formwork was constructed and adapted, casting was then performed in a single lift using metal formwork made of half-shells:

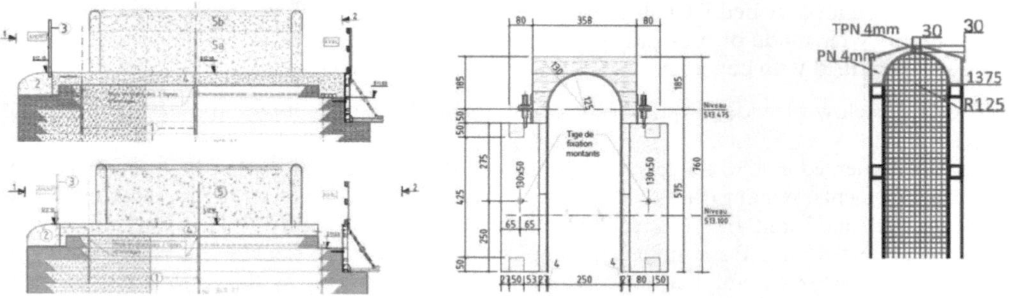

Figure 10. In-place casting of the labyrinth walls – Up/Left: initial CCN method; Down/Right: final CCN method.

The advantages and disadvantages of the two methods are summarized below:

Table 6. Nachtigal labyrinth cycles parameters according to (Crookston, 2013a).

Method	Advantage	Disadvantage
Single lift	More rigid formwork and better alignment, greater durability	No adjustment possible during installation, no adaptability to the situation (position and alignment of expectations and reinforcement), less control over the spacing of formwork faces and their offsets.
Two lifts	Allows finer adjustment of the altitude of the second lift, allows teams to take more care in concreting the second lift with the rounded crest	Less rigid formwork inducing geometry and alignment defects, adds a concrete joint

In general, the main construction defects observed were as follows:

- Non-compliance with crest altitude contract tolerance along the 4917 ml of sill: The crest altitude varied between -5.7 cm and +6.5 cm, with a target of 513.60 m NGC. Forty-four percent of control measurements were outside the stringent (and challenging) contract tolerance (+/- 1 cm). See Section 5 for the observations during first spilling. These defects are attributed, in addition to the quality of the topographic controls and the lack of detailed altitude adjustment methods mentioned above, to the accumulation of wiping products on the crest of the structure and defects in air evacuation at the crest,
- Formwork and concreting defects: The wooden formworks were too rudimentary and deteriorated quickly, with a maximum of five concreting cycles. These formworks, which did not allow for precise altitude adjustments, limited the ability to control the geometry of the structures. The defects were exacerbated by challenges in managing the workability of self-compacting concrete and the arrangement of certain construction features (chimneys, vents, etc.). As a result, extensive repair work was required, including re-grouting and bush hammering.

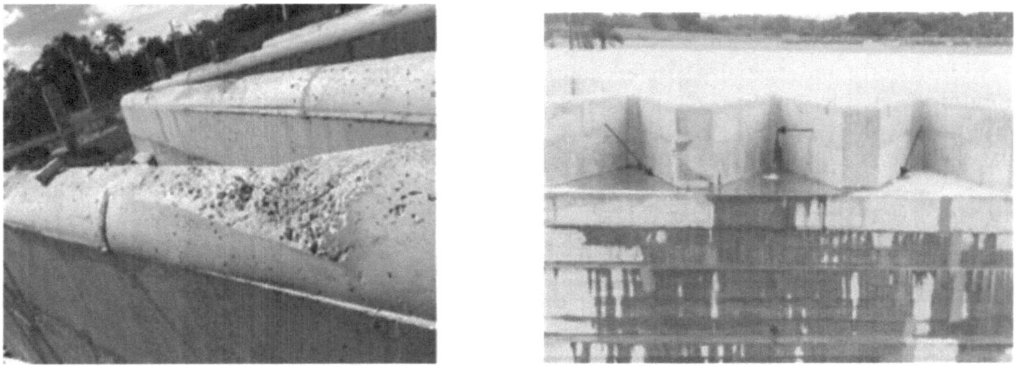

Figure 11. Concreting defects on labyrinth sill and wall sealing defects.

- Localized sealing defects of the slabs and labyrinth walls: These localized defects, which required significant repair efforts by the contractor, were primarily due to the quality of concrete joint treatments, sealing issues at the junctions between formwork and vertical walls, and problems with the concrete formulation. There were also considerable difficulties in vibrating the formwork feet due to the density of the reinforcement.

5 OBSERVATIONS DURING FIRST SPILLING

The reservoir size allows to force a first spilling during the first impounding with a Sanaga inflow rate of 651 m^3/s during the dry season (August 2023). The reservoir water level reached 513.75 m NGC (FSL + 25 cm; labyrinth crest + 15 cm), with 125 m^3/s discharging through the spillway gates and 526 m^3/s spilling over the labyrinth (specific flow rate to 0.38 $m^3/s/ml$).

The spillover phase revealed that the spilling initiation was not fully synchronous along the long 4917 ml of the crest. Indeed, the right bank discharges first, in accordance with the topographic crest survey (3699 topographic control points). The elevation recorded reached on average the precision +0.4 cm relative to the target value of 513.6 m NGC. From a discharge perspective, the reasoning is acceptable on average, as the control density is fairly consistent. The discharge is ensured between 99% and 100% of the theoretical value.

The hydraulic behavior during this first spilling was consistent with the scaled model (TYPE 1 behaviour). The spilling water flow first impacted the face at the third step (508.5 m NGC) from the top, then flowed into the dissipation basin in the area of the step near the basin foot.

Figure 12. First spilling in August 2023 (0.38 $m^3/s/ml$).

No instability and partially aerated nappe behavior, nor any noise characteristic of nappe vibration, were identified on the labyrinths. However, during the first spilling, the H_T/P rate was limited to 0.1. The theoretical and observed behavior was a clinging of the water nappe to the downstream face of the labyrinth's wall. Therefore, the phenomenon nappe instability and partially aerated was not observed during the impounding, as according to section 3.6, this phenomenon typically occurs with higher water head.

6 DISCUSSIONS AND CONCLUSIONS

The decision to use a general labyrinth spillway at the crest of the Nachtigal Dam was driven by the project owner's desire to meet several key objectives:

- High Hydraulic Safety: Despite the site's remote location, the project prioritizes hydraulic safety. This is achieved through the spillway's passive operation (free overflow spillway), ensuring reliability without the need for complex mechanisms.
- Optimization of the Operating Level: The design aims to optimize the operating level under normal conditions. This is achieved by maintaining a low overflow water depth, even during major flood events.
- Environmental and Social Considerations: The project also complies with strict environmental and social criteria, particularly by ensuring the seasonal replenishment of the

natural channels on the right bank of the Sanaga River. This replenishment is effective as soon as the labyrinth spillway begins to overflow.
- Cost Optimization: The design optimizes both construction and operational costs. The spillway's simplicity, industrialized implementation, and the relatively low quantity of concrete required for its construction all contribute to cost savings.

This choice, made following a detailed multi-criteria analysis during the Preliminary Design phase, remains the most appropriate even at the conclusion of the detailed design and construction phases.

However, the detailed design of the structure and its adaptation to the actual site conditions, as well as the typical cross-section of the RCC dam, required significant adjustments compared to the theoretical characteristics found in the literature. In particular, the upstream hydraulic approach conditions had to be studied with the utmost precision, as they have a significant impact on the facility's capacity to evacuate water under all load conditions. It is crucial to recall that the hydraulic design during the early stages of project development should include a margin of safety, particularly when the design relies on literature where test conditions may not fully represent site-specific conditions. Additionally, when a scale model is used within the framework of an EPC contract, the project schedule is often tight, leaving little opportunity for significant modifications once the contract has been finalized and construction is set to begin shortly after.

Additionally, it should be noted that the energy dissipation conditions on the downstream steps could have been studied more thoroughly. In fact, the optimization of the dam's typical cross-section, which was related to stability studies and its high L/H ratio, resulted in a downstream slope of 0.5/1, which is hydraulically inefficient. A minimum slope of 0.7/1, regardless of the structure's stability conditions, seems to be a more appropriate project criterion.

Finally, the execution challenges related to the substantial length of the structure (including formwork with a high reuse rate, significant crest topographical precision constraints, and the complicated concreting of the labyrinth feet due to the thin vertical walls) and the goal of perfect execution for the parts exposed to flows required the company to extensively anticipate needs and methods to ensure quality execution that met expectations. Optimization opportunities include the use of more efficient metal formwork for better altitude alignment and air evacuation, as well as the systematic use of self-compacting concrete or even fiber-reinforced concrete.

In conclusion, despite the challenges in implementation, mainly due to the unusual dimensions of the structure, the construction of an optimized labyrinth weir at the crest of the Nachtigal dam proved to be an excellent solution. It was perfectly suited to the various objectives identified by the project owner and the specific local characteristics of the project. Actual observations during the first spill and after two years of operation have confirmed the designers' decision.

REFERENCES

B.M. Crookston & B.P. Tullis 2013a. Hydraulic Design and Analysis of Labyrinth Weirs. I: Discharge Relationships, ASCE.

B.M. Crookston & B.P. Tullis 2013b. Hydraulic Design and Analysis of Labyrinth Weirs. II: Nappe aeration, instability and vibration, ASCE.

M. Lodomez et al. 2019. Nappe Oscillations on Free Overfall Structures: Size Scale Effects, ASCE.

Miller 1994. Discharge Characteristics/Hydraulic Structures Design Manual, A.A. Balkema Publishers.

Labyrinth and Piano Key Weirs IV – Erpicum, Crookston & Paxson (Eds)

High-altitude labyrinth weir design and construction

N. Miller and M. Provencher
GEI Consultants, Inc, Denver, Colorado, USA

ABSTRACT: GEI worked with a Colorado district to design and construct a labyrinth spillway at a high-altitude reservoir west of Denver. A labyrinth weir configuration was chosen to increase discharge capacity with minimal adjustment to the existing spillway footprint and to pass the Inflow Design Flood with the higher water surface. Empirical design methods were used to develop the labyrinth weir geometry. The spillway has three cycles and is 28.0 m wide and 33.8 m long. The modifications increased the reservoir normal water surface elevation 3 m, increasing reservoir storage capacity by 149.7 ML. The use of a labyrinth weir resulted in minimal modification to the existing embankment dam. This paper presents a comparison of the recent computational fluid dynamics modeling and the spreadsheet-based results used in design. The site presented challenges including significant ice loading on the weir structure, complex rock conditions requiring seepage control measures, and unique construction techniques.

1 INTRODUCTION

This paper presents the design and construction of a labyrinth spillway project located high in the Colorado mountains. The purpose of the project was to add 149.7 ML of storage to the existing reservoir storage of 317.0 ML, for a total of 466.7 ML. The additional storage required the reservoir normal water surface elevation to increase by 3 m. To accommodate the higher reservoir water surface without significantly modifying the existing dam, a more hydraulically efficient 3.0 m high, three-cycle reinforced concrete labyrinth spillway was constructed. The new labyrinth spillway has a total width of approximately 28 m over a length of 33.8 m. The reservoir is located at high altitude, at approximately 2560.3 m, and is used as a primary raw water storage facility. The dam is a zoned earth dam with a structural height of about 26.8 m impounding approximately 317.0 ML at the normal pool and approximately 716.7 ML at the dam crest. The hydraulic height is approximately 18.3 m. The dam was originally constructed as a homogeneous earthfill dam around 1924. In 1992, the dam was modified, raising the crest approximately 7.9 m and adding internal drainage features to the embankment. In its current condition, the dam crest is about 158.5 m long with an average crest width of 7.6 m. The upstream slope is approximately 2.5H:1V (Horizontal to Vertical) and the downstream slope is about 1.75H:1V.

The existing spillway was excavated into the rock slope of the left abutment. The service spillway was founded on this rock cut and had a simple 0.3 m wide, 24.4 m long control section. The excavated rock channel side slopes were constructed at 1H:4V with a 3.0 m wide bench cut into the left side wall. The existing rock channel approach was approximately flat from the control structure for about 61.0 m upstream. Beyond the control structure, the spillway chute had an approximately 25% slope, curving towards the northeast and discharging into a natural rock stilling basin. The stilling basin is located approximately 39.6 m downstream of the toe of the dam and consists of a rock-lined basin that is approximately 18.3 m wide and 30.5 m long. The spillway modifications maintained the existing approach while discharging into the existing stilling basin.

DOI: 10.1201/9781003655541-25

This paper summarizes the geotechnical, hydrologic, hydraulic, and structural elements of the design and the challenges encountered during construction.

2 SITE GEOLOGY

The bedrock at the spillway consists of Precambrian basement rocks, and geological mapping indicates that the dominant units are amphibolite and feldspar gneiss. The rock mass at the existing spillway cut is composed of Precambrian metamorphic rocks which are strongly foliated with alternating layers of feldspar, schist, and amphibolite, and localized pegmatite intrusion generally less than 0.6 m thick. Foliations are generally 0.6-0.9 m thick, but thicknesses vary from less than 25.4 mm to 1.0 m. Foliation planes dip steeply downstream (north) and are generally perpendicular to the axis of the spillway. The existing rock cuts have maintained steep faces since originally excavated, with observed failures mostly limited to localized toppling of small rock blocks. No faults or shear zones are observable in the existing spillway walls. The rock face on the west side of the spillway is jagged with a rough surface, while the east rock face is in general much less jagged as a result of controlled blasting methods.

Subsurface conditions at the spillway were investigated by advancing two boreholes. These boreholes were advanced from the left abutment, above the existing spillway channel. Boreholes were inclined to characterize subsurface bedrock conditions beneath the existing spillway floor. A truck-mounted drill rig was used to collect HQ size rock core samples. Each borehole encountered Precambrian gneiss of the Idaho Springs Formation with isolated pegmatites with thicknesses varying from 25.4 mm to 0.6 m. Gneiss encountered was generally black and white mottled, medium grained, slightly to moderately weathered, slightly to intensely fractured, and soft to hard. Isolated zones of completely decomposed iron-stained rock were encountered. Clay infilling up to 25.4 mm thick was found on fractures. Rock Quality Designations (RQD) varied from 0 to 100 with a general trend of increasing with depth and had an average of 42 for all drives. Permeability values of the bedrock were obtained by water pressure testing using a single inflatable packer to isolate the test zone. Permeability values ranged from $1.5\text{x}10^{-4}$ to $9.7\text{x}10^{-5}$ cm/sec for all tests.

3 DESIGN

The design included geotechnical engineering, hydrology and hydraulic engineering, and structural engineering. Each design component is summarized below.

3.1 *Geotechnical design*

Seepage and slope stability analyses were performed to verify the dam performance is within allowable criteria under the raised reservoir water surface conditions. Seepage analyses were performed to estimate the phreatic surface through the existing embankment dam and seepage quantities for evaluating the capacity of the existing chimney, blanket, and toe drains. Seepage analyses were performed using the two-dimensional finite element method program SEEP/W, Version 8.14.1.10087. The typical maximum cross-section was selected for analysis. The seepage results indicate that the embankment internal filter and drainage system is effectively controlling seepage at the current normal pool and is adequately sized for the proposed normal and maximum pools. However, seepage losses will increase by a factor of 2 to 3 times in the elevated pool. The existing drainpipe has the capacity to convey these increased seepage flows while meeting design criteria for drain capacity.

Slope stability analyses were performed using SLOPE/W, Version 8.14.1.10087, limit equilibrium computer program and the Morgenstern-Price limit equilibrium method. The maximum embankment cross-section was analyzed. For the steady-state seepage condition, the reservoir was analyzed at the proposed normal pool and maximum pool under fully developed steady-state seepage conditions obtained from the seepage analyses. The results of the slope stability analyses showed that for all loading conditions analyzed the calculated factor of safety met or exceeded the minimum required factor of safety.

3.2 *Hydrologic design*

The reservoir has a gross watershed area of 16.8 km^2. The basin was modeled as a single sub-basin and no additional sub-basins were delineated within the watershed. The Inflow Design Flood (IDF) for the dam and spillway were developed using the Probable Maximum Precipitation (PMP) estimates from the methods in Hydrometeorological Report No. 55A (HMR 55A). The critical PMP is the general-storm PMP (all-season or seasonal) or the local-storm (thunderstorm) PMP, depending on which corresponding event produces the maximum reservoir elevation and is therefore the IDF. The general-storm PMP has a 72-hour duration and was estimated to have a total rainfall depth of 66.7 cm. The local-storm PMP has a 6-hour duration and was estimated to have a total depth of 23.0 cm.

The US Army Corps of Engineers (USACE) Hydrologic Engineering Center's - Hydrologic Modeling System (HEC-HMS) Version 3.5 computer model was used to estimate the IDF inflows and outflow hydrographs for the watershed and reservoir. Level-pool routing was used to route the IDF through the reservoir. Based on the modeling results, the critical IDF for the reservoir was produced by the HMR 55A local storm. The IDF results in a peak inflow of 630.1 cms, and a total peak discharge of 585.6 cms. The unmodified spillway was able to pass the IDF with 0.3 m of freeboard below the dam crest.

3.3 *Hydraulic design*

The new spillway was designed to accommodate the IDF without significant modification to the existing dam. The crest of the new weir will be 3.0 m higher to provide for increased storage while still passing the recommended IDF. A stage-discharge rating curve, which depicts reservoir elevation versus spillway discharge, was developed for the proposed spillway. The trapezoidal labyrinth weir rating curve was developed using the methods described in Crookston and Tullis (2013). This method was developed for the hydraulic design and analysis of labyrinth weirs based upon experimental results of physical modeling. The design method provides a robust, empirically derived framework for the hydraulic design, emphasizing the use of discharge coefficients, geometric considerations, and awareness of nappe behavior to optimize their performance for spillway applications. The results of this design method were compared to previously published labyrinth weir head-discharge data including Falvey (2003).

Using successive approximations in an Excel spreadsheet, the raised spillway was designed to be a three-cycle labyrinth weir. Each cycle will be 9.1 m wide and 33.2 m long. The angle of side legs was 6.04 degrees, and the wall height was 3.0 m. The total spillway width was increased to 27.4 m, which required widening the existing spillway by 3.0 m. A summary of the labyrinth weir design input parameters is provided in Table 1. Routing the IDF through the raised spillway resulted in a modest increase of the peak outflow from 585.6 cms to 611.3 cms and increased the peak reservoir water surface elevation by less than 3.0 cm. To provide a minimum of 0.3 m of freeboard between the maximum routed water surface elevation and the dam crest, the dam crest was raised by 7.6 cm. The resulting freeboard is 0.4 m.

Table 1. Labyrinth spillway design parameters.

Parameter	Value	Parameter	Value
No. of Cycles	3	A/w	0.12
Spillway Width (m)	27.4	P/tw	6.0
Wall Angle (degree)	6.04	Length of Apron (m)	33.2
Crest Shape	Quarter Round	Apex Width (m)	1.1
Maximum Discharge (cms)	611.3	Wall Thickness at Crest (cm)	55.9
Lde/lc	0.40	Wall Thickness at Base (cm)	66.0
Ht/P	0.98	Floor Slab Thickness (cm)	61.0
w/P	3.0	Wall Height (m)	3.0

Figure 1 depicts the raised labyrinth spillway configuration in an isometric view. The abutments are colored green, and the spillway slab and walls are gray, with foundation material in tan.

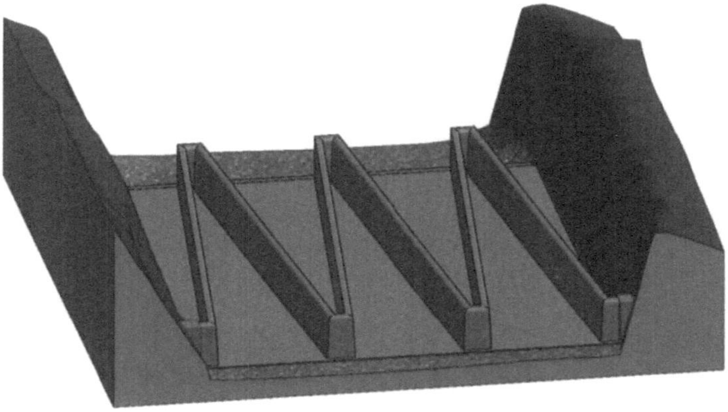

Figure 1. Labyrinth spillway configuration isometric view.

3.3.1 *Computational fluid dynamic modeling*
An additional hydraulic analysis of the labyrinth spillway was performed to obtain a better understanding of the spillway crest discharge capacity and to compare newer numerical modeling techniques to the iterative spreadsheet approach discussed above. CFD simulations were performed using FLOW-3D HYDRO software, Version 23.2.1.1. The CFD model included a partial reservoir, the labyrinth crest and slab, and a portion of the downstream channel. The model extended approximately 83.8 m upstream of the labyrinth structure to approximately 30.5 m downstream of the end of the labyrinth structure.

The model used mesh cell spacings of 0.6 m for the reservoir and transitioned to 0.3 m, 15.2 cm, and 7.6 cm as it approached the labyrinth crest. Downstream of the labyrinth, the mesh cell size transitioned from 7.6 cm to 152 cm. The total number of cells in the model was about 124 million. All meshes use the standard FLOW-3D hexahedral (cubic) mesh to resolve fluid and solid objects.

To develop the terrain for the CFD model, the original survey used for the project drawings was converted to a three-dimensional solid. Additionally, record drawings of the spillway were used to develop geometry for the labyrinth structure. This information was supplemented by photographs taken during and after the construction of the structure. Terrain data and spillway geometries were used to develop the mesh-generated geometry in the CFD model.

Two CFD modeling scenarios were developed to evaluate the spillway discharge capacity and hydraulic characteristics of the spillway flow and include 1) a steady reservoir water surface elevation at El. 2566.1 m, the design water surface elevation, and 2) steady-state discharge 608.8 cms, the design spillway flow. Additional CFD modeling runs were used to evaluate the model's sensitivity to the selected modeling parameters including air entrainment, turbulence modeling, surface roughness values, and minimum mesh cell sizes of 0.3 m, 15.2 cm, and 7.6 cm. To model the turbulent flow, the Renormalized Group (RNG) turbulence model was used. Model simulation durations were approximately 3 days using a 32-core computer processor.

3.3.2 *Computational fluid dynamic results*
The CFD modeling evaluated two scenarios to estimate the discharge capacity of the labyrinth spillway. Scenario 1 represents the estimated spillway capacity based on the spreadsheet design, where the water surface elevation in the reservoir is at the design elevation of 2566.1 m. At this reservoir elevation, the modeled spillway discharge is about 555.0 cms. In comparison, the spreadsheet design tool predicts a discharge capacity of approximately 608.8 cms at the same water surface elevation. The CFD model results represent a reduction of about 8.8% in estimated discharge capacity.

The observed difference in discharge capacity is likely caused by two primary factors: energy losses in the approach channel upstream of the spillway structure and a reduction in flow due to abutment effects, primarily on the right side of the approach channel, which are not accounted for in the spreadsheet calculations. The spreadsheet tool assumes idealized flow conditions with a uniform velocity distribution and negligible approach energy losses. However, the CFD model provides a more detailed representation of the actual hydraulic conditions at the site.

As flow approaches the spillway, it passes through an approach channel approximately 61.0 m long. Energy losses within the approach channel contribute to reduced spillway discharges at the design water surface elevation. A plan view of the velocity results is shown in Figure 2.

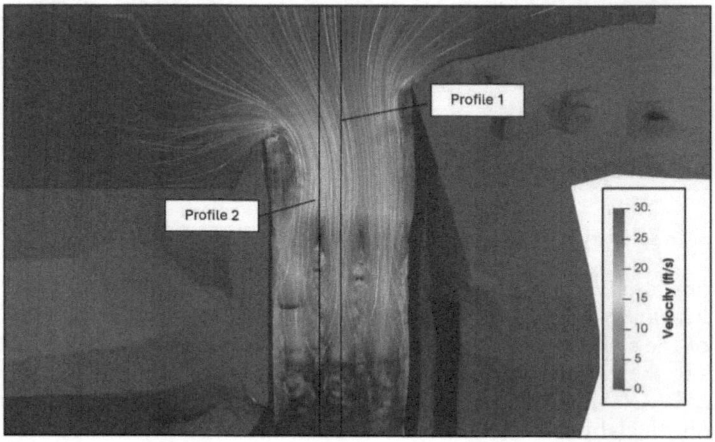

Figure 2. Labyrinth spillway plan view velocity results. Note profile 1 and profile 2 lines.

The geometry of the abutments also impacts the flow through the approach channel and the spillway. As shown in Figure 2, flow around the right abutment of the spillway approach channel results in flow contraction and an area of recirculation at the upstream end of the approach channel. The effect is also seen at the left abutment but is less pronounced. This reduces the effective width of the approach channel and subsequently reduces the discharge through the spillway. Profile views through the upstream and downstream ends of the labyrinth cycle are shown in Figure 3.

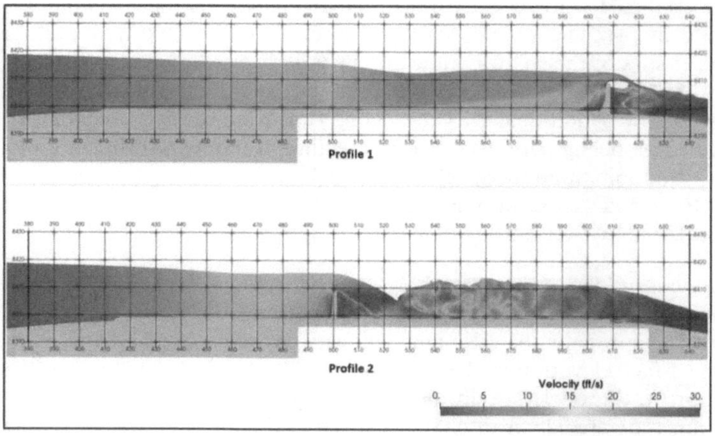

Figure 3. Labyrinth spillway velocity results profile views.

Scenario 2 was developed to evaluate the water surface elevation required to achieve the design discharge of 608.8 cms. The CFD model for this scenario estimates that the reservoir level must increase to about 2566.3 m to achieve the target flow rate. This result suggests that an increase of approximately 15.2 cm in water level is required to compensate for approach channel energy losses and maintain the design discharge. However, this 15.2 cm increase in water surface elevation reduces the available freeboard during the PMF event from 36.6 cm to 21.3 cm. The results highlight the influence of approach flow conditions on spillway capacity and indicate that real-world discharge may be lower than spreadsheet-based estimates under non-ideal approach conditions.

Sensitivity analyses were conducted to evaluate the impact of various modeling parameters on the estimated spillway discharge. The analyses included variations in surface roughness (both higher and lower than the baseline), including air entrainment using FLOW-3D default values, modifications to mesh cell size, and a change in turbulence modeling. These variations were analyzed by comparing the resulting flow over the spillway at the design water surface elevation of 2566.1 m. The sensitivity analysis results demonstrate that variations in roughness, air entrainment, turbulence model selection, and mesh cell size had a minimal impact on the estimated spillway discharge. The largest deviation from the base case was observed with the high roughness scenario (-1.0%) and the low roughness scenario (+1.0%), while changes to mesh cell size and air entrainment produced variations within ±0.9%. Applying a different turbulence model showed no difference from the base case, reinforcing the reliability of the selected model. The results indicate that the selected modeling parameters are appropriate for estimating the flow rate over the labyrinth spillway, as none of the tested variations resulted in a significant change in the estimated discharge.

3.4 *Structural design*

The spillway consists of labyrinth weir walls, abutment walls, a flat slab, and upstream and downstream cutoff walls. Both the upstream and downstream ends of the spillway include subsurface cutoff walls. The reinforced concrete elements were designed in accordance with the requirements of the American Concrete Institute (ACI) "Building Code Requirements for Structural Concrete – ACI 318" (2011), and guidance from the USACE's "Strength Design of Reinforced Concrete Hydraulic Structures" (1992). The concrete design compressive strength for the labyrinth walls and slab was 41.4 MPa at 28 days and all other conventional concrete design compressive strength was 31.0 MPa at 28 days. Reinforcing steel was grade 60 (413.7 MPa yield strength), uncoated. The higher strength concrete was specified for the labyrinth structure to better accommodate potential shear forces generated by winter ice loading without increasing the weir thickness which would impact hydraulic characteristics. The ice load of 145.9 kN/m was applied 0.3 m from the top of the weir wall, assuming an ice thickness of 0.6 m. This ice load is assumed to be developed by thermal expansion of a massive ice sheet rather than localized ice impact loads.

Stability analysis of the spillway section was conducted for sliding and overturning. The structural design was generally controlled by the usual condition including ice load. Sliding stability required factor of safety was 3.0 and the foundation was required to maintain 100% base compression per state requirements. For Usual Conditions with lateral loads developed by full reservoir head and ice against the labyrinth wall and vertical gravity loads due to the weight of water behind the wall, the weight of concrete, uplift below the slab and resisting anchor loads, the calculated sliding factor of safety was estimated to be 5.4, exceeding the required value.

Labyrinth walls were designed for three loading conditions: Normal operating conditions with ice, the IDF flood, and normal operating conditions with the maximum credible earthquake. The normal operating conditions plus ice loading case controlled the design. The weir wall varied in thickness from 66.0 cm at the base to 50.8 cm at the top of the wall. The weir walls are connected to the rock walls by 1.5 m wide abutment walls matching the height of the weir wall. The spillway slab was designed for load transfer from the wall sections. The slab was designed with construction joints in three locations along the length and two locations across the width of the spillway. The design required two 0.9 m deep cutoff walls below the slab acting as shear keys. Twenty-one rock anchors, placed at 3.7 m on center in 3 rows, were installed to help resist buoyancy effects and to add to the overall stability of the spillway structure.

4 SPILLWAY CONSTRUCTION

The storage within the reservoir was lowered during the winter of 2015 and spring of 2016 to accommodate construction of the spillway weir. The reservoir was drawn down 1.4 m, which corresponds with the maximum depth of excavation for the upstream cutoff wall. The lowered reservoir water level was maintained using the low-level outlet works.

The excavation for the new spillway was into rock and required controlled blasting. The excavation slopes were set at 1H:4V, which was consistent with the existing spillway excavation. The maximum excavation on the left abutment was about 6.2 m. The maximum excavation on the right abutment was about 9.9 m. Rock excavation by controlled blasting was performed in two areas. First, a test blast was conducted along the access road south of the right abutment to remove a bedrock knob that allowed for construction of the elevated access road. A second production blast was conducted along the right lateral wall of the existing spillway channel to widen the channel by approximately 3.0 m and allow construction of the new labyrinth spillway. Drilling began at the test blast holes along the access road. A pre-split line was drilled on 76.2 cm centers to control the limit of the cut into the rock outcrop. Production holes were spaced approximately 1.5 m on center and drill hole depth ranged from about 1.5 to 3.7 m. Prior to blasting, the Contractor cleared the blasting exclusion zone, and four seismographs were set up to monitor ground acceleration and vibrations. The seismographs were set up in pre-approved strategic locations, including the dam crest at the right abutment, dam crest near the outlet works controls, upstream dam face near the right abutment, and along the access road approximately 15.2 m from the blast. Blast monitoring results were reviewed immediately following the test blast and were determined to be below the project specified vibration thresholds. Blast rubble was processed immediately following the blast to re-open access to the dam crest for vehicle traffic. The production blast along the right spillway wall was performed the following week. Blast rubble was processed over the next week and spillway construction began. Excavation of the upstream cutoff wall occurred next followed by the foundation and wall grouting program.

Seepage through the rock mass beneath the spillway was anticipated to be minimal due to the low permeability values observed during water pressure testing. However, rock excavation and blasting had the potential to result in a disturbed rock surface and could widen existing joints and possibly form new joints. Any resulting discontinuities at the rock surface were planned to be treated with dental concrete and slush grout. Additional seepage protection at the spillway floor was provided by the cutoff wall on the upstream side of the spillway slab. A grout curtain was placed along the upstream toe of the spillway slab and at the spillway abutments to reduce deeper seepage through the bedrock. The alignment of the grout holes was selected to target specific discontinuities identified during the exposure mapping. This formed a continuous line of grout holes up to the top of spillway elevation and minimized seepage around the right lateral side of the structure.

4.1 *Cold weather concrete construction*

When concrete was placed in cold weather (below 4.4 deg. C) the Contractor implemented their cold weather placement plan in accordance with the project specifications. The cold weather placement provisions included protecting the subgrade from frost using heating, monitoring air temperatures during placement, protecting placed concrete with tarps/blankets after placement, and applying heat as necessary to maintain the ambient air temperature above 4.4 deg. C for at least 7 days. No adjustments were made to address low oxygen conditions. The Contractor also collected extra cylinders for field curing under the heated blankets adjacent to each placement and subsequent lab testing to verify field curing performance.

4.2 *Achieving the designed wall shape*

The quarter round upstream edge of the labyrinth wall was important to developing the designed flow characteristics over the top of the labyrinth walls. Irregularities or roughness in the quarter round occurring during construction would be expected to result in irregular wearing patterns on the top of the concrete wall as well as the potential for reductions in reservoir storage. A test section was required to prove that the Contractor was able to construct the top of the wall section to the design tolerances and finish. To minimize inconsistencies in the rounded shape, the Contractor elected to use a form made using pipe sections rather than hand forming the curved portion. Test section construction and results of the test section are shown in Figure 4. Where honeycombing was observed below the pipe form, Dayton Superior Sure Bond ™ J58 epoxy adhesive was used to patch holes and form a smooth surface. The completed wall section using J58 for patching is also shown in Figure 4.

Figure 4. Labyrinth spillway crest test section and finished quarter round crest as constructed.

5 CONCLUSIONS

This paper documents the design and construction activities associated with the raised labyrinth spillway weir crest. The paper compares the empirically based hydraulic design calculations to modern CFD modeling techniques and describes the differences observed due to non-ideal approach conditions and abutment effects. The geotechnical and structural engineering design analyses required for raising the reservoir level with a labyrinth structure were also briefly summarized. Additionally, the construction challenges were discussed, including rock excavation by blasting, grout curtain construction, concrete placement and wall corner treatments. GEI worked closely with the Client, the Contractor, and the regulators to address construction challenges and implement engineering solutions with minimal impact to the project budget and schedule.

REFERENCES

American Concrete Institute (ACI), 2011. Building Code Requirements for Structural Concrete (ACI 318-11). Farmington Hills, MI.

Crookston, B. M. and J. P. Tullis, 2013. "Hydraulic Design and Analysis of Labyrinth Weirs," Journal of Irrigation and Drainage Engineering, 139:363–377.

Falvey, H. T., 2003. "Hydraulic Design of Labyrinth Weirs", ASCE Press, Reston, VA.

Flow Science. 2019. "FLOW-3D v12.0 User Manual", 2019.

US Army Corps of Engineers (USACE), 1992. Strength Design for Reinforced Concrete Hydraulic Structures. Engineering Manual 1110-2-2104.

US Bureau of Reclamation (USBR), 1989. "Flood Hydrology Manual." United States Government Printing Office, Denver, CO.

US Bureau of Reclamation (USBR), 1988. "Hydrometeorological Report No. 55A. Probably Maximum Precipitation Estimate – United States Between the Continental Divide and the 103rd Meridian." Silver Springs, Maryland. June.

Labyrinth and Piano Key Weirs IV – Erpicum, Crookston & Paxson (Eds)

A stainless-steel piano key weir at the Oule Dam spillway (France)

Y. Fournié, H. Tralli and J. Alende
Société Hydroélectrique du Midi, Toulouse, France

S. Erpicum
UEE - HECE, Liège University, Liège, Belgium

ABSTRACT: The spillway of the Oule dam (Pyrénées, France) was successfully modified in 2018, with the discharge capacity increased in accordance with the updated hydrological analysis and French regulations. In order to achieve a significant increase in total discharge capacity, the dam operator (Société Hydroélectrique du Midi - ENGIE) re-designed the whole spillway. A 1 m high type-A piano key weir solution was selected and implemented to replace the existing ogee crest. This solution is unique since it is made of stainless steel, precast, transported by helicopter on site and assembled by bolting. This paper focuses on the design and the building method, which was adapted to the complex geometry of the structure and site constraints. Information about design assumptions and installation are also presented (on-site procedure, duration, costs).

1 INTRODUCTION

The Oule dam, located in the Pyrénées mountains (France), is part of a multipurpose hydropower scheme operated by SHEM (Société Hydroélectrique du Midi - ENGIE), the French 3rd largest hydropower producer. It is a concrete gravity dam commissioned in 1923 and raised in 1950. The discharge capacity of the initial spillway (Fig. 1 & 2), composed of a lateral ogee crest followed by a very narrow and complex stepped chute, was limited to 10 m^3/s that corresponds to chute channel overflowing.

The spillway was undersized compared to the updated hydrology and design flood (1000 year return period with a peak discharge of 103 m^3/s). In order to significantly increase the discharge capacity, a new spillway layout was required. Several rehabilitation solutions have been sized and analysed with various scaled models, including a new upstream piano key weir (PKW) and downstream stepped channel (Erpicum et al, 2017). The upgrading works were carried out during three summer seasons from 2016 to 2018. This paper focuses on the uncommon design of the PKW, which is unique from several perspectives.

2 DESIGN

2.1 *Overall presentation and site constraints*

The new Oule spillway is equipped with a side PKW on the right bank of the reservoir, followed by a collecting channel and a 170m long straight stepped chute with varying slope (Figure 1).

The design of the new weir addressed various constraints:

DOI: 10.1201/9781003655541-26

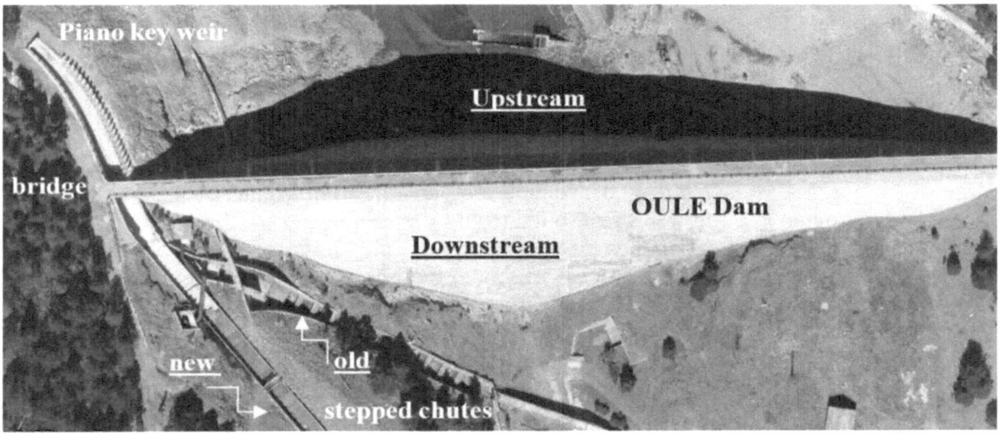

Figure 1. Aerial view of the dam site during a reservoir drawdown and the spillway located on the right bank (left of photo).

Figure 2. View of the new downstream straight stepped chute (left) and upstream reservoir with right bank weir (right).

- The normal reservoir elevation was to be kept constant as before construction, because the volume is secured for water release and agriculture requirements during summer.
- The section of the collecting channel could not be modified on its downstream end due to the presence of an existing bridge (Figure 3). This creates submergence of the upstream section of the weir at high heads operation.

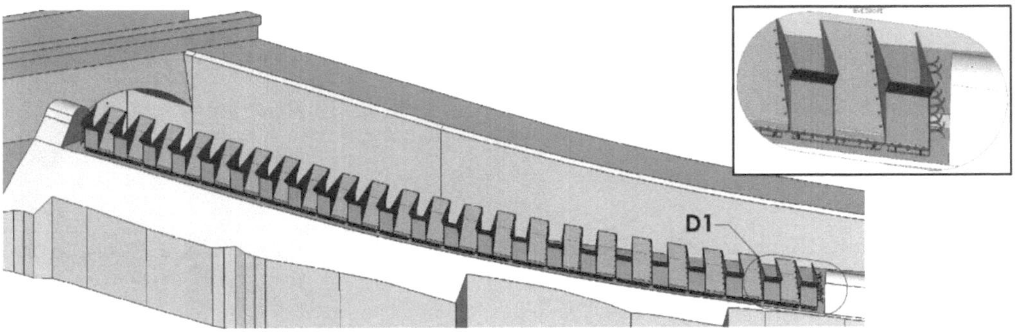

Figure 3. Global 3D view of the PKW.

- The high-altitude environment and isolated site create difficult access and weather conditions with ice. Modifying the existing arrangements with passive structures was preferred to gated systems, in order to avoid complex and expensive maintenance works.

Multiple rehabilitation designs were studied, and the inclined PKW solution was considered to be the optimal alternative for the following reasons:

- One of its main advantages is to drastically increase the discharge capacity at low heads with the lengthening of the weir crest, maximising design flood peak discharge attenuation by the reservoir routing. This benefits from the specific shape of the flood hydrograph which shows a sharp peak discharge. Consequently, the PKW enables the release of the design flood without exceeding the available 1m head between the maximum reservoir elevation and the normal reservoir elevation. Also, the spillway maximum outflow discharge is reduced to 72 m^3/s compared to the inflow peak.
- The inlet and outlet keys 30° inclination towards downstream reduces submergence effects by improving flow acceleration in the collecting channel.
- The very compact geometry and use of steel material allowed a very low weight structure, easy to transport and assemble on site, requiring limited civil works on the existing weir.

The solution selection process with the main steps of the hydraulic studies are detailed in Erpicum et al (2017). First, various geometries of type-A PKW were compared, by means of analytical calculations: from small structure made of steel (0.75 m high) to larger structure made of concrete (3m high). The outflow discharge study was then carried out in two phases with different physical models (Figure 3) of the selected solution (1m high steel structure). The hydraulic analysis begun with a 1:5 model representing a limited section of the PKW in a simplified channel configuration to be free of scale effects. A second global 1:25 model was built with the existing right bank environment (collecting channel, bridge, dam) to take into account the partial submergence at high heads, the effect of alveoli inclination and approach flow conditions. These studies could confirm the 72 m^3/s outflow discharge required to safely evacuate the 1000 year flood return period.

Figure 4. 1:5 model with 0.4$m^3/s/m$ equivalent discharge (left); 1:25 full 3D model with 0.3 $m^3/s/m$ equivalent (center); overflow of the prototype, unknown discharge (right).

2.2 *Main characteristics*

The PKW includes 22 units 1.23m wide and 1.23 m high (Figure 5). Each unit is linked to the other by fasteners (bolts). It is composed of symmetric type-A steel PKW with side walls inclined by 30° towards downstream (Figure 6).

The PKW modules occupy a length of 33m atop the original lateral weir. The upstream part of the existing spillway remains with a 7.9 m long linear ogee crested weir. The base of the PKW

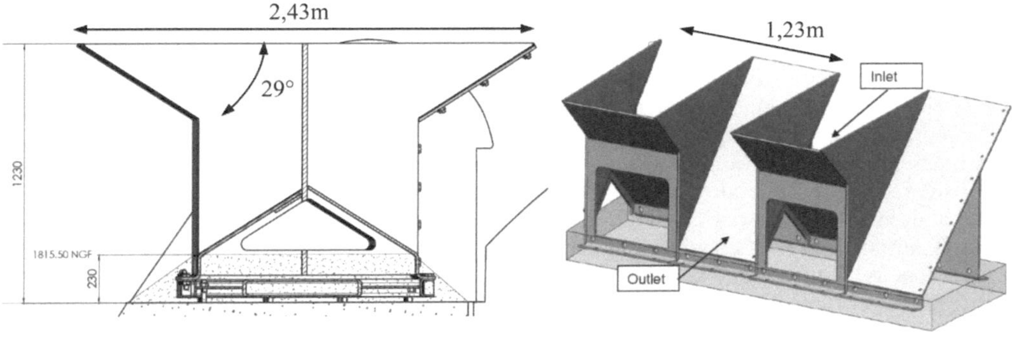

Figure 5. Drawings of the structure (cross-section and 3D view of 2 modules).

structure is linked to a steel frame embedded in concrete. Therefore, the visible PKW height (P) is only 1m. Two types of frames have been designed to adapt to the curvature of the existing weir (Figure 6). They are anchored in the concrete of the partially demolished existing structure.

The PKW is made with 12 mm thick steel plates. Some stainless 304L grade steel was selected for the laminated sheets to avoid future maintenance needs of repetitive protective

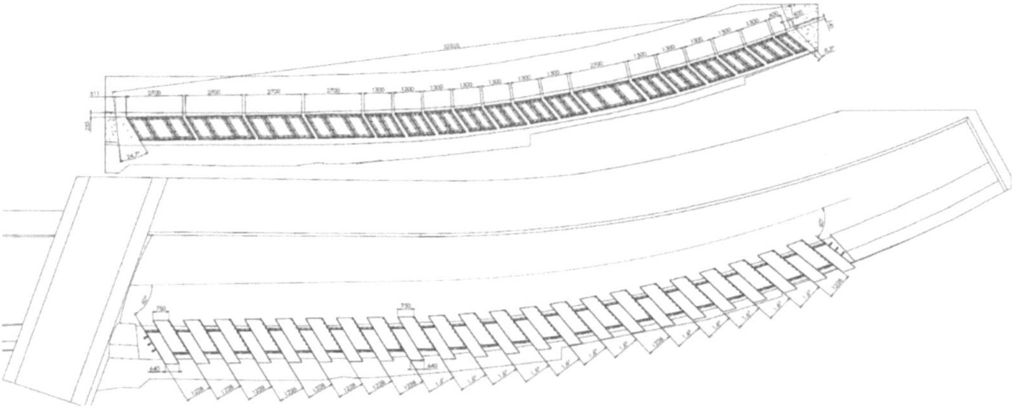

Figure 6. PKW layout (down) and set up of the frames on the concrete levelled weir (up).

corrosion painting. S235J2 steel was used for intermediate lower support angles. The downstream base of the structure comprises recesses to minimize the quantity of steel and create a better aeration of the flow nappes (Figure 5).

2.3 *Calculation assumptions and main results*

The following design criteria and loads have been considered for this project:

- Structure's lifespan: 75 years minimum.
- Characteristics of material: stainless steel 304L with a limited admissible elastic stress of 209MPa.
- European standards are used for calculations: Eurocode 0 (basis of structural design), 1 (actions on structures), 3 (design of steel structures), 8 (earthquake resistance).

- Maximum Water Level: 1m high overflow on the structure which corresponds to the 72 m^3/s total required discharge capacity for the spillway. Uplift water pressure was applied at the base with triangular diagram.
- Extreme Water Level: 3,38 m overflow on the structure. The downstream channel being submerged, the force is applied also on the outlet part.
- Air temperature variation: between -30°C and +70°C. The keying up of the structures is considered to be done at 15°C. Some more realistic temperatures values were applied for the stainless steel. It was considered that the bottom area (of a length about 100mm at least) cannot be at 70°C, while the same sheet embedded in the concrete is at 15°C. This area acts as a buffer zone allowing the temperature to decrease from 70°C to 15°C. Therefore the average temperature of the steel plates is estimated at (70°C + 20°C)/2 = 45°C (with 20°C being the ambient air temperature). For negative temperatures, the reasoning cannot be the same, as the negative temperature is brought by the ambient air that touches almost all surfaces. Additionally, there is no possible warming (by conduction) and the exposure duration to negative temperatures is longer (duration of the night in winter). Therefore, thermal exchanges (and consequently cooling) are much faster. The entire mass of stainless steel can thus reach -30°C. These assumptions result in a temperature positive load of 45°C - 15°C = +30°C, and -30°C used on the model for the stainless steel without waterlevel.

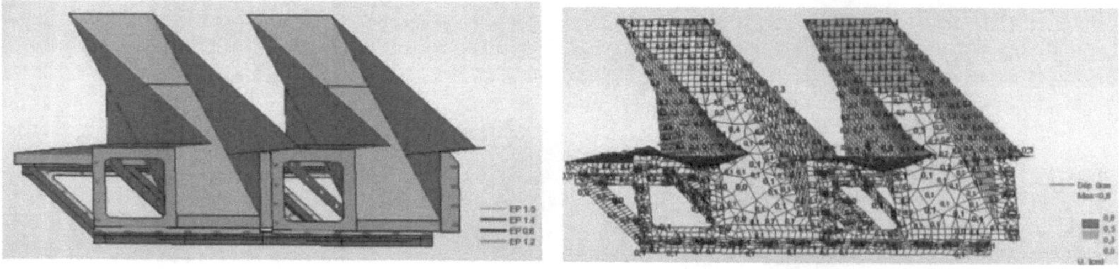

Figure 7. View of FE model Autodesk Robot Structural Analysis (left), maximum displacement (right).

- Ice load: DIN 19704-1 (2014-11) standard was used. The ice pressure generates a horizontal surface force of 150 kPa applied over a thickness of 0.30m. These effects are cumulative with the hydrostatic pressure and applied at normal water level.

The Finite Element model analysed the most critical configuration: 2 bolted modules placed on a unique frame (Figure 7).

The maximum deformation of the structure (0,8cm see Figure 6) at Serviceability Limit State (SLS) is generated by ice thrust. The most critical combinations governing Ultimate Limit State (ULS) include ice load and thermal gradient. They result in local over-stresses concentrated in areas with holes and are not representative of the structure behaviour. The other static and seismic load cases were validated. The stresses in the fasteners between modules was also justified. There are no stresses exceeding the allowable limits in the welded areas.

3 MANUFACTURING

The modules are assembled in the workshop by welding. An overview of the docking is shown on the following Welding Map (Figure 8).

Manufacturing monitoring consisted of:

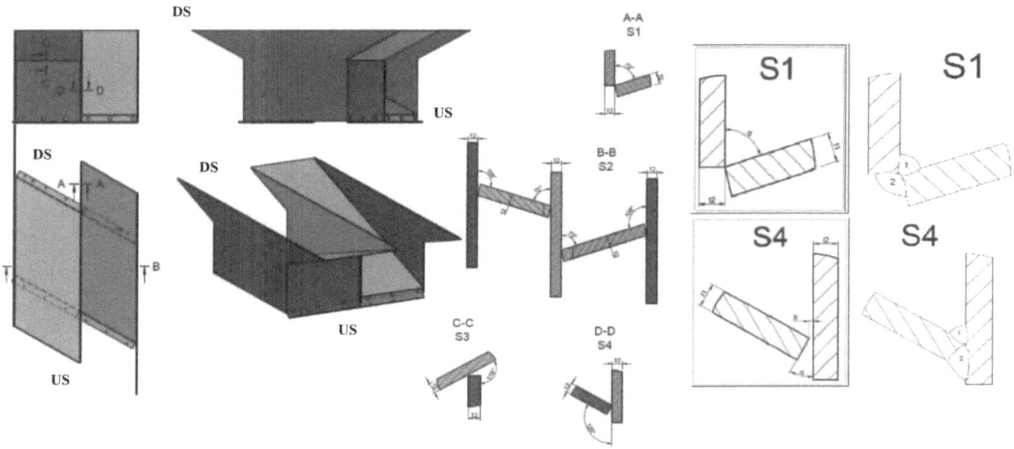

Figure 8. View of the welding map.

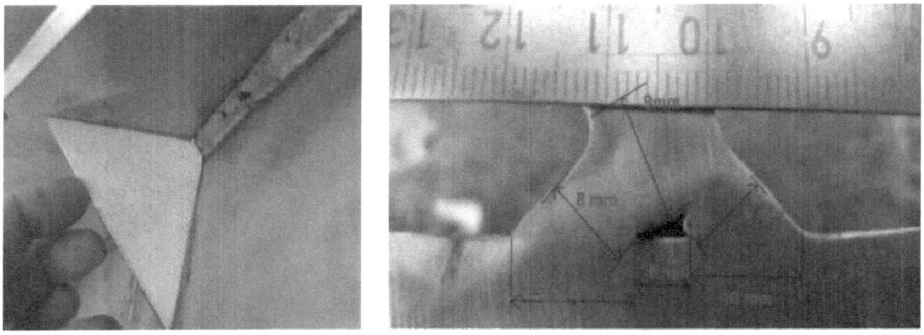

Figure 9. Geometrical control of welding using template (left) and example of weld (right).

- Qualification of welders and welding procedures
- Weld control; these are controlled 100% visual (VT) + 10% surface (MT)
- Dimensional control of welds using templates (Figure 9)

A blank assembly was carried out in the factory to ensure the correct assembly and compliance with the requested dimensions (Figure 10). The total weight of the stainless steel plates is estimated to 14,5 tonnes.

Because of the innovative PKW solution, studies have taken much more time than initially foreseen. Many complications were encountered in designing the structure:

- Difficulty to find a competent contractor to get a satisfactory detailed design and suitable manufacturing methods.
- Evolution of design criteria that led to an increase in the thickness of stainless steel sheets from 9 to 12 mm;
- Manufacturing and mounting process constraints of the PKW units;
- Design of the expansion joints between modules. Several alternatives have been studied (Figure 10). The simplest, most robust solution that does not require future maintenance was finally chosen, compared to those that were too complex to build and maintain.

Figure 10. Photograph from the manufacturing shop: blank assembly (left) one PKW unit (right).

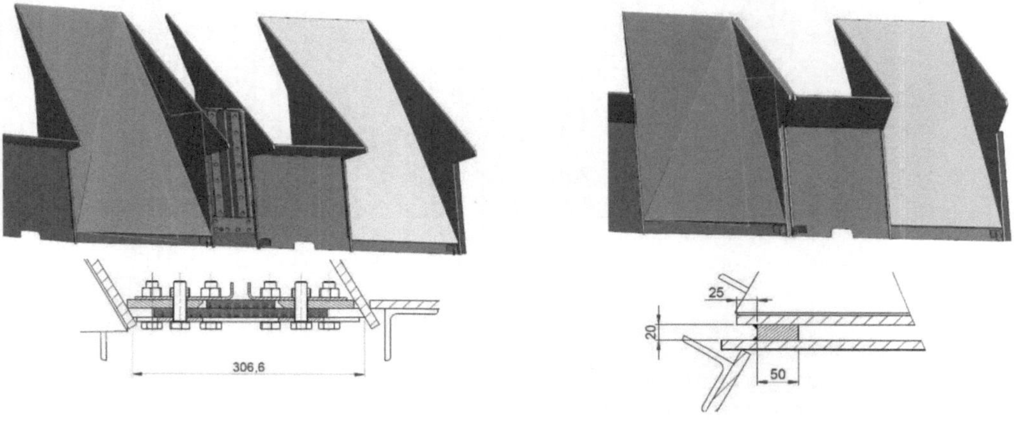

Figure 11. 2 types of dilatation joints studied and finally not selected.

4 ASSEMBLY AND ON-SITE INSTALLATION

On-site assembly was carried out during only 4 weeks by means of relevant procedures described in the following steps.

- The existing concrete weir was lowered by 1.2m using concrete cutting and chipping with an hydraulic breaker.
- The frames were supplied first. Helicopters were used to transport them one at a time from the closest location accessible by truck (Figure 12). Once installed, they were adjusted in position and height, then anchored to the concrete support with threaded rods and chemical sealing. 16 anchored were used for each unit (diameter 24 mm upstream and 16 mm intermediary and downstream).

Figure 12. View of the PKW frames (left), module placement by helicopter (center) and position of anchors (right).

- The PKW units were designed to weigh less than 800 kg, in order to be transported by helicopter. Welded lifting rings ensured stability and balance during helicopter transport.
- The 22 modules were positioned on the frames and manually adjusted using centering rods with o-ring placed at the junction between 2 modules to facilitate the assembly (Figure 13).
- Each unit was then bolted with the frame and with its side module. In order to prevent leakage, a relevant joint connexion was chosen as shown on the following sketch (Figure 14). This solution has several advantages: simple sealing with standard joint, the joint is replaceable and protected from direct sunlight.
- The base of the PKW was then sealed with compensated shrinkage sealing mortar (Clavex 708 MB). It was carried out in 5 stages to limit the amount of daily poured mortar ($2.5m^3$), for a total of around $12m^3$. A helicopter-transportable silo was used to store the Clavex on-site, which was then poured into a booster pump and conveyed to the sealing areas.

Figure 13. Detail of the assembling part between module and frame.

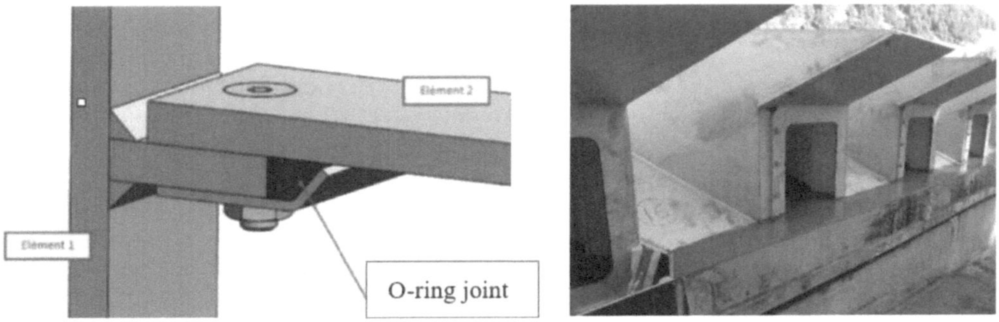

Figure 14. Detail of the joint for connection between modules (left) and modules embedded in concrete base (right).

5 COSTS & FEEDBACK

The main works were carried out by two local companies: NGE Fondations (contractor in charge of all civil works: demolition, anchoring, sealing, general installations) and ACTI (subcontractor for the manufacture and installation of the PKW).

The total budget for the weir rehabilitation works with this unique PKW solution was 410 000€ (2018 final cost). This amount is divided between (i.) detailed design and building studies (11%), (ii.) civil works and PKW installation (18%), (iii.) material supply and manufacturing (70%). To be noted that the value issued from the 2016 call for tender was much less (270 000€). The important increase was due to unforeseen PKW design complexity, the increase of stainless steel thickness during the project and the change of transportation method (use of an helicopter instead of the cable crane which was dismantled due to delay constraints).

The use of Clavex sealing mortar is not compatible with the thermal expansion of the modules, which induced some cracks at the surface. This was expected and some leakages between the stainless steel PKW and concrete junction are observed (Figure15). This problem should be addressed for new projects of this type.

It was observed that large floating bodies such as tree trunks would get trapped during spills in the structure when water levels went down and these had to be removed manually (Figure 15). This was due to the low water head (maximum observed water overflow to date is 7 m^3/s) and the narrowness of the keys. The structure's performance is considered satisfactory after 7 years of operation and no deterioration of the steel PKW has been noted.

Figure 15. PKW assembled on site (left), in operation (center) and small leakages observed (right).

6 CONCLUSION

The new steel PKW was able to meet the increased capacity requirements of the project. Its design was complex due to multiple site constraints (geometry, difficult access, stain-steel material). Therefore, it was chosen to focus on practical on-site assembly and maintenance. However, for future projects, it is recommended to find more relevant solutions to ensure the correct sealing of the interface between the steel module and the civil engineering support.

REFERENCES

Erpicum, S., Archambeau, P., Dewals, B., Pirotton, M., Tralli, H., & Alende, J. 2017. A piano key weir to improve the discharge capacity of the Oule Dam spillway (France). Labyrinth and Piano key weirs III – PKW 2017: 195–204. Leiden: CRC Press.

Author Index